Photoeffects at Semiconductor–Electrolyte Interfaces

Photoeffects at Semiconductor-Electrolyte Interfaces

Arthur J. Nozik, EDITOR

Solar Energy Research Institute

Based on a symposium

sponsored by the Division

of Colloid and Surface Chemistry

at the 179th Meeting of the

American Chemical Society,

Houston, Texas,

March 25–26, 1980.

ACS SYMPOSIUM SERIES **146**

AMERICAN CHEMICAL SOCIETY

WASHINGTON, D. C.   1981

Library of Congress CIP Data

Photoeffects at semiconductor–electrolyte interfaces.
 (ACS symposium series; 146 ISSN 0097-6156)

 Includes bibliographies and index.

 1. Photochemistry—Congresses. 2. Photoelectricity—
Congresses. 3. Semiconductors—Congresses.
 I. Nozik, Arthur J., 1936–   . II. American Chemi-
cal Society. Division of Colloid and Surface Chemistry.
III. Series: American Chemical Society. ACS sympo-
sium series; 146.

QD701.P47   541.3′5   80-27773
ISBN 0-8412-0604-X   ASCMC8   146 1–416 1981

ACS Symposium Series

M. Joan Comstock, *Series Editor*

FOREWORD

The ACS SYMPOSIUM SERIES was founded in 1974 to provide a medium for publishing symposia quickly in book form. The format of the Series parallels that of the continuing ADVANCES IN CHEMISTRY SERIES except that in order to save time the papers are not typeset but are reproduced as they are submitted by the authors in camera-ready form. Papers are reviewed under the supervision of the Editors with the assistance of the Series Advisory Board and are selected to maintain the integrity of the symposia; however, verbatim reproductions of previously published papers are not accepted. Both reviews and reports of research are acceptable since symposia may embrace both types of presentation.

CONTENTS

PREFACE

This volume, based on the symposium "Photoeffects at Semiconductor–Electrolyte Interfaces," consists of 25 invited and contributed papers. Although the emphasis of the symposium was on the more basic aspects of research in photoelectrochemistry, the covered topics included applied research on photoelectrochemical cells. This is natural since it is clear that the driving force for the intense current interest and activity in photoelectrochemistry is the potential development of photoelectrochemical cells for solar energy conversion. These versatile cells can be designed either to produce electricity (electrochemical photovoltaic cells) or to produce fuels and chemicals (photoelectrosynthetic cells).

The first 12 chapters of this volume are concerned with the vital subject of the effects of surface properties on photoelectrochemical behavior. This includes work on the effects of the chemical modification of semiconductor electrode surfaces either through molecular derivatization or ionic adsorption; the effects of surface structural defects and surface electronic states on photo-induced charge transfer across semiconductor–electrolyte interfaces; the kinetics of competing chemical reactions on semiconductor electrode surfaces; catalytic effects on semiconductor surfaces; and the problems of photocorrosion of semiconductor electrodes. Chapters 13–15 deal with new electrode materials (oxide semiconductors) and structures (protective layers and interfacial chlorophyll layers). Chapters 16 and 17 relate to the basic energetics of the semiconductor–electrolyte interface (potential distribution and the effects of charge inversion leading to band-edge unpinning), while Chapters 18–22 present results on new chemical systems and phenomena associated with photoelectrochemistry. These latter include luminescence studies, surfactant assemblies, a new model based on the effects of ionic desorption, studies of carbanion oxidation on semiconductor surfaces, and the behavior of molten-salt electrolytes. Finally, the volume concludes with three chapters on electrochemical photovoltaic cell devices dealing with models for current–voltage characteristics, stability performance, and solid electrolytes.

The exceptional interest and ferment in photoelectrochemistry has been manifested in 1980 by the appearance of at least five major conferences, symposia, or workshops in the field with international participation, including the present symposium. It is apparent from these meetings, as well as from the burgeoning amount of published literature, that photoelectrochemistry is a vital, productive, and stimulating field of science in which much significant and exciting scientific progress can be expected for a long time. Although very rapid progress has also been made in the applications of photoelectrochemistry, it is still too early to predict what impact photoelectrochemistry will have on practical solar energy conversion systems. We can be confident, however, that basic research in photoelectrochemistry will continue to produce important new knowledge, and that attractive potential applications to our energy problems will continue to provide the impetus for the work.

Solar Energy Research Institute   ARTHUR J. NOZIK
Golden, Colorado 80401
October 3, 1980

1

The Influence of Surface Orientation and Crystal Imperfections on Photoelectrochemical Reactions at Semiconductor Electrodes

HEINZ GERISCHER

Fritz-Haber-Institut der Max-Planck-Gesellschaft, Faradayweg 4-6,
D-1000 Berlin 33, West Germany

It is well known that the surface orientation of crystals and imperfections in the surface, like grain boundaries or dislocations, affect largely the reaction rates at electrodes made of metals or semiconductors. Such effects are most pronounced in those reactions where atoms leave their position in a crystal lattice or have to be incorporated into such one. These processes are connected with activation barriers which are particularly high for semiconductors where the chemical bonds between the components of the crystal lattice are highly directed and localized. If we consider photoelectrochemical reactions at semiconductors we have additionally to discuss the influence of these factors on light absorption and its consequences.

Factors which control the yield of photocurrents

Photocurrents and photovoltages are induced by the generation of excessive mobile charge carriers. In a semiconductor these are electron hole pairs generated by light absorption. The size of the effects is largely dependent on how far the recombination between electrons and holes can be prevented. An efficient separation of electron hole pairs occurs only in the space charge layer beneath the semiconductor/electrolyte contact. Large efficiencies can be reached if this space charge layer forms a high enough energy barrier for the two charge carriers to encounter each other. Such a situation is found in a depletion layer of n-type or p-type semiconductors or in an inversion layer (1,2). Here we shall not consider insulating materials where one can use high external electric fields to obtain charge separation (3).

Assuming that we have such a situation favorable for charge separation, we have to consider what factors influencing the efficiency of charge separation in an illuminated semiconductor electrode are affected by crystal orientation or crystal imperfections. Five such factors are listed in the following table:

0097-6156/81/0146-0001$05.00/0

(1) Schottky barrier height: $\Delta\varphi_{sc}$
(2) Schottky barrier extension: W
(3) rate of surface recombination: r_s
(4) mean diffusion length of minorities: L
(5) penetration depth of light: $\dfrac{1}{\alpha}$
     with α = absorptivity

The quantum yield of the photocurrent for an electrode illuminated from the front side can be calculated from a simple model described by Gärtner (4) provided some simplifying assumptions are applicable. This model is shown in Figure 1. If surface recombination can be neglected, the quantum yield ϕ is obtained as

$$\phi = 1 - \frac{1}{1 + \alpha L}\,\exp\,(-\alpha W) \tag{1}$$

The equation above can be approximated if the width of the space charge layer is small compared with the penetration depth of the light $1/\alpha$ by

$$\phi \approx \frac{L + W}{1/\alpha + L} \quad \text{if } \alpha \cdot W \ll 1 \tag{2}$$

The width of the space charge layer depends on the height of the Schottky barrier according to

$$W = \left(\frac{2\varepsilon\,\varepsilon_o\,\Delta\varphi_{sc}}{e_o \cdot N} \right)^{1/2} \tag{3}$$

where N is the donor or acceptor concentration, ε, ε_o and e_o have their usual meaning of permittivity and elementary electric charge. These equations contain four parameters of Table 1 and indicate how the quantum yield is affected by these factors. The surface recombination rate is only important if the barrier height is low and can otherwise be neglected. This was assumed in the derivation of Equation (1) which requires a high enough barrier height.

The main effect of crystal orientation is caused by different barrier heights on different crystal faces. It is well known that Volta-potential differences are dependent on crystal orientation because the surface dipole differs for different faces. In the case of a semiconductor electrode this means that the flat band potential which can be determined experimentally (5,6,7) depends on surface orientation. Consequently, the band bending at the same position of the Fermi level in the bulk of the semiconductor, i.e. at the same electrode potential, differs for different faces. Figure 2 gives a schematic picture for such differences.

Besides the effects of different surface dipoles, the concentration and energy position of surface states depend also largely on surface orientation with the result that the electric excess charge in surface states can be very different on different surfaces. This is indicated in Figure 3 by a comparison between the flat band situation and the situation at equal electrode potential for different surfaces. Case (a) is a surface free of surface

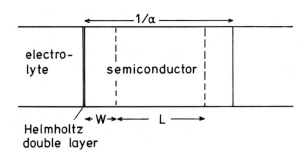

Figure 1. *Geometric parameters characterizing the semiconductor–electrolyte contact*

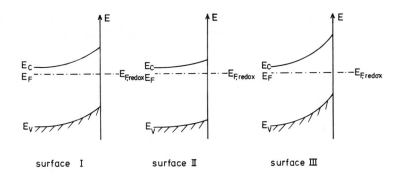

Figure 2. *Energy scheme for different surfaces of a semiconductor electrode at equilibrium with the same redox system*

states, case (b) a surface with acceptor states and case (c) a surface with donor states both having energy levels close to the conduction band. Since the excess charge in these surface states changes, with varying polarization, the band bending depends in a different way on the voltage applied.

Besides such electronic surface states which can interact either with electrons in the bulk of the semiconductor or with a redox system in the electrolyte, we have to consider another type of excess charge at the surface. This stems from adsorbed ions or from ionic groups attached to the surface of the semiconductor. This is well known from the pH dependence of the flat band potential of semiconducting oxides (8) or the dependence of the flat band potential of sulfides on the sulfide concentration in solution (9). Since surfaces of different orientation will interact differently with such ionic charge, this again will affect the photoelectrochemical processes via the different barrier heights at different surface orientation.

Electric charge in surface states or ionic groups on the surface will particularly depend on imperfections found in the surface. Imperfections in the bulk will affect the extension of a Schottky barrier since such imperfections can form traps for electric charge. Such a case has been studied for example at zinc oxide (10).

The rate of surface recombination is only important at low barrier height. Since, however, the barrier height depends on both factors, orientation and imperfections, there are different potential ranges, where this effect has to be taken into account.

The mean diffusion length of the minorities is drastically dependent on crystal imperfections. It may also be dependent on surface orientation, if they have an anisotropic mobility. Optical anisotropy is necessary for seeing an influence of crystal orientation on the penetration depth of the light while crystal imperfections will only affect the penetration depth in a range of wave lengths where light absorption in the crystal is very weak. The photocurrents are very small in this case which will not be discussed here.

We conclude from this discussion that a very complex correlation between structure and photoelectrochemical behavior is to be expected and it will often be difficult to decide what may be the main influence. The following examples are selected under the aspect to demonstrate some effects of surface orientation and crystal imperfections in systems where they are very pronounced. Materials with a large anisotropy of the crystal properties are the best candidates for this purpose. Therefore semiconductors with layer structure which have been introduced into photoelectrochemical studies by Tributsch (11,12,13) are predominantly used as examples.

Photoredox reactions at different surfaces
of layered semiconductors

Molybdenum selenide and molybdenum sulfide electrodes shall demonstrate the influence of different surface orientation on photoredox reactions. Figure 4 represents the crystal structure of MoS_2 where one can immediately see that on the van der Waals-surface exclusively sulfur atoms are exposed to the contact with the electrolyte. They should behave very differently from any other surface where the molybdenum atoms are not so well shielded from interaction with the electrolyte solution.

Photocurrent voltage curves have been studied with molybdenum selenide crystals of different orientation and different pretreatment. Figure 5 represents results for three typical surfaces of n-type $MoSe_2$ (14). An electrode with a very smooth surface cleaved parallel to the van der Waals-plane shows a very low dark current in contact with the KI containing electrolyte since iodide cannot directly inject electrons into the conduction band and can only be oxidized by holes. At a bias positive from the flat band potential U_{fb} where a depletion layer is formed a photocurrent can be observed as shown in this Figure. This photocurrent reaches a saturation at a potential about 300 mV more positive than U_{fb} when surface recombination becomes negligible.

An electrode with another surface structure which was not cleaved but used as grown appeared to contain faces of other orientation. Its current voltage behavior is seen in curve 2 of Figure 5. This electrode had a similar blocking character in the dark as the first one. The most prominent difference is that the onset of the photocurrent is shifted to more anodic potentials and is somewhat less steep than at the first electrode. The surface structures of these two electrodes are shown in the microscopic pictures of Figure 6 (15). It appears that the second electrode still contains a large portion of van der Waals-planes with some surfaces of other orientation in between. Although the latter ones are still inactive in the dark, they have a different dipole or differ in adsorption of charged ions from solution. Therefore the flat band potential is shifted into anodic direction.

A drastically different behavior was found if the electrode was prepared by a mechanical cut normal to the van der Waals-surface. The dark current increases steeply above a critical anodic potential as is seen in curve 3 of Figure 5. Only a small photocurrent can be observed which quickly becomes indistinguishable from the dark current, if the electrode potential is further increased anodically. This shows that the recombination rate is much higher at this kind of surface and that electron injection into the conduction band is now catalyzed by surface states generated by the formation of steps and other surface imperfections (16).

Figure 7 shows another set of examples for a molybdenum sulfide electrode (15). Current voltage curve No. 1 gives the result for a smooth electrode surface cleaved along the van der Waals-

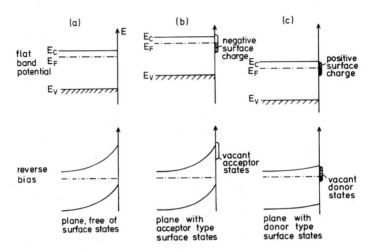

Figure 3.   Energy correlations for three different surfaces of an n-type semiconductor at the flat band potential (upper row) *and at equal anodic bias* (lower row): *(a) plane free of surface states; (b) plane with surface states of acceptor character; (c) plane with surface states of donor character*

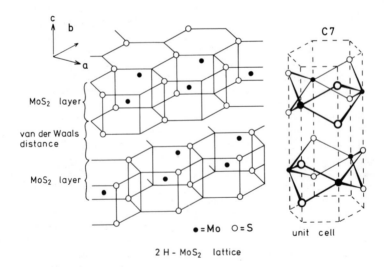

Figure 4.   Model of a 2 H-crystal lattice for MoS₂

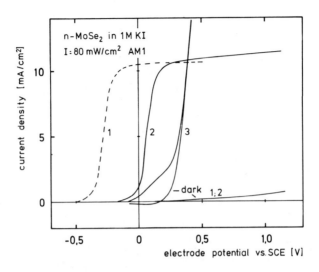

Figure 5. Current–voltage curves in the dark and under illumination (solar-like light with 80 mW/cm²) of MoSe₂ electrodes with different surface structure (14): (1) ⊥-c face, cleaved (van der Waals face); (2) ∥-c face, as grown; (3) ∥-c face, mechanically cut; (electrolyte: 1M KCl + 0.05M KI)

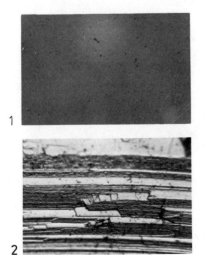

Figure 6.   Pictures of the surfaces used in Experiments 1 and 2 of Figure 5

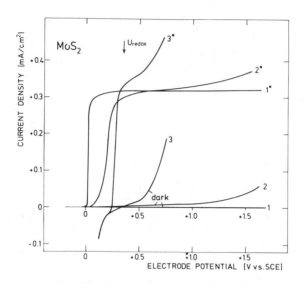

Figure 7. Current–voltage curves in the dark and under illumination (680-nm light, 1 mV/cm²) of a MoS₂ electrode of ⊥-c orientation at different states of surface perfection (15): (1) freshly cleaved; (2) after 25-min photocorrosion (solar light of AM 1) at 1 V_{SCE} in 1M KCl solution; (3) after 115-min photocorrosion, same conditions as (2); electrolyte: 1M KCl + 0.05M KI

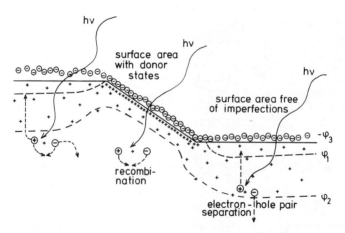

Figure 8. Model for charge distribution and course of the electric potential at a semiconductor with two different adjacent surface areas in contact with an electrolyte

plane. The dark current is again very small and the photocurrent
reaches saturation very quickly. With continuous exposure to photocorrosion in the absence of the redox system the electrode surface
becomes more and more less perfect. An increasing number of steps
is formed on the surface and the dark current increases as shown
in curves No. 2 and 3. The onset of the photocurrent is simultaneously hifted to more anodic potentials. However, in this example,
the saturation photocurrent seems practically not to decrease.

In order to explain these effects one has to assume that the
electric double layer structure is different at different faces
(16). While the van der Waals-face should not contain surface
states which can accumulate excess electric charge, other faces or
surfaces with structural defects form surface states with such a
property. Therefore, depending on the sign of the charge in these
surface states, the space charge underneath the surface will, at
a given position of the Fermi level in the bulk, be larger or
smaller in these regions than underneath the van der Waals-planes.
Figure 8 shows a model where the excess charge on the surface is
positive and therefore the extension of the depletion layer in the
n-type semiconductor is smaller in this region beneath the surface
than in the other regions. Recombination is enhanced at the defective parts of the surface while the photocurrents are unaffected
on the other parts as is indicated in this picture.

If the surface states would pick up negative charge, we would
have the opposite situation with a more extended space charge layer beneath the surface areas containing such defects. This would
shift the onset of the photocurrent to more negative potentials
and could improve the photocurrent yield. It appears that such a
situation can also be found in some cases.

Photocurrent and anisotropy of light absorption

Anisotropic crystals have a light absorption coefficient depending on the direction of the light wave and its polarization
(17). This again can be demonstrated in electrochemical experiments with layered crystals (18). Some results obtained with gallium selenide crystals are shown in the following Figures.

Gallium selenide, Ga_2Se_2, is a layered material where the
Ga_2-molecules are enclosed between two selenium layers. The
structure is shown in Figure 9. The anisotropy of the optical absorption has been studied with very thin crystals (19). It was
found that the absorption coefficient for light polarized normal
to the layers is higher than for light polarized parallel to the
layers. Because we were not able to prepare smooth surfaces oriented parallel to the c-axis, we studied the photocurrents obtained
under illumination with polarized light at various angles of incidence. Since s-polarized light has the electrical vector parallel to the van der Waals-planes independent of the angle of incidence, the photoresponse of this light could be used for a nor-
malization of the photocurrent spectra obtained. This was neces-

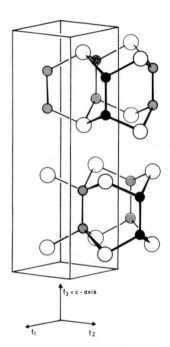

Figure 9.   Structure model of a Ga$_2$Se$_2$ crystal

sary because with varying angle of incidence the diffracted light
beam entering the crystal varies direction and the penetration
depth of light must be corrected for this. What is more important
for a comparison, the illuminated area of the electrode varies
with varying angle of incidence in a not fully controllable
manner. These differences can, however, be excluded if one meas-
ures the relative size of the photocurrents for s- and p-polarized
light at every wave length and relates these values to the photocurrent spectrum obtained at normal incidence where no difference
between s- and p-polarized light was found in accordance with the
theoretical expectation. A further correction has to be made for
the difference of the reflectivities for s- and p-polarized light.

A set of results obtained in such experiments is shown in
Figure 10. One sees that the light absorption with p-polarized
light leads to a drastic increase of the photocurrents in the whole
spectral range. This corresponds to the larger absorption coefficient for light with p-polarization. Since such differences can
only be seen in the photocurrents if the quantum yield is small
– cf Equation (1) and (2) – the diffusion length of electrons, L_n,
must be small compared to the penetration depth of the light $1/\alpha$.

Figure 11 gives another example performed in the same way with
a gallium selenide crystal of another (lower) doping concentration. The pronounced differences from the previous Figure indicate
clearly that it is not only the different absorption coefficient
which controls the height of the photocurrents. The diffusion
length of the minority carriers normal to the surface is of equal
importance for the quantum yield as Equation (1) and (2) have
shown. Therefore the relation between the yields for s- and p-polarized light will strongly depend on the life time of the minority
carriers. Both spectra should coincide if $L_n \gg 1/\alpha_{\parallel}$ and $1/\alpha_{\perp}$.
This can partly explain the differences found for the two elec-
trodes. The somewhat preliminary results shown here demonstrate
that large effects of this kind can be found and one has to take
into account the anisotropy of light absorption in the study of
photoeffects at semiconductor electrodes.

The role of crystal imperfections

Crystal imperfections play an enormous role in semiconductor
electrochemistry. Examples for imperfections in the surface have
already been given in the previous section. Imperfections in the
bulk mainly influence recombination. This is always seen in the
decreased yields after mechanical surface polishing where a great
number of crystal defects has been generated. Such crystals show
large photocurrents only after careful etching until all mechanical
defects have been removed from the boundary layer in which the photocurrents are generated. An example is given in Figure 12 for the
photocurrents observed at a zinc oxide electrode. It has been shown
that mechanically formed defects in zinc oxide act as traps for
holes and as recombination centers (10). Their presence in the po-

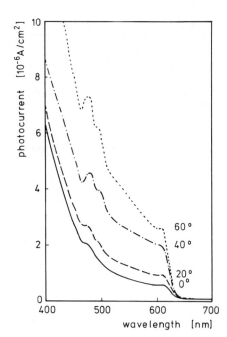

Figure 10.   Photocurrent spectra for p-polarized light at different angles of incidence, normalized to s-polarized light, represented by the spectrum at 0° (GaSe electrode with $N_A = 7 \times 10^{16}$ cm⁻³ at $V_{SCE} = -0.7$ V; electrolyte: 1M H_2SO_4)

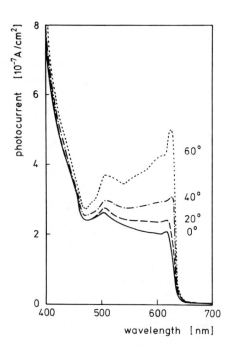

Figure 11.   The same as in Figure 10 for a GaSe electrode with $N_A = 3 \times 10^{14}$ cm⁻³

sitively charged state decreases the extension of the space charge layer. Therefore the photocurrent is reduced in two ways, by enhanced recombination and by a decrease of the electron hole pair separation efficiency of the space charge layer. Similar results have been observed with CdS (20) and GaAs(21).

One more example for the role of imperfections in the bulk and on the surface shall be demonstrated here. This is their influence on recombination luminescence which can be quite indicative for the type of defect which is involved. Luminescence can be studied by illumination of the crystal with light of shorter wave length as widely done in solid state investigations. In the case of an electrochemical system one can observe luminescence also due to the injection of minorities into the surface. This has been done with semiconductors of small band gap very early (22,23). Semiconductors with a wider band gap have first been studied by Beckmann and Memming (24) who could observe luminescence at GaP from recombination via surface states.

We have applied this technique to zinc oxide where hole injection can be reached with very oxidizing redox species (25). Figure 13 shows the result of such experiments. The luminescence observed covers two spectral ranges. One can be attributed to interband recombination and corresponds to the light absorption edge of zinc oxide. There is a tail at longer wave lengths which can come either from recombination via surface states or via bulk states having energy levels within the gap. Mechanical polishing of the surface changes the spectral distribution of the luminescence light drastically as shown in this Figure. The interband luminescence is largely reduced but not the tail at longer wave lengths. This can be taken as an indication that the longer wave length range of luninescence stems from surface state recombination since the mechanical polishing will certainly decrease the life time of minority carriers in the bulk and enhance radiationless recombination. On the other hand, the same treatment might increase the number of surface states due to the formation of differently oriented surface areas or of surface defects. It appears reasonable that the luminescence via surface states cannot be quenched by the increased concentration of defects in the bulk and may even increase due to a higher concentration of such states. More systematic investigations are necessary to analyze the details of such processes. However, this is a good example showing how sensitive photoeffects in semiconductors respond to structural changes of the surface and in the boundary layer underneath the interface.

Conclusions

Surface orientation and imperfections of the surface or the bulk expose very drastical influences on the photoelectrochemical behavior of semiconductor electrodes. This is very important for all applications of such systems, for example for the conversion

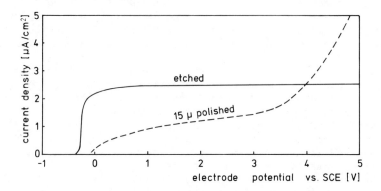

Figure 12. Photocurrent–voltage curves for a ZnO electrode after two different surface pretreatments (electrolyte: 1M KCl)

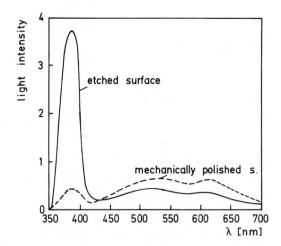

Figure 13. Luminescence from a ZnO electrode into which holes are injected by SO_4^- radicals (25) (spectra for two different surface pretreatments)

of solar energy (14,26,27). It appears that high efficiencies can only be reached with materials and surfaces where all the effects which favor recombination between electrons and holes can be minimized. No surface states should be present which can pick up charge of the same sign as that of the minority carriers since this would reduce the band bending under illumination by an increased voltage drop in the Helmholtz double layer and would diminish the photovoltage output accessible from such a contact. Besides this, these factors have also a very large influence on the corrosion reactions which could not be discussed here although they are extremely dependent on surface imperfections (16). The optical anisotropy of most semiconductor materials, which has been neglected nearly completely in electrochemical experiments until now, must also be taken into account.

Acknowledgement:

I want to thank Mrs. M. Lübke for performing the experiments with GaSe.

Literature Cited

1. e.g. Milnes, A. G.; Feucht, D. L. "Heterojunction and Metal Semiconductor Junctions", Academic Press: New York – London, 1972.

2. Gerischer, H., in "Light-Induced Charge Separation in Biology and Chemistry", Eds. H. Gerischer and J. J. Katz; Dahlem Konferenzen: Berlin, 1979; p. 61.

3. cf Gerischer, H.; Willig, F. "Topics in Current Chemistry", Vol. 61; Springer-Verlag: Berlin, 1976; p. 31-84.

4. Gärtner, W. W. Phys. Rev., 1959, 116, 84.

5. Myamlin, V. A.; Pleskov, Yu. V. "Electrochemistry of Semiconductors"; Plenum Press: New York, 1967.

6. Gerischer, H., in "Physical Chemistry", Vol. IX A: "Electrochemistry"; Eds. H. Eyring, D. Henderson, W. Jost; Academic Press: New York, 1970; p. 463-542.

7. Van den Berghe, R. A. L.; Cordon, F.; Gomes, W. P. Surf. Sci., 1973, 39, 368.

8. Lohmann, F. Ber. Bunsenges. Phys. Chem., 1966, 70, 87, 428.

9. Inoue, T.; Watanabe, T.; Fujishima, A.; Honda, K. Bull. Chem. Soc. Japan, 1979, 52, 1243.

10. Gerischer, H.; Hein, F.; Lübke, M.; Meyer, E.; Pettinger, B.; Schöppel, H. Ber. Bunsenges. Phys. Chem., 1973, 77, 284.

11. Tributsch, H., Z. Naturforsch., 1977, 32a, 972.

12. Tributsch, H., Ber. Bunsenges. Phys. Chem., 1978, 82, 169.

13. Tributsch, H., J. Electrochem. Soc., 1978, 125, 1086.

14. Gobrecht, J., Thesis, Technical University Berlin, 1979.

15. Kautek, W., Thesis, Technical University Berlin, 1980.

16. Kautek, W.; Gerischer, H.; Tributsch, H. Ber. Bunsenges. Phys. Chem., 1979, 83, 1000.

17. e.g. Greenaway, D. L.; Harbeke, G. "Optical Properties and Band Structure of Semiconductors", Pergamon Press: Oxford 1968.

18. Gerischer, H.; Gobrecht, J.; Turner, J. Ber. Bunsenges. Phys. Chem., 1980, 84, 596–601.

19. Akhundov, G. A.; Musaev, S. A.; Bakhyshev, A. E.; Musaev, L. G. Sov. Phys. Semicond., 1975, 9, 95.

20. Van den Berghe, R. A. L.; Gomes, W. P.; Cordon, F. Ber. Bunsenges. Phys. Chem., 1973, 77, 291.

21. Van Meirhaage, R. L.; Cordon, F.; Gomes, W. P. Ber. Bunsenges. Phys. Chem., 1979, 83, 236.

22. Gobrecht, H.; Bender, J.; Blaser, R.; Hein, F.; Schaldach, M.; Wagemann, H. G. Phys. Lett., 1965, 16, 132.

23. Bernhard, J.; Handler, P. Surf. Sci., 1973, 40, 141.

24. Beckmann, K. H.; Memming, R. J. Electrochem. Soc., 1969, 116, 368.

25. Pettinger, B; Schöppel, H. R.; Gerischer, H. Ber. Bunsenges. Phys. Chem., 1976, 80, 849.

26. Heller, A.; Chang, K. C.; Miller B. J. Electrochem. Soc., 1977, 124, 697.

27. Lewerenz, H. J.; Heller, A.; DiSalvo, F. J. J. Am. Chem. Soc., 1980, 102, 1877.

RECEIVED October 23, 1980.

Carrier Recombination at Steps in Surfaces of Layered Compound Photoelectrodes

H. J. LEWERENZ, A. HELLER, H. J. LEAMY, and S. D. FERRIS

Bell Laboratories, Murray Hill, NJ 07974

The performance of $n-WSe_2$ and $n-MoSe_2$ photoanodes differing in surface morphology has been investigated. A correlation between the smoothness of the surface as determined by scanning electron microscopy and solar conversion efficiency was found. With a smooth WSe_2 photoanode, a solar-to-electrical conversion efficiency of $\sim 5\%$ is reached in the $n-WSe_2/2MKI-0.05MI_2/C$ cell. Charge collection electron (EBIC) microscopy on $p-WSe_2$ shows that steps on the surface of layered semiconductors are recombination sites. The deleterious effect of steps is explained by deflection of minority carriers towards recombination sites at the edges of steps by an electric field component which parallels the layers.

Layered compound transition metal dichalcogenides gained recent interest as electrode material in semiconductor liquid junction solar cells (1-5). These materials show improved stability when used as photoanodes even in oxidizing solutions such as relatively non-toxic air stable cells, which do not require hermetic seals. The improvement in stability to photocorrosion has been attributed to the fact that excitation of the layered dichalcogenides involves transitions between hybridized metal d-bands, leaving the bonding of the illuminated semiconductors relatively unaffected (2,3). Review of the metal-metal distances which determine the d-d-interaction and hence also the d-d-splitting, suggests that WSe_2, $MoSe_2$ and WS_2 are likely to be particularly stable when used as photoanodes. Because of uncertainties even in the most basic data for these materials none can be ruled out or preferred in photoelectrochemical solar cells. Thus, for example, band gaps ranging from 1.16eV (6) to 1.56eV (7) have been reported for WSe_2. The differences in the data may be due to the methods of preparations of the different crystals (8,9). With respect to semiconductor liquid junction solar cells, the unknown parameters include flat band positions, photoelectrode kinetics and compatibility with various redox couples (10). Substantial differences in the performance of the various layered dichalcogenides in solar cells have been reported. Molybdenum dichalcogenides appeared to have higher short circuit currents, open circuit voltages and fill factors (11). The first investigation of $p-WSe_2$ revealed losses, of unknown origin, due to carrier recombination (12). The best solar conversion efficiency reached with this material was of 2% (under laboratory conditions).

This paper analyzes the causes of the losses in $n-WSe_2$ and $n-MoSe_2$ photoanodes and proposes ways to reduce these. Tungsten and molybdenum diselenide crystallize in the molybdenite structure with the space group symmetry D_{6h}^4-P_6/mmc (13, 14). The arrangement of the metal atoms with respect to each other is that of a close packed hexagonal layer. Each of these layers is surrounded by two adjacent close-packed planes containing the anions, so that the cations occupy trigonal prismatic sites. Strong covalent bonding is assumed within the layers (15, 16, 17) whereas the interlayer attractive forces are predominantly characterized by van der Waals interactions. This particular bonding results in the remarkable anisotropic properties of layered semiconductors. The peculiar features of the band structures of the transition layered metal compounds include the following (16):

(i) Substantially split bonding and antibonding orbitals result from the strong covalent bond between metal and chalcogen s and p orbitals. The resulting energy separation of the respective bands is relatively large (16).

(ii) The interaction between the metal d- and chalcogen sp-orbitals is weaker and causes the d-bands to be located within the metal-chalcogen sp-gap.

(iii) The d_{z^2} and the $d_{x^2-y^2}$, d_{xy} type subbands are strongly hybridized. The hybridization produces an energy gap within the d-bands. Theoretical calculations as well as experimental data show that the highest occupied band has d_{z^2}-character (15, 16, 18) whereas the four other d-subbands are unoccupied. The $n-WSe_2$ band structure is presented schematically in Figure 1. A band gap of 1.35eV is assumed (8).

Experimental

Single crystals of $n-WSe_2$ and $n-MoSe_2$ were grown using chlorine transport; for $p-WSe_2$, $TeCl_4$ was used as transport agent (19). The doping of these samples was approximately the same ($\sim 8 \times 10^{16}cm^{-3}$). The area of the crystals varied between $0.2cm^2$ and $0.5cm^2$ and their thickness from 0.1mm to 0.5mm. To avoid differences in doping or composition all samples studied were selected from the same batch of crystals. The mounting procedure of the WSe_2 and $MoSe_2$ photoanodes has been described earlier (20). The electrochemical experiments were performed in the normal three-electrode potentiostatic configuration. All potentials are referred to that of the I^-/I_3^- redox couple. Except in solar efficiency measurements, which were performed in sunlight, a 150W tungsten iodine lamp (Oriel Corp.) was used. All solutions were prepared from deionized water with analytical grade chemicals.

The Correlation Between Surface Morphology and Solar Cell Performance of WSe_2 and $MoSe_2$ Photoanodes

Initial inspection of the as-grown crystals, revealed substantial variations in shape, thickness and crystal morphology. Examination with a light microscope showed large differences in surface topography, which were particularly pronounced in WSe_2 samples. The main features, found on the surfaces were:

(a) crack-like steps

(b) deep ruptures

(c) heaps of small, thin crystallites lying on the surface

(d) very small terraces associated with growth spirals

(e) smooth but curved (concave or convex) surfaces

(f) smooth planar single crystalline surfaces of hexagonal shape

In order to study the influence of surface morphology on the photoanode performance, samples differing in surface structure have been selected. Electron micrographs of representative areas of three different $n-WSe_2$ electrodes are shown in Fig. 2. Substantial differences in surface structure ranging from nearly smooth (Fig. 2a) through moderately structured (Fig. 2b) to extremely high structured (Fig. 2c) are noted.

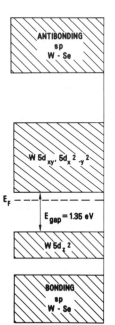

Figure 1. Schematic of the band structure of n-WSe₂ assuming a band gap of 1.35 eV (8)

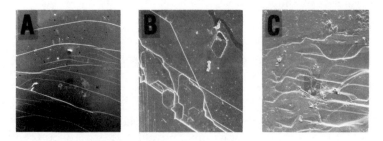

Figure 2. Electron micrographs of representative areas of three n-WSe₂ samples

In order to obtain a semiquantitative estimate of the respective amount of the non-horizontal surface area on each sample, we counted the height and length of the dislocations, which were mostly step-like. This was done by (a) taking electron micrographs at two different angles, thus obtaining a stereoscopic view of the surface and (b) by estimating step-heights with "defocusing" under a light microscope. Since the surface structure is to a large extent semi-macroscopic, we only considered step heights larger than $1\mu m$. The non-horizontal surface area was estimated to be 5% of the horizontal area in case of sample A (Fig. 2a) $\sim$20% for sample B (Fig. 2b) and $\sim$80% in case of sample C (Fig. 2c) which had particularly high steps. The current voltage curves corresponding to these samples, and measured in the $n-WSe_2/2MKI-50mMI_2/C$ cell are shown in Fig. 3. Drastic differences in short circuit current, open circuit voltage and fill factors are found. The smoothest electrode, sample A, shows the highest values whereas the most structured sample C shows the lowest ones. Sample A exhibits an open circuit voltage of 0.63V and a fill factor of 0.37 with the maximum power point at 0.35V. The fill factors of the approximately four times more structured sample B is 0.25, the open circuit voltage is 0.5V and the short circuit current is down by a factor of 2.5. The fill factor of sample C is only 0.23, the open circuit voltage is 0.39V and the short circuit current is one third of that of sample A.

Practically the same correlation is found on $n-MoSe_2$ electrodes. The electron micrographs of representative areas of two samples, differing in surface morphology, are shown in Fig. 4. Sample A, although not completely smooth, reveals considerably less surface structure than sample B. The resulting solar cell performance of these electrodes is shown in Fig. 5. The smoother photoanode (Fig. 4a) shows substantially higher values of fill factor, open circuit voltage and short circuit current than the more structured one (Fig. 4b).

After the results established a correlation between surface structure and cell performance in $n-WSe_2$ and $n-MoSe_2$ electrodes, crystals with smooth surfaces were selected to build semiconductor liquid junction solar cells of increased efficiency. The electron micrographs of these $n-WSe_2$ samples, labeled D, E and H are shown in Fig. 6. Their surfaces reveal almost no structure. Current voltage curves of these and additional samples are displayed in Fig. 7. With the best these anodes, labelled D-H, have open circuit voltages approaching 0.6V and exhibit fill factors as high as 0.55. The short circuit current of sample F, with some surface structure, is slightly lower than that of the other electrodes.

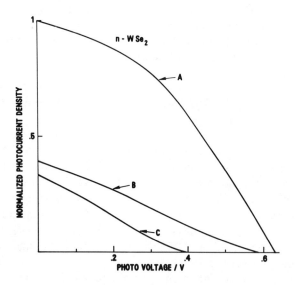

Figure 3.   Current–voltage characteristics of the samples shown in Figure 2 in 2M KI–0.05M I_2 solution (the solution was exposed to air)

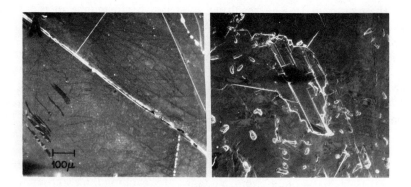

Figure 4.   Electron micrographs of representative areas of two n- *$MoSe_2$ samples*

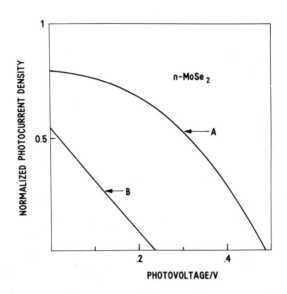

Figure 5. Current–voltage characteristics of the samples shown in Figure 4 in 2M KI–0.5M I₂ solution (air)

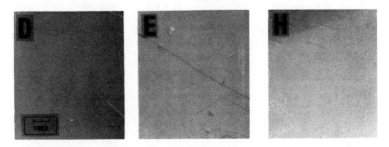

Figure 6. Electron micrographs of representative areas of three smooth n-WSe₂ samples

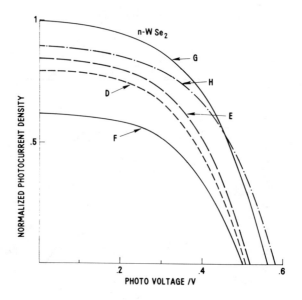

Figure 7.   Current–voltage characteristics of the samples shown in Figure 6 (D, E, H).   Also shown are the current–voltage characteristics of additional smooth samples F and G; solution: 2M KI–0.05M I_2 (air).

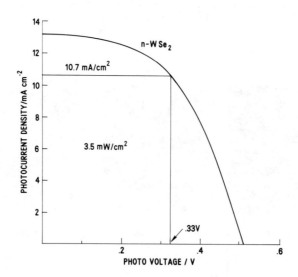

Figure 8.   Current–voltage characteristic of photoanode D (see also Figures 6 and 7) under 92.5 mW/cm² insolation in the n-WSe_2/2M KI–0.05M I_2/C cell

In order to get an estimate of the solar-to-electrical conversion efficiency on layered compounds, sample D has been measured in sunlight. The result, obtained at $92.5 mW/cm^2$ insolation is shown in Fig. 8. The maximum power point is at 0.33V and $10.7 mA/cm^2$, with a resulting solar conversion efficiency of 3.7%. As is evident from Fig. 7, some samples show better overall performance than sample D. The best of these, sample G, the surface of which was accidentally damaged before being measured in the sun, had a maximum power output which exceeded that of sample D by a factor of 1.4 bringing the estimated solar conversion efficiency to 5.2%.

The existence of recombination sites at or near the surface of a semiconductor is known to affect the short circuit current spectra in three respects: first, the current efficiency decreases at all wavelengths; second, the loss being greatest at short wavelengths; third, under intense illumination (with a laser beam) the quantum efficiency declines further (21,22). In Fig. 9, the photocurrent spectra of two parts of the same $n-WSe_2$ electrode are shown. Spectra measured on a relatively smooth part of the sample are displayed in Fig. 9a. The spectra of a highly structured part are shown in Fig. 9b. It is seen that the current from the structured surface is lower for all wavelengths and a drastic drop of current efficiency under intense illumination ("laser on") is noted. No such behavior is observed in Fig. 9a on the smoother part of the sample. The spectra with the "laser on" in Fig. 9 were obtained at an irradiance of $2 mW/cm^2$ at 6328Å with a He-Ne laser. Comparison of the current efficiencies of the smoother and the highly structured parts of the crystal shows a decline of the current by a factor of 4 under low level illumination through a monochromator. Additional illumination with laser light causes a further decrease by a factor of 5, thus reducing the quantum efficiency to 5% of that of the smooth part of the sample.

Identification of Recombination Sites by EBIC

A powerful method for the characterization of the origin of losses on structured surfaces is provided by the EBIC (Electron Beam Induced Current) technique also known as charge collection microscopy (23, 24).

A typical EBIC experiment is shown schematically in Fig. 10. A Schottky barrier is formed between a semiconductor and a metal. An electron beam impinges through the metal and creates electron hole pairs in the depletion region of the semiconductor. The charge carriers are separated by the electric field in the space charge region and are detected as collected current I_c in the external circuit. In the presence of recom-

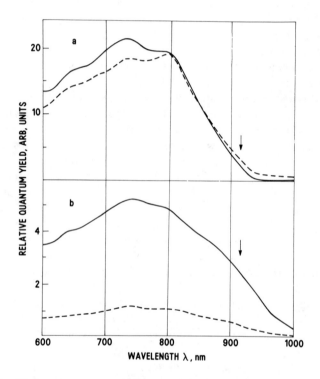

Figure 9.   Photocurrent spectra of two parts of the same n-WSe_2 *electrode: (a) relatively smooth part and (b) heavily structured part. The arrows indicate the assumed band edge; (———) the relative quantum yields at very low levels of illumination; (– – –) the relative quantum yields when the sample is illuminated also by a superimposed laser beam of 2 mW/cm² at 6328 Å. The spectra are corrected for the response of the system.*

bination sites, I_c is decreased, and by scanning the electron beam across the surface, one obtains information on the location and effectiveness of recombination sites.

We utilized this technique on a structured $p-WSe_2$ sample (25) (good quality Schottky barriers with n-type materials were difficult to form). The Schottky junction was made by evaporation of 500Å Al onto the $p-WSe_2$ surface (26). An ohmic back contact was obtained by evaporating 300Å of Zn followed by 1000Å of Au onto the back of the sample. Figs. 2 and 4 reveal that the surfaces of layered compounds are often terraced. We have therefore investigated the charge collection efficiency on a part of the WSe_2 sample which showed very fine steps under the scanning electron microscope. Fig. 11a is an electron micrograph of a typical stepped surface and Fig. 11b shows the corresponding EBIC-picture. In the latter, high charge collection efficiency is indicated by dark areas and poor efficiency by light areas. Fig. 11 shows very poor collection efficiency at and near the edges of steps and provides direct proof that steps act as recombination sites. The effect of steps on the collection efficiency can be more quantitatively displayed if I_c is recorded during a single lateral scan of the electron beam. Fig. 12 shows a typical result, obtained on another part of the sample. Clearly, the collection current I_c drops whenever the electron beam scans across a step, demonstrating further the recombination at these.

Discussion

The results in Section 3 and 4 demonstrate that steps on the surface of layered semiconductors are recombination sites, and hence predominantly responsible for the poor cell performance of structured electrodes. We therefore proceed in examining the role of steps in layered compounds.

First, we consider the ideal case of an atomically smooth sample as shown in Fig. 13 for WSe_2. Here, we have assumed an average absorption coefficient of $1 \cdot 10^5 cm^{-1}$ in the energy range above 1.5 eV (7) hence the light intensity I_o drops to I_o/e within 1000Å. With a static dielectric function ϵ (perpendicular to the layer structure) of 9(27), a barrier height of 0.6V and a carrier concentration of $8 \cdot 10^{16} cm^{-3}$, a space charge layer thickness of 860 Å is obtained in the depletion approximation. Thus within the accuracy of the above approximation, the absorption length L_α and the depletion layer width W have very similar values (as shown in Fig. 13) indicating that most of the incoming light is absorbed in the space charge region.

The absorption of light results in creation of electron hole pairs, predominantly in the depletion region, and is then followed, as usual, by transport of minority carriers to the surface. In layered compounds, however, mobility within the layers is high, while mobility perpendicular

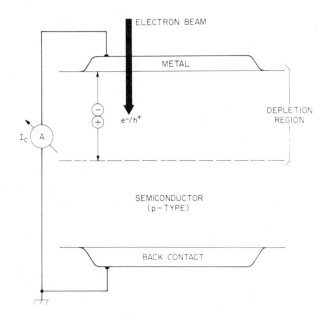

Figure 10. Schematic of an EBIC experiment with a p-type semiconductor as photoactive part (I_c denotes the charge collection current)

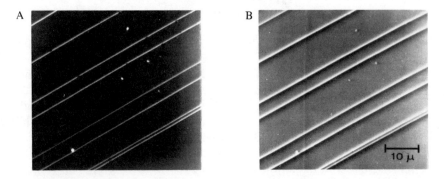

Figure 11. (a) Electron micrograph of a stepped part of a p-WSe$_2$ photocathode; (b) EBIC picture of the same part of the p-WSe$_2$ sample; the charge collection efficiency is lowest for light and highest for dark areas.

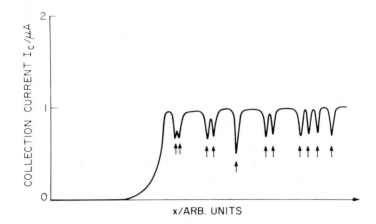

Figure 12. Collection current of an EBIC experiment obtained by scanning the electron beam across a stepped surface of a p-WSe₂ sample (x-direction); beam voltage 20 kV, beam current 6.5 × 10⁻¹⁰ A. The arrows indicate the position of the steps.

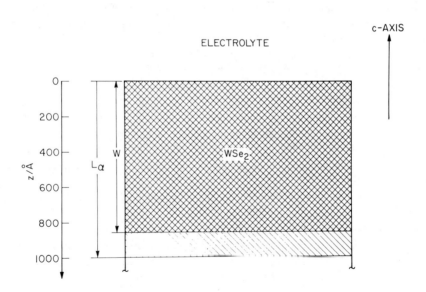

Figure 13. Absorption length L_α and depletion layer width W of an ideally smooth WSe₂ electrode in contact with an electrolyte, assuming $\epsilon_\perp = 9$ and $n_c = 8 \times 10^6$ cm⁻³

to the layers is low. The charge transport in the perpendicular direction is related to randomly distributed stacking faults, which interconnect the layers, introducing an extrinsic conduction mechanism (28). The presence of stacking faults leaves the translational invariance parallel to the layers unaffected whereas it destroys translational invariance perpendicular to them. Thus, layered semiconductors may be viewed as one dimensional disordered systems with extrinsic conduction paths parallel to the c-axis.

The pronounced anisotropy of the electrical conductivity in layered compounds (8,15) suggests that the charge carriers move, on their way to the surface, predominantly within the layers, i.e. parallel to the main surface as shown in Fig. 14 (in which the relative path of the charge carriers within the layers is compressed). The random character of interlayer charge transport due to extrinsic conduction leads to a variety of possible paths, two of which are represented in Fig. 14.

The presence of a step introduces two significant changes. First, in contrast to a smooth surface which does not exhibit unsaturated bonds in the direction perpendicular to the surface, strong covalent bonds are exposed to the electrolyte at the edge of a step. These are likely to chemisorb species from the environment and introduce surface states within the band gap (29). These are recognized recombination sites and shunts in solar cells, known to reduce fill factors, current collection efficiencies and open circuit voltages (30, 31).

Second, the exposure of a step to the conductive electrolyte results in a space charge region parallel to the layers. The depth of this region is given by the differences in the static dielectric constant parallel and perpendicular to the layers. For WSe_2, $\epsilon_\| = 16$ and $\epsilon_\perp = 9(27)$, hence in case of homogenous doping, the width of the depletion region parallel to the layered structure, $W_\|$, is increased approximately by a factor of $\sqrt{2}$ compared with $W_\perp = 860\text{Å}$. In case of $MoSe_2$, this anisotropy in W is somewhat larger, since $\epsilon_\perp = 4.9$ and $\epsilon_\| = 20(32)$, but also not particularly substantial.

We will consider two physically different situations: (a) the height d of the step is comparable with the extension of the depletion layer (d ~W) and (b) the step is much higher than the depletion width (d >> W). In the first case, charge carriers are generated in a region of the sample in which two electric fields exist: one parallel, the other perpendicular to the layers. Minority carriers, created within the depletion region $W_\|$ will drift towards the edge of the step due to the acting electric field and the highly anisotropic conductivity. A step can thus be regarded as a collector of minority carriers which otherwise would have reached the surface that parallels the layers in the semiconductor. Such deflection of carriers is shown schematically in Fig. 15. Once the minor-

ELECTROLYTE

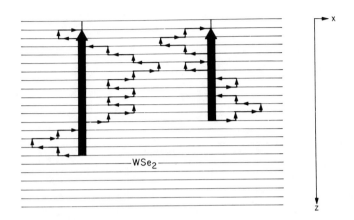

Figure 14.   Schematic of trajectories of minority carriers for an ideally smooth surface in contact with an electrolyte. The random character of the interlayer charge transport is also indicated (small arrows). The average minority carrier motion is given by the large arrows.

ELECTROLYTE

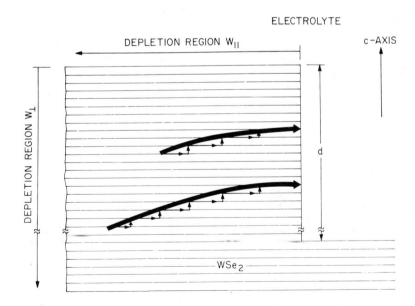

Figure 15.   Schematic of trajectories of minority carriers for a surface with a step exposed to the electrolyte (the step height d is assumed to equalize approximately the depletion layer width)

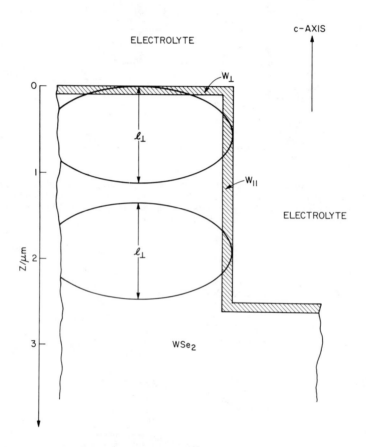

Figure 16.   Schematic of the influence of steps on diffusion processes in case d $\gg$ W. *The lined areas indicate the extension of the depletion layer parallel or perpendicular to the layered structure,* $W_{||}$ *and* $W\perp$, *respectively, and* $1_\perp$ *denotes the minority carrier diffusion length perpendicular to the layered structure (the horizontal radius of the ellipses is compressed somewhat).*

ity carriers reach the edge of a step, most of them will undergo recombination due to the high density of surface states.

In case d $>> W$, an electric field and a space charge layer parallel to the layer structure is present in a region, in which light absorption is negligible and no field perpendicular to the layers exists. However, the efficiency of solar cells is a function of the minority carrier diffusion length (33, 34). In WSe_2, evidence has been obtained, that this length $l_\perp$ is in the order of 10^{-4} cm.(6, 25) The presence of a step associated with an electric field parallel to the layers leads to a deflection of diffusing minority carriers towards the edge where recombination occurs as schematically indicated in Fig. 16. The diffusion coefficient D is proportional to the carrier mobility μ, which is highly anisotropic in layered semiconductors thus the diffusion profile is elliptic as shown in the figure. Hence, a step will also reduce the effective minority carrier diffusion length and lifetime, causing further deterioration in solar cell performance.

In summary, it has been demonstrated that surface morphology is critically important in determining the performance of solar cells with layered compound semiconductors. Steps on structured surfaces of transition metal dichalcogenides have been identified as carrier recombination sites. The region defined by the depth of the space charge layer parallel to the van der Waals planes can be considered as essentially "dead" in the sense that its photoresponse is negligible. As the "step model" predicts, marked improvement in solar cell performance is found on samples with smooth surfaces.

Two conclusions concerning future application of layered compounds in solar energy converting devices can be derived from the present investigation:

(i) the growth of large area smooth samples is highly desirable

(ii) a method to reduce the disadvantageous effects of steps on surfaces of layered semiconductors is needed

If neither of these goals can be realized, layered semiconductors may not become useful electrode material in either semiconductor liquid junction or Schottky junction devices. Fortunately, evidence is already being obtained that the negative effects due to steps can be at least temporarily and partially alleviated (35, 36). Future development of chemical methods to inhibit deflection of minority carriers to the edges of steps and to reduce the high recombination rates at steps may open the way for the use of polycrystalline layered chalcogenide semiconductors in solar cell devices.

REFERENCES

(1) Gerischer, H. *Electroanal. Chem. and Interfac. Electrochem.* (1975), *58*, 263.

(2) Tributsch, H. *Z. Naturforsch.* (1977), *32a*, 972.

(3) Tributsch, H. *Ber. Buns. Ges. Phys.Chem.* (1977), *81*, 361.

(4) Tributsch, H. *Ber. Buns. Ges. Phys. Chem.* (1978), *82*, 169.

(5) Tributsch, H. *J. Electrochem. Soc.* (1978), *125*, 1086.

(6) Kautek, W., Gerischer, H., Tributsch, H. *J. Electrochem. Soc.* in press.

(7) Frindt, R. F., *J. Phys. Chem. Solids,* (1963), *24*, 1107.

(8) Upadhyayula, L. C., Loferski, J. J., Wold., A., Giriat, W., Kershaw, R., *J. Appl. Phys.* (1968), *39*, (10), 4736.

(9) Levy, F., Schmid, Ph. Berger, H. *Phil. Mag.* (1976), *34*, 1129.

(10) Menezes, S., DiSalvo, F. J. Miller, B., *J. Electrochem. Soc.* (1978), *125*, 2085.

(11) Gobrecht, J., Tributsch, H., Gerischer, H., J. Electrochem. Soc. (1978), *125*, 2085.

(12) Gobrecht, J., Gerischer, H., Tributsch, H. *Ber. Buns. Ges. Phys. Chem.,* (1978), *82*, 1331.

(13) Kershaw, R., Vlasse, M., Wold, A., *Inorganic Chem.* (1967), *6*, (8), 1599.

(14) Monzack, J., Richamn, M. H., *Metallography* (1972), *5*, 279.

(15) Wilson, J. A., Yoffe, A. D., *Adv. Phys.* (1969), *18*, 193.

(16) Mattheis, L. F., *Phys. Rev. B* (1973), *8*, (8), 3729.

(17) White, R. M., Lucovsky, G. *Sol. St. Comm.,* (1972), *11*, 1369.

(18) Title, R. S., Shafer, M. W., *Phys. Rev. Lett.* (1972), *28*, (13), 808.

(19) Schaefer, H. in "Chemical Treansport Reactions", Academic Press, (1964).

(20) Lewerenz, H. J., Heller, A., DiSalvo, F. J., *J. Am. Chem. Soc.,* (1980), *102*, 1877.

(21) Heller, A., Chang, K. C., Miller, B., *J. Am. Chem. Soc.* (1978), *100*, 684.

(22) Chang, K. C., Heller, A., Miller, B., *J. Electrochem. Soc.* (1977), *124*, (5), 697.

(23) Leamy, H. J., Kimerling, L. C., Ferris, S. D., in "Scanning Electron Microscopy" (1976), Part IV, Vol. 1, ed. Johari, O., (IIT Res. Inst., Chicago Ill), p. 529.

(24) Wu, C. J., Wittry, D. B. *J. Appl. Phys.* (1978), *49*, 2827.

(25) Lewerenz, H. J., Ferris, S. D., Leamy, H. J., Doherty, C. J., to be published.

(26) Clemen, C., Saldana, X. I., Munz, P., Bucher, E., *Phys. Stat. Sol.* (1978), A47, 437.

(27) Zeppenfeld, K. *Opt. Comm.* (1970), *1*, 377.

(28) Fivaz, R., Schmid, Ph. E., in "Physics and Chemistry of Materials with Layered Structures", Vol. 4, ed. P. A. Lee, D. Reidel, Holland (1976).

(29) Ahmed, S. M., Gerischer, H., *Electrochimica Acta,* (1979), *24*, 705.

(30) Parkinson, B. A., Heller, A., Miller, B., *J. Appl. Phys.* (1978), *33*, 521.

(31) Parkinson, B. A., Heller, A., Miller, B., *J. Electrochem. Soc.,* (1979), *126*, 954.

(32) Neville, R. A., Evans, B. L. *Phys. Stat. Sol.* (1976), *B73*, 597.

(33) Gosh, A. K., Maruska, H. P. *J. Electrochem. Soc.* (1977), *124*, 1516.

(34) Graff, K., Fischer, H. *Top. Appl. Phys.* (1979), *31*, 173.

(35) Parkinson, B. A., Furtak, T. E., Canfield, D., Kam, K., Kline, G., to be published.

(36) Lewerenz, H. J., Heller, A., Gallagher, P. K., Menezes, S., Miller, B., to be published.

RECEIVED October 23, 1980.

Rate of Reduction of Photogenerated, Surface-Confined Ferricenium by Solution Reductants

Derivatized *n*-Type Silicon Photoanode-Based Photoelectrochemical Cells

NATHAN S. LEWIS and MARK S. WRIGHTON[1]

Department of Chemistry, Massachusetts Institute of Technology, Cambridge, MA 02139

N-type semiconductors can be used as photoanodes in electrochemical cells (1, 2, 3), but photoanodic decomposition of the photoelectrode often competes with the desired anodic process (1, 4, 5). When photoanodic decomposition of the electrode does compete, the utility of the photoelectrochemical device is limited by the photoelectrode decomposition. In a number of instances redox additives, A, have proven to be photooxidized at n-type semiconductors with essentially 100% current efficiency (1, 2, 3, 6, 7, 8, 9). Research in this laboratory has shown that immobilization of A onto the photo-anode surface may be an approach to stabilization of the photoanode when the desired chemistry is photooxidation of a solution species B, where oxidation of B is not able to directly compete with the anodic decomposition of the "naked" (non-derivatized) photoanode (10, 11, 12). Photoanodes derivatized with a redox reagent A can effect oxidation of solution species B according to the sequence represented by equations (1) – (3) (10–15).

$$\text{n-type semiconductor} \xrightarrow{\text{h}\nu} \text{e}^- + \text{h}^+ \tag{1}$$

$$A_{\text{surf.}} + \text{h}^+ \xrightarrow{k_e} A^+_{\text{surf.}} \tag{2}$$

$$A^+_{\text{surf.}} + B \xrightarrow{k_{et}} A_{\text{surf.}} + B^+ \tag{3}$$

Thus, $A_{\text{surf.}}$ is oxidized by the photogenerated h^+, and the photogenerated $A^+_{\text{surf.}}$ in turn oxidizes B. By such a mechanism, the photooxidation of B is possible for wavelengths of light that will create e^- – h^+ pairs in the n-type semiconductor ($\geq E_g$) and for processes where the chemistry represented by equation (3) is spontaneous in a thermodynamic sense. The $(A^+/A)_{\text{surf.}}$ reagent system must also result in a suppression of the anodic decomposition, equation (4), in order to achieve

$$\text{n-type semiconductor} + \text{h}^+ \xrightarrow[\text{solvent}]{} \text{decomposition products} \tag{4}$$

[1] Address correspondence to this author.

0097-6156/81/0146-0037$05.00/0

a durable interface for photooxidation when the naked electrode undergoes photoanodic decomposition in the presence of B.

Large values of k_{et} are needed to achieve photoanode-based cells having high efficiency. Measurement of k_{et} can be made directly, since derivatized photoanodes are two stimuli response systems (10–16). Oxidation of $A_{surf.}$ requires $\geq E_g$ light and some electrode potential, E_f. However, reduction of $A_{surf.}^+$ on the n-type semiconductor only requires a sufficiently negative potential. This two stimuli response (light and potential) allows evaluation of the rate constant k_{et} by measuring the time dependence of the $A_{surf.}^+$ concentration in the dark in the presence of B (16). The $A_{surf.}^+$ concentration can be monitored by either a negative sweep or step from the positive potential needed in the photogeneration of $A_{surf.}^+$, equation (5), and

$$ e^- + A_{surf.}^+ \xrightarrow{\quad k_5 \quad} A_{surf.} \tag{5} $$

integrating the charge passed in the reduction of $A_{surf.}^+$. Separation of the charge passed corresponding to reduction of $A_{surf.}^+$ from that associated with reduction of B^+ can be accomplished by moving B^+ away from the photoanode by stirring. The key fact is that even though the positive limit may be more positive than $E^o(A^+/A)_{surf.}$, regeneration of $A_{surf.}^+$ after reaction according to equation (3), will not occur in the dark. Thus, the consumption of $A_{surf.}^+$ by B is detectable by the decline in $A_{surf.}^+$ concentration (16) measured by cathodic current associated with its reduction after a reaction time t_i at specified conditions. Direct monitoring of $A_{surf.}^+$ concentration in this sense is not possible on a reversible electrode, such as Pt or Au, since the $(A^+/A)_{surf.}$ ratio depends only on E_f. If $A_{surf.}^+$ does effect chemistry according to equation (3), the $A_{surf.}^+$ is regenerated to an extent that depends only on E_f (16). However, indirect procedures for evaluating k_{et} do exist for reversible electrodes, particularly when the oxidation of B at the naked electrode occurs at a negligible rate compared to the rate at an electrode derivatized with the $(A^+/A)_{surf.}$ system (17).

In photoelectrochemical cells high efficiency depends on having a high quantum yield for electron flow, Φ_e, at all light intensities to be used. If k_{et} is too small, Φ_e may be less than unity because back electron transfer, equation (5), can compete when E_f is sufficiently negative. Negative values of E_f are desirable, since the extent to which $A_{surf.} \to A_{surf.}^+$ can be driven uphill with light, and here $B \to B^+$, depends on E_f (1). Further, if k_{et} is too small, direct $e^- - h^+$ recombination, equation (6),

$$ e^- - h^+ \xrightarrow{\quad k_6 \quad} \text{heat and/or light} \tag{6} $$

may occur when the $A_{surf.} \to A_{surf.}^+$ conversion is complete. When k_{et} is large, back electron transfer and recombination can still be competitive if the concentration of B is too low. If B^+ is present back reaction, equation (7), can contribute to

$$ A_{surf.} + B^+ \xrightarrow{\quad k_7 \quad} A_{surf.}^+ + B \tag{7} $$

a low value of Φ_e. Light intensity is a consideration for two reasons. First, the rate of $e^- - h^+$ recombination is a "bimolecular" process whereas the other e^- or h^+ processes are "unimolecular"; Φ_e might be lower at higher light intensity. Second, when B is being consumed according to equation (3) it can only be replenished at the interface at a mass transport controlled rate; the excitation rate can exceed the mass transport rate resulting in a low steady-state value of Φ_e.

In this article we wish to amplify on our previous studies (10-16) of derivatized photoanode surfaces by reporting new results related to the measurement of k_{et} for n-type Si photoanodes derivatized with (1,1'-ferrocenediyl)dichlorosilane, I. We report that a number of solution reductants B can be oxidized

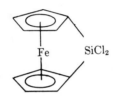

I
~

by the photogenerated, surface-confined ferricenium with a value of k_{et} that exceeds 6×10^8 cm^3 mol^{-1} s^{-1} at 298 K. Larger values of k_{et} cannot be measured owing to difficulties associated with mass transport controlled rates. This would correspond to a homogeneous bimolecular rate constant of $>6 \times 10^5$ M^{-1} s^{-1}. In practical terms this means that k_{et} is large enough to yield a good value of Φ_e at solar irradiation intensities and at generally accessible concentrations of B. However, the extent to which the oxidation of B can be driven uphill, E_V, is generally modest (0.4 - 0.5 V at open-circuit) compared to $E_g = 1.1$ eV for Si. Small values of E_V give low overall optical energy conversion efficiency.

Background and Working Hypothesis

N-type Si derivatized with I is believed to have the interface structure and energetics represented by Scheme I (11). Taking $E^o(FeCp_2^{+/0})_{surf.}$ to be +0.43 V vs. SCE in EtOH from measurements for Au or Pt electrodes derivatized with I, (18, 19) E_V for the $(FeCp_2^0)_{surf.} \rightarrow (FeCp_2^+)_{surf.}$ oxidation can be up to ~0.60 V (10). That is, the value of E_f of the photoanode where the $(FeCp_2^{+/0})_{surf.}$ ratio is one is ~0.60 V more negative than for Au or Pt in the best cases. More typically, the value of E_V is 0.3-0.4 V as shown in Figure 1 for electrodes (Pt vs. illuminated n-Si) characterized by cyclic voltammetry.

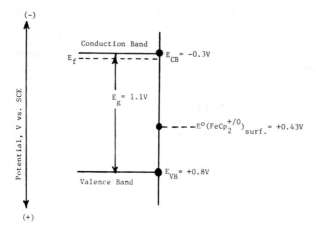

<u>Scheme I</u>   Interface energetics for an n-type Si photoanode at the
          flat-band condition showing the formal potential for a
          surface-confined ferricenium/ferrocene reagent relative
          to the position of the top of the valence band, E_{VB}, and
          the bottom of the conduction band, E_{CB}, at the interface
          between the Si substrate and the redox/electrolyte system.
          Interface energetics apply to an EtOH/0.1 $\underline{M}$ [n-Bu$_4$N]ClO$_4$
          electrolyte system.

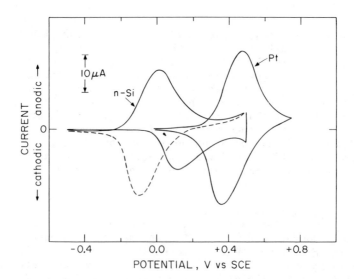

*Figure 1.   Typical comparison of cyclic voltammetry (100 mV/s) for Pt vs. n-type
Si in CH$_3$CN/0.1M [n-Bu$_4$N]ClO$_4$ derivatized with (1,1'-ferrocenediyl)dichloro-
silane.*

The Pt exhibits a reversible wave in the dark whereas the n-type Si exhibits no oxidation
current unless illuminated with $\geq E_g$ light (632.8 nm, $\sim$ 50 mW/cm^2). The photoanodic
peak is more negative than the anodic peak on Pt, reflecting the extent to which ferrocene
$\rightarrow$ ferricenium can be driven uphill (380 mV in this case). For n-Si, (– – –) represents the
reduction when the light is not switched off at the +0.6 V positive limit; (———) on the
cathodic sweep corresponds to the dark reduction.

The main point is that at a photoanode potential of ~0.4 V vs. SCE the $(FeCp_2^{+/0})_{surf.}$ ratio is typically >10 and the available oxidant is $(FeCp_2^+)_{surf.}$. Thus, any material B that is oxidizable with $(FeCp_2^+)$ in solution should be oxidizable with the n-type Si photoanode derivatized with I (10-16). N-type Si derivatized with I can be used in H_2O/electrolyte solution, unlike the naked n-type Si that is rapidly passivated by photoanodic growth of an oxide layer on the surface (10, 11, 12, 16).

As a guide to understanding the heterogeneous oxidation of B by $(FeCp_2^+)_{surf.}$, we adopt the theory of Marcus (20, 21). However, we underscore the fact that redox reaction of the surface-confined redox system, like the solution system, is accompanied by solvation changes (16). Heterogeneous electron transfer at a naked electrode need not involve an electrode solvation term. Qualitatively, Marcus predicts that k_{et} will be large when the (B^+/B) and $FeCp_2^{+/0})$ self-exchange are fast and when the driving force is large.

Results

Solvent Dependence of I⁻ Oxidation

In an earlier study, (16), we found the rate law represented by equation (8) to be appropriate for B=I⁻ in EtOH or H_2O solvent.

$$\text{Rate} \quad = \quad k_{et} \quad [(FeCp_2^+)_{surf.}][B] \tag{8}$$

Units Analysis: Rate, $molcm^{-2}s^{-1}$; k_{et}, $cm^3mol^{-1}s^{-1}$; $[(FeCp_2^+)_{surf.}]$, $molcm^{-2}$; $[B]$, $molcm^{-3}$

We assume the same rate law to generally apply when B is a one-electron reductant. The ability to prove the rate law for B=I⁻ stems from a rather low value of k_{et}, Table I. Interest in the photochemical oxidation of I⁻ to I_3^- for energy storage purposes and the ~10-fold difference in k_{et} in EtOH vs. H_2O prompted us to determine k_{et} for B=I⁻ for several other solvents. Values of k_{et}, $E_{1/4}(I^-_{oxdn})$, and $E^0(FeCp_2^{+/0})_{surf.}$ in the various solvents used are given in Table I.

The $E_{1/4}(I^-_{oxdn.})$ values are data from the literature (22) and our own measurements, and the $E^0(FeCp_2^{+/0})_{surf.}$ are from the cyclic voltammetry peak positions for Pt electrodes functionalized with I in the various solvents. Together, these data provide information concerning the driving force for the oxidation of I⁻ by $(FeCp_2^+)_{surf.}$ in the various solvents.

Values of k_{et} were determined as previously reported for H_2O or EtOH solvent (16). The derivatized electrode is first characterized by cyclic voltammetry in solvent/electrolyte solution without added I⁻. The value of k_{et} is then determined from the time dependence of the surface-concentration of $(FeCp_2^+)_{surf.}$ in the presence of variable I⁻ concentration and as a function of solvent. The $(FeCp_2^+)_{surf.}$ is generated in a linear potential sweep from -0.6 to +0.5 V vs. SCE while the

Table I.   Reduction of Surface-Confined Ferricenium by Iodide
in Various Solvents at 298 K.

Potential, V vs. SCE

Solvent/Electrolyte	$E^0(FeCp_2^{+/0})^a$	$E_{\frac{1}{4}}(I^-)^b$	k_{et}, $cm^3 mol^{-1} s^{-1 c}$
EtOH /[$\underline{n}$-Bu$_4$N]ClO$_4$	0.43	0.42	$3 \times 10^{4\,d}$
H$_2$O (pH=1.3)/NaClO$_4$	0.35	0.30	$1 \times 10^{3\,d}$
Glacial Acetic Acid/– [$\underline{n}$-Bu$_4$N]ClO$_4$	0.42	0.42	3×10^4
EtOH/Toluene (1/1)/– [$\underline{n}$-Bu$_4$N]ClO$_4$	0.44	0.44	3×10^4
CH$_2$Cl$_2$/[$\underline{n}$-Bu$_4$N]ClO$_4$	0.48	0.40	6×10^4
CH$_3$CN/[$\underline{n}$-Bu$_4$N]ClO$_4$	0.50	0.32	$>6 \times 10^{8\,e}$

[a] Formal potential for surface-confined $(FeCp^{+/0})$ as determined by
slow sweep cyclic voltammetry for Pt electrodes derivatized with
(1,1'-ferrocenediyl)dichlorosilane.

[b] Data are quarter wave potentials for I^- oxidation reported in
ref. 22 and measured at Pt, see Experimental.

[c] Heterogeneous electron transfer rate constant, see equation (8)
in text, for I^- oxidation determined as in ref. 16.

[d] Data from ref. 16.

[e] Assuming $[(FeCp_2^+)_{surf.}] = 10^{-10}$ mol/cm^2.

electrode is illuminated at a sufficiently high 632.8 nm light
intensity (~50 mW/cm^2) to insure that the $(FeCp_2^+/0)_{surf.}$ ratio
at +0.5 V vs. SCE is >10/1. The light is then switched off at
the positive limit. Concentration of $(FeCp_2^+)_{surf.}$ vs. time, t_i,
in the dark is then determined by holding the potential at +0.5 V
vs. SCE for a time, t_i, and then sweeping the potential at
>300 mV/s to -0.6 V vs. SCE while monitoring cathodic current
corresponding to $(FeCp_2^+)_{surf.}$ reduction. Another way to vary
reaction time, t_i, is to simply use no delay at the positive
limit and to vary the sweep rate on the negative scan. This
method has been used as our routine method of determining k_{et} (16).
Equivalent data were also obtained by measuring $[(FeCp_2^+)_{surf.}]$
by doing a potential step from +0.5 V to -0.6 V vs. SCE and
integrating current. Solutions are stirred to remove oxidized
product, I_3^-, from the vicinity of the electrode.

Plots of $\ln\{[FeCp_2^+]_{t=0}/[FeCp_2^+]_{t=t_i}\}$ vs. t_i are linear with
a slope = $k_{obsvd}[I^-]$. Consistently, the slopes are directly
proportional to $[I^-]$. Since $k_{obsvd} = k_{et}[(FeCp_2^+ \text{ surf.}]$, the
value of k_{et} is given by dividing k_{obsvd} by surface coverage. By
holding the electrode at +0.5 V vs. SCE and irradiating with a
sufficiently high light intensity to insure that excitation rate
is not rate limiting, the steady-state photocurrent should be
predictable using equation (8). Indeed, using the surface
coverage, typically ~10^{-9} mol/cm^2, and the k_{et}'s given in Table I
calculated and observed steady-state photocurrents are in good
agreement.

For CH$_3$CN solvent we can only place a lower limit on the
value of k_{et}. Using a derivatized rotating disk photoelectrode
we find that the steady-state photocurrent at +0.5 V vs. SCE is
directly proportional to $\omega^{\frac{1}{2}}$, the square-root of the rotation
velocity. Such a dependence is expected when the limiting current
is controlled by mass transport (23, 24, 25) and not by the
electron transfer rate (i.e. k_{et} is very large). For CH$_3$CN
solvent, the steady-state photocurrent is independent of coverage
of $[(FeCp_2^+)_{surf.}]$, as expected for a mass transport controlled
rate. But for every other solvent system investigated we find
values of k_{et} that are small compared to what would be expected
for mass transport limited currents; steady-state currents do not
depend on ω or whether the solution is stirred when the current is
not mass transport limited.

Unfortunately, the large value of k_{et} associated with I^-
oxidation in CH$_3$CN is accompanied by an unstable interface. For
reasons that we do not presently understand, the n-type Si
electrodes derivatized with I are not durable enough in CH$_3$CN
solution to sustain I^- oxidation for prolonged periods of time.
However, the electrodes do survive long enough to establish that
the rate of I^- oxidation is limited by mass transport. At the
highest ω from our 2000 r.p.m. motor our strictly linear plots of
limiting current vs. $\omega^{\frac{1}{2}}$ establish that the heterogeneous rate
constant, $k_{et}[(FeCp_2^+)_{surf.}]$, is >0.06 cm/sec (16). Thus, if
$[(FeCp_2^+)_{surf.}]$ is ~10^{-10}mol/cm^2, which approximates a 'monolayer' of
reagent exposed to the solution, k_{et} is $\geq 6 \times 10^8$ cm^{3}mol^{-1}s^{-1}.

Reduction of Surface-Ferricenium by One-Electron Reductants

In our earlier work, we showed that a number of reagents could be oxidized according to equations (1) – (3) where k_{et} is so large that the rate is limited by mass transport (16). Thus, as for B = I⁻ in CH_3CN, k_{et} is $\geq 6 \times 10^8$ $cm^3mol^{-1}s^{-1}$ for an effective surface coverage of ~10^{-10} mol/cm^2. The reagents previously studied (16), B = $Fe(\eta^5-C_5H_5)_2$, $Fe(\eta^5-C_5H_4Me)_2$, $Fe(\eta^5-C_5H_5)(\eta^5-C_5H_4Ph)$, $Fe(\eta^5-indenyl)_2$, and $[Fe(CO)(\eta^5-C_5H_5)]_4$, are all one-electron reductants that have fast self-exchange rates and should, therefore, reduce $(FeCp_2^+)_{surf.}$ rapidly. Table II lists some of our earlier data along with information for several other systems that we have now determined to have large values of k_{et}.

Figure 2 shows the sort of direct evidence that shows that $(FeCp_2^+)_{surf.}$ can oxidize solution reductants. The figure shows cyclic voltammograms for the derivatized electrode in the stirred EtOH/0.1 M n-Bu₄N ClO₄ electrolyte solution first in the absence of B and then in the presence of B = $Co(bipy)_3^{2+}$ at 3.0 mM. The photocurrent at the positive limit of +0.3 V vs. SCE is directly proportional to $\omega^{\frac{1}{2}}$ in the presence of 3.0 mM $Co(bipy)_3^{2+}$. The cathodic current associated with $(FeCp_2^+)_{surf.} \rightarrow (FeCp_2^0)_{surf.}$ on the negative sweep in the dark from +0.3 V vs. SCE is completely absent in the presence of 3.0 mM $Co(bipy)_3^{2+}$ at the scan rate used. The lack of a return wave for the $(FeCp_2^+)_{surf.} \rightarrow (FeCp_2^0)_{surf.}$ reduction directly evidences complete reduction of the $(FeCp_2^+)_{surf.}$ by the $Co(bipy)_3^{2+}$. Linear plots of limiting current vs. $\omega^{\frac{1}{2}}$ establish k_{et} to be $\geq 6 \times 10^8$ $cm^3mol^{-1}s^{-1}$. Representative data for equilibration of $(FeCp_2^+)_{surf.}$ with solution ferrocene are shown in Figure 3 where the limiting current varies linearly with solution ferrocene concentration at a fixed ω and varies linearly with $\omega^{\frac{1}{2}}$ for a fixed solution ferrocene concentration. Such data have been obtained for all of the couples listed in Table II and for I⁻ in CH_3CN.

The measurement of the mass transport rate constant by monitoring $[(FeCp_2^+)_{surf.}]$ vs. t_i in the presence of the various fast reductants generally gives a value that does not accord well with the steady-state photocurrents. This situation results even though the steady-state photocurrent density for rotating, derivatized disk electrodes and the current density at a rotating reversible disk electrode (e.g. Pt) are nearly the same for all reagents when corrected for minor differences in diffusion constants. A representative situation is shown in Figure 4 where cyclic voltammetry of an n-type Si photoanode, derivatized with I, is shown for several situations. Included are data for B = ferrocene and 1,1'-dimethylferrocene under identical conditions. These two solution reductants result in identical steady-state photocurrents at +0.5 V vs. SCE, but as shown in Figure 4b vs. 4c the 0.5 mM 1,1'-dimethylferrocene consumes more of the $[(FeCp_2^+)_{surf.}]$ than does the 0.5 mM ferrocene under identical conditions.

Table II. Rate Constants for Reduction of Surface-Ferricenium by Various Reductants at 298 K.

Reductant, B	$E^{\circ}(B^{+}/B)$, V vs. SCE	Solvent	k_{et}, $cm^3 mol^{-1} s^{-1}$ [a]
$Co(bpy)_3^{2+}$	+0.37	EtOH	$>6 \times 10^8$
$Fe(CN)_6^{4-}$	+0.20	H_2O	$>6 \times 10^8$
$Ru(NH_3)_6^{2+}$	-0.20	H_2O	$>6 \times 10^8$
$(Me_2dtc)^{-}$	Irrev. Oxdn @+0.2 @ Pt	EtOH	$>6 \times 10^8$
$Fe(\eta^5-C_5H_5)_2$	+0.45	EtOH	$>6 \times 10^8$
	+0.46	CH_2Cl_2	$>6 \times 10^8$
	+0.22	sulfolane	$>6 \times 10^8$
$Fe(\eta^5-C_5H_4Me)_2$	+0.36	EtOH	$>6 \times 10^8$
$Fe(\eta^5-C_5H_5)(\eta^5-C_5H_4Ph)$	+0.46	EtOH	$>6 \times 10^8$
$Fe(\eta^5-indenyl)_2$	+0.17	CH_2Cl_2	$>6 \times 10^8$
$[Fe(CO)(\eta^5-C_5H_5)]_4$	+0.34	CH_2Cl_2	$>6 \times 10^8$

[a] See equation (8) of text. All k_{et}'s here are lower limits, since observed rate is mass transport limited (e.g. Figure 3), see text.

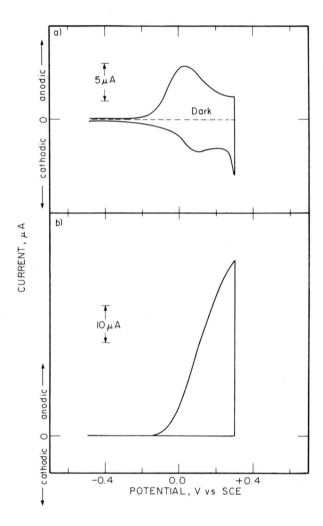

Figure 2. Cyclic voltammetric (100 mV/s) characterization of an n-type Si electrode.

(a) Derivatized with (1,1'-ferrocenediyl)dichlorosilane (5 × 10⁻⁹ mol/cm²) in stirred EtOH/0.1M [n-Bu₄N]ClO₄. (– – –) the dark current; (——) the effect of illumination (632.8 nm, ~ 50 mW/cm²) from −0.5 V to +0.3 V. The light is switched off at +0.3 V showing that photogenerated ferricenium can be reduced in the dark on the negative sweep. (b) Same as in (a) except 3.0 mM Co(bipy)₃Cl₂ is in the stirred electrolyte solution. Note enhanced photoanodic current indicating Co(bipy)₃²⁺ → Co(bipy)₃³⁺ and the lack of a return wave for ferricenium → ferrocene showing that Co(bipy)₃³⁺ formation occurs via surface ferricenium + Co(bipy)₃²⁺ → surface ferrocene + Co(bipy)₃³⁺.

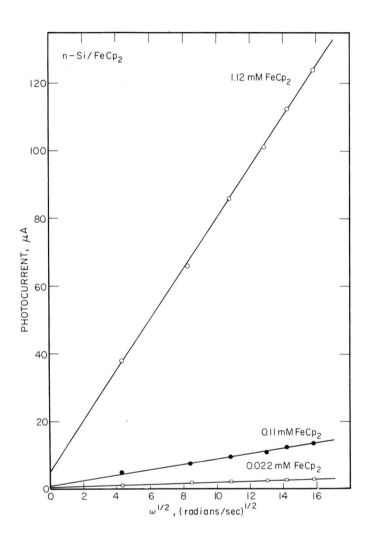

Figure 3.   Photocurrent vs. $\omega^{1/2}$ at +0.5 V vs. SCE from an n-type Si disk derivatized with 5×10^{-9} mol/cm² of ferricenium/ferrocene from (1,1'-ferrocenediyl)dichlorosilane.

The solution is EtOH/0.1M [n-Bu₄N]ClO₄/ferrocene and illumination is at 632.8 nm, ~50 W/cm². The strict linearity of the plots shows that oxidation of solution ferrocene is mass transport–limited for all ω used and for each ferrocene concentration used.

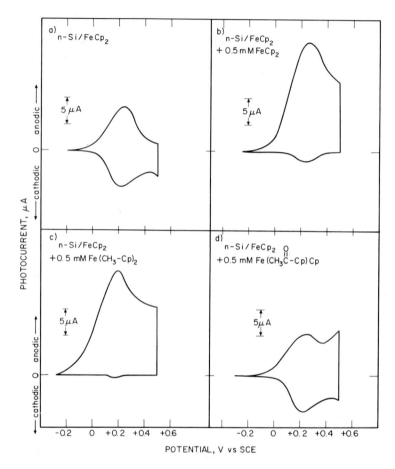

Figure 4.   Cyclic voltammetry (100 mV/s) in stirred EtOH/0.1M [n-Bu₄N]ClO₄ solutions for n-type Si derivatized with (1,1'-ferrocenediyl)dichlorosilane (5×10^9 mol/cm²) with illumination at 632.8 nm, ~ 50 mW/cm² from negative initial potential to the positive limit at +0.5 V.

Light is switched off at +0.5 V vs. SCE for the cathodic sweep. In (a) there is no added reductant; (b), (c), and (d) contain 0.5mM ferrocene, 1,1'-dimethylferrocene, and acetyl-ferrocene, respectively. Acetylferrocene does not attenuate the surface ferricenium → surface ferrocene wave since it is not a sufficiently powerful reductant. Ferrocene and 1,1'-dimethylferrocene both attenuate the surface ferricenium → surface ferrocene wave. But 1,1'-dimethylferrocene is more effective under identical conditions despite the fact that the same, mass transport–limited, steady-state photocurrent is found for these two reductants. These data suggest that after the light is switched off the reduction of surface ferricenium is controlled partially by mass transport and partly by the electron transfer rate (see text).

Since the 1,1'-dimethylferrocene consumes more of the $(FeCp_2^+)_{surf.}$ we conclude that it is a faster reductant than ferrocene. Consistently, the ratio of cathodic peak areas for the reductants examined is strictly maintained for a variety of scan rates, pulse times t_i, concentrations of reductant, and stirring rates. Using a similar technique we order the rates for several other reagents:

$$Fe(\eta^5\text{-indenyl})_2 > [Fe(CO)(\eta^5\text{-}C_5H_5)]_4 \sim Fe(\eta^5\text{-}C_5H_4Me)_2 >$$

$$Fe(\eta^5\text{-}C_5H_5)_2 \sim phenylferrocene \gg acetylferrocene$$

The acetylferrocene does not consume the $(FeCp_2^+)_{surf.}$, Figure 4d, because the reaction is not thermodynamically spontaneous. The conflict in our data is that steady-state photocurrents for the fast reductants is the same, but the $(FeCp_2^+)_{surf.}$ can be consumed at different rates in the dark for the various fast reductants. The data demand the conclusion that reduction of $Fe(Cp_2^+)_{surf.}$ can become limited partially by mass transport and partially by k_{et} at some point in the reaction (large fractional consumption of $(FeCp_2^+)_{surf.}$) in the dark. This seems reasonable in view of the expected rate law, equation (8), the declining $[(FeCp_2^+)_{surf.}]$ and a constant mass transport rate of fresh solution reductant.

Discussion

Data given in Table I for mediated I^- oxidation rate, k_{et}, and for energetics of the process $E^o(FeCp_2^{+/0})_{surf.}$ vs. $E_{1/4}(I^-)$ seem to suggest that both solvent and driving force for reaction can influence k_{et}. In CH_3CN where the driving force is greatest, we find large values for k_{et}. However, it does not seem that 0.18 V (CH_3CN) vs. 0.08 V (CH_2Cl_2) would account for a factor of 10^4 in rate, and we conclude that medium effects can contribute to rate variation as well. In support of this conclusion we note that H_2O yields a rate constant about a factor of 10 lower than the other solvents (EtOH, glacial acetic acid, EtOH/toluene) where the driving force is even slightly smaller. A complication is that self-exchange rates are not known for all of the media used. This precludes a detailed interpretation of the data within the framework of Marcus theory (20, 21).

With respect to solar energy conversion and photoelectro-chemical synthesis involving $I^- \rightarrow I_3^-$ oxidation, it is noteworthy that significant variation in k_{et} can be brought about by variation of the electrolyte solution. The data suggest that CH_3CN could be a good solvent in terms of the measured value of k_{et}, but the durability of the interface in CH_3CN is too poor. Perhaps a mixture of CH_3CN/H_2O would prove workable.

Two points of interest regarding derivatized electrodes emerge from the data for mediated I^- oxidation. First, it is noteworthy that ferricenium in H_2O will not oxidize I^-; the thermodynamics are such that the reaction is not spontaneous (26). However, we do find that $(FeCp_2^+)_{surf.}$ from derivatization with $\underset{\sim}{I}$ does spontaneously oxidize, albeit slowly, I^- in H_2O. Apparently, the surface-confined system has a sufficiently more positive potential that I^- can be oxidized. Since monosubstituted ferrocenes do not have a potential identical to that for ferrocene itself, the surface reagent may be enough different that its oxidizing power can be greater. It may also be that, despite the general similarity in E^O's for surface-bound and solution species (27), the difference in E^O brought about by binding the reagent to the surface is sufficient to change the direction of spontaneity for a given redox process. Second, the direct proportionality of steady-state photocurrent to $[(FeCp_2^+)_{surf.}]$ for k_{et} < (mass transport limited) indicates that all surface-bound material is I^- accessible. This is consistent with the ability of small anions to penetrate throughout the redox polymer. Large anions have been electro-statically bound to polycationic material confined to an electrode surface (28), and such bound anions slowly exchange with smaller counterions. We do not find evidence for slow I^- or I_3^- ion diffusion in and out of polymer on electrodes derivatized with $\underset{\sim}{I}$. Larger counterions, however, do effect the electrochemical behavior as previously noted (19).

As shown in Table II, there are a number of reagents that will rapidly reduce $(FeCp_2^+)_{surf.}$. The value of $k_{et} \geq$ 6 x 10^8 $cm^3mol^{-1}s^{-1}$ comes from the observation for I^- in CH_3CN and all reductants in Table II that photocurrent density is directly proportional to $\omega^{\frac{1}{2}}$ for rotating disk electrodes. The highest ω allows the conclusion that $k_{obsvd} \geq 0.06$ cm/s. Dividing by a coverage of $\sim 10^{-10}$ mol/cm^2 gives $k_{et} \geq$ 6 x 10^8 cm^3 $mol^{-1}s^{-1}$. We are not certain that 10^{-10} mol/cm^2 is the proper coverage to use; we take this value because the data in fact show limiting photocurrent to be independent of coverage in the range $\sim$6 x 10^{-10} to 1 x 10^{-8} mol/cm^2. The $\sim 10^{-10}$ mol/cm^2 is the likely coverage that would correspond to a monolayer of reagent derived from a molecule as large as $\underset{\sim}{I}$. But some caution should be exercised in interpreting k_{et} that is assigned from k_{obsvd}. In our earlier work (16) we used a coverage of $\sim 10^{-9}$ mol/cm^2 to represent the appropriate monolayer coverage but we believe this to be too high, since $<10^{-9}$ mol/cm^2 will, experimentally, result in the same steady-state photocurrent.

The value of $k_{et} \geq$ 6 x 10^8 $cm^3mol^{-1}s^{-1}$ in the usual units for a bimolecular rate constant for homogeneous solution is $\geq$ 6 x 10^5 $M^{-1}s^{-1}$. The ferrocene self-exchange constant is 5 x 10^6 $M^{-1}s^{-1}$ (29). Various cross reactions of substituted ferrocenes and ferricenium derivatives have bimolecular rate constants that exceed 10^6 $M^{-1}s^{-1}$ where the equilibrium constant exceeds unity (30). Further, in the cross reactions, the rate constants varied by almost two orders of magnitude for a change in driving force of ~ 0.25 V (30). Thus, the data in Table II relating to the ferrocene-like molecules is reasonable.

The cross reaction rates and self-exchange rates predict $k_{et} > 6 \times 10^5$ $\underline{M}^{-1}s^{-1}$, provided that the surface-confined reagent has the same activity as the solution species. While we have not ordered k_{et}'s quantitatively, the order of k_{et}'s for the ferrocene-like reagents does appear to depend in a systematic way on the driving force for reaction.

$Co(bipy)_3^{3+/2+}$ self-exchange is fairly slow (31), and we felt that this species might reduce $(FeCp_2^+)_{surf.}$ slowly enough that k_{et} could actually be measured. However, it too has too large a k_{et} to measure. The dimethyldithiocarbamate, Me_2dtc^-, is irreversibly oxidized by $(FeCp_2^+)_{surf.}$ and again the rate constant is too large to measure. The product(s) from this reaction have not yet been identified. Ferrocene in sulfolane is reported to have a low rate constant for heterogeneous exchange (32); we examined this system with the hope of being able to measure k_{et}, but found that ferrocene again has too large a k_{et} to measure. The aqueous $Ru(NH_3)_6^{2+}$ and $Fe(CN)_6^{4-}$ show that water soluble redox reagents can be found that give large values of k_{et}. Indeed, the similarity of $E°$ from I_3^-/I^- and $Fe(CN)_6^{3-/4-}$ and the $>10^5$ change in rate constant shows that the I^- oxidation has a kinetic barrier in H_2O. We could attribute this kinetic difficulty solely to the two-electron process to form I_3^- vs. the one-electron oxidation of $Fe(CN)_6^{4-}$. However, this explanation is not entirely satisfactory in view of the results for I^- in CH_3CN. The $Fe(CN)_6^{3-/4-}$ system would seemingly be a good couple for a photoelectrochemical cell for the conversion of light to electricity but here long term durability is again a serious problem: $Fe(CN)_6^{3-/4-}$ is photosensitive (33) and the long term durability of electrodes derivatized with I in the presence of high concentrations of $Fe(CN)_6^{3-/4-}$ has not been demonstrated.

Several of the reversible redox couples could, in fact be used in a cell for electricity generation. However, the output photovoltage, E_V, associated with these n-type Si/ferricenium/-ferrocene photoanodes is only in the 0.3-0.4 V range at open-circuit. Such is likely too low to be useful in practical schemes for solar energy conversion. The approach of derivatization, though, may be applied to other photoanode materials to realize improved efficiency and durability.

Summary

Photogenerated, surface-confined ferricenium can be reduced by a number of reductants including I^-, $Co(bipy)_3^{2+}$, ferrocene, 1,1'-dimethylferrocene, phenylferrocene, $Fe(\eta^5\text{-indenyl})_2$, $[Fe(CO)(\eta^5\text{-}C_5H_5)]_4$, $Ru(NH_3)_6^{2+}$, $Fe(CN)_6^{4-}$ and dimethyldithio-carbamate. With the exception of I^-, all generally reduce the surface-confined ferricenium with an observed heterogeneous rate constant of >0.06 cm/s which corresponds to a bimolecular rate constant $> 6 \times 10^5$ $\underline{M}^{-1}s^{-1}$, assuming that a monolayer ($\sim 10^{-10}$ mol/cm^2) of the surface-ferricenium participates in the reaction. While good kinetics for I^- oxidation can be obtained in CH_3CN, durability of the photoelectrode is not good enough to promise long term operation. For electrode/solvent/electrolyte/redox couple combinations that are durable and where good kinetics obtain, low output photovoltage remains a problem for n-type Si electrodes derivatized with ferrocene reagents.

Acknowledgements

We thank the United States Department of Energy, Office of Basic Energy Sciences, Division of Chemical Science for support of this research. N.S.L. acknowledges support from the John and Fannie Hertz Foundation 1977–present, and M.S.W. as a Dreyfus Teacher-Scholar Grant recipient, 1975–1980. We acknowledge the assistance of Dr. A.B. Bocarsly in some aspects of this work.

Experimental

Chemicals

Ferrocene (Aldrich Chemical Co.) was purified by sublimation. Bisindenyliron (34), phenylferrocene (35), $[(\eta^5\text{-}C_5H_5)Fe(CO)]_4$ (36) $Co(bipy)_3Cl_2 \cdot 7H_2O$ (37), and sodium dimethyldithiocarbamate (38) were prepared by literature methods. 1,1'-Dimethylferrocene (Polysciences), acetylferrocene, and 1,1'-diacetylferrocene (Aldrich) were purified by chromatography on alumina with hexane as eluant. $K_4Fe(CN)_6$ was used as received (Mallinckrodt), as was anhydrous $NaClO_4$ (G. Frederick Smith). $Ru(NH_3)_6^{2+}$ was conveniently prepared by electrochemical reduction at -0.3 V vs. SCE at a Hg pool electrode of $Ru(NH_3)_6Cl_3$ (Alfa Ventron) in pH = 4.0 $HClO_4/0.1$ M $NaClO_4$. The solution was then acidified to pH = 2.0 with $HClO_4$ just before use, and manipulated under Ar. Polarographic grade $[\underline{n}\text{-}Bu_4N]ClO_4$ (Southwestern Analytical Chemicals) was dried at 353 K for 24 hours and stored in a dessicator until use. $[\underline{n}\text{-}Bu_4N]I$ (Eastman) was recrystallized twice from absolute EtOH. Absolute EtOH, spectroquality isooctane, CH_2Cl_2, CH_3CN, and toluene were used as received, as was reagent grade acetic acid and sulfolane. All aqueous solutions were prepared from doubly distilled deionized water.

Electrodes

Single-crystal Sb-doped, polished n-type Si wafers ((111) face exposed) were obtained from General Diode Co., Framingham, MA. The wafers were 0.25 mm thick and had resistivities of 4-5 ohm-cm. Electrodes were fashioned as previously reported (12). Ohmic contacts were achieved by rubbing Ga-In eutectic onto the back side of the electrode after a 48% HF etch and H_2O rinse. The electrode was attached to a Cu wire with Ag epoxy, and the Cu wire was passed through a 4 mm Pyrex tube. The electrode surface was defined by insulating all other surfaces with ordinary epoxy, yielding electrodes of areas 3-15 mm^2.

Circles of 5 mm were cut ultrasonically from the original Si wafers for use as rotating disk electrodes. Mounting was carried out as above in a 6 mm o.d. capillary tube, except that the Cu wire was replaced by a Hg contact through the capillary. Ordinary epoxy insulation was used sparingly and care was taken to maintain as flat a Si electrode surface as possible. For calibration purposes, rotating Pt disks from circles of Pt foil were mounted in exactly the same manner.

Rotating disk electrodes were mounted vertically and stirred by a variable speed motor from Polysciences, Inc. Rotation velocities were calibrated by two methods: (1) a slitted piece of cardboard was mounted on the disk shaft and the time response of a photodiode was recorded on an oscilloscope , and (2) plots of the limiting current as a function of $K_4Fe(CN)_6$ concentration in 2 $\underline{M}$ KCl yielded straight lines whose slope is $\omega^{\frac{1}{2}}$ (D for $Fe(CN)_6^{4-}$ is 6.3 x 10^{-6} cm^2/sec) (39). Agreement between the two methods was better than 10%, and the motor was found to be extremely stable over long periods of time.

Before use, all Si surfaces were etched in concentrated HF and rinsed with distilled H_2O. Electrodes to be derivatized were then immersed in 10 $\underline{M}$ NaOH for 60 seconds, washed with H_2O followed by acetone, and then air dried. Derivatization was accomplished by exposing the pretreated electrodes to dry, degassed isooctane solutions of (1,1'-ferrocenediyl)-dichlorosilane for 2-4 hours (19). The electrodes were then rinsed with isooctane followed by EtOH.

Electrochemistry

All experiments were performed in single compartment Pyrex cells equipped with a saturated calomel reference electrode (SCE), Pt wire counterelectrode, and the appropriate working electrode. Irradiation was supplied by a beam expanded He-Ne laser of ~50 mW/cm^2 (5 mW total) at 632.8 nm. Laser intensities were measured using a Tektronix J16 digital radiometer equipped with a J6502 probe, and were adjusted to desired illumination levels by use of Corning transmission filters. Cyclic voltammetry measurements were obtained with a Princeton Applied Research Model 173 potentiostat equipped with a Model 179 digital coulometer and driven by a Model 175 voltage programmer. Potential step experiments were performed using the same apparatus, and the coulometer readings were verified by oscilloscopic current-time traces on a Tektronix 564B storage oscilloscope with Type 2B67 time base plug-in. Cyclic voltammetry traces were recorded on a Houston Instrument Model 2000 X-Y recorder and current-time plots were obtained using a Hewlett-Packard strip chart recorder.

The supporting electrolytes were 0.1 $\underline{M}$ [n-Bu$_4$N]ClO$_4$ for CH_3CN, EtOH, CH_2Cl_2, EtOH/toluene (1:1 v/v). sulfolane, and glacial acetic acid solvents, and 1.0 $\underline{M}$ NaClO$_4$/0.01-0.1 $\underline{M}$ HClO$_4$ for H_2O. $E_{\frac{1}{4}}$ values for I^- oxidation were obtained at Pt

rotating disk electrodes at 600 r.p.m.; excellent agreement
with literature values (22), where available, was obtained.
Potential values for surface-attached material were obtained
by derivatizing Pt and Au surfaces as previously described
(18, 19), and taking $E^{o}(FeCp_2^{+/0})_{surf.}$ to be the arithmetic
mean of the anodic and cathodic peak positions; generally
peak-to-peak separations were less than 10 mV at a 20 mV/s
scan rate.

Kinetic Measurements

For measurement of k_{et}, derivatized electrodes were cycled
in a solution of appropriate solvent and electrolyte until stable
cyclic voltammetric parameters were obtained at 100 mV/sec
scan rate. Cyclic voltammograms at several scan rates were
recorded with illumination for the anodic portion of the scan,
but the light was blocked at the anodic limit by a solenoid
driven by the trigger output of the PAR 175 voltage programmer.
Stock solutions of reductant were prepared as needed and
aliquots injected into the Pyrex electrochemical cell immediately
prior to use. Cyclic voltammetry data was then collected for
the same set of scan rates and illumination conditions in the
presence of solution reductant to obtain kinetic data. The
electrodes were rinsed with solvent and checked for decay in the
absence of reductant between every kinetic measurement. At least
four different concentrations of reductant were used for each set
of data points. From this data, cathodic currents associated
with the reduction of surface-attached ferricenium were
integrated manually to determine the time and concentration
dependence of the extent of consumption of surface-confined
ferricenium. The reaction time associated with a cyclic voltammetric sweep was chosen as the period from the anodic limit to
the peak cathodic current in the cyclic voltammogram. Potential
step experiments were performed by scanning anodically at 500 mV/sec
to the anodic limit (+0.5 V vs. SCE), holding at this limit in the
dark for a time t_i, and then pulsing cathodically back to -0.6 V
vs. SCE where the reduction was observed.

Literature Cited

1.   Wrighton, M.S., Acc. Chem. Res., 1979, 12, 303.
2.   Nozik, A.J., Ann. Rev. Phys. Chem., 1978, 29, 189.
3.   Bard, A.J., Science, 1980, 207, 139.
4.   Gerischer, H. J. Electroanal. Chem., 1977, 82, 133.
5.   Bard, A.J.; Wrighton, M.S., J. Electrochem. Soc., 1977, 124,
      1706.
6.   Ellis, A.B.; Kaiser, S.W.; Wrighton, M.S., J. Am. Chem. Soc.,
      1976, 98, 1635, 6418, and 6855.
7.   Hodes, G., Nature (London), 1980, 285, 29.
8.   Parkinson, B.A.; Heller, A.; Miller, B., Appl. Phys. Lett.,
      1978, 33, 521.
9.   Nakatani, K.; Matsudaira, S.; Tsubomura, H., J. Electrochem.
      Soc., 1978, 125, 406.

10. Wrighton, M.S.; Bocarsly, A.B.; Bolts, J.M.; Bradley, M.G.;
     Fischer, A.B.; Lewis, N.S.; Palazzotto, M.C.; Walton, E.G.,
     Adv. Chem. Ser., 1980, 184, 269.
11. Bocarsly, A.B.; Walton, E.G.; Wrighton, M.S., J. Am. Chem.
     Soc., 1980, 102, 3390.
12. Bolts, J.M.; Bocarsly, A.B.; Palazzotto, M.C.; Walton, E.G.;
     Lewis, N.S.; Wrighton, M.S., J. Am. Chem. Soc., 1979, 101,
     1378.
13. Bolts, J.M.; Wrighton, M.S., J. Am. Chem. Soc., 1979, 101,
     6179.
14. Bocarsly, A.B.; Walton, E.G.; Bradley, M.G.; Wrighton, M.S.,
     J. Electroanal. Chem., 1979, 100, 283.
15. Bolts, J.M.; Wrighton, M.S., J. Am. Chem. Soc., 1978, 100,
     5257.
16. Lewis, N.S.; Bocarsly, A.B.; Wrighton, M.S., J. Phys. Chem.,
     1980, 84, 2033.
17. Murray, R.W., Acc. Chem. Res., 1980, 13, 135 and references
     therein.
18. Wrighton, M.S.; Austin, R.G.; Bocarsly, A.B.; Bolts, J.M.;
     Haas, O.; Legg, K.D.; Nadjo, J.; Palazzotto, M. C.,
     J. Electroanal. Chem., 1978, 87, 429.
19. Wrighton, M.S.; Palazzotto, M.C.; Bocarsly, A.B.; Bolts, J.M.
     Fischer, A.B.; Nadjo, L., J. Am. Chem. Soc., 1978, 100, 7264.
20. Marcus, R.A., Ann. Rev. Phys. Chem., 1964, 15, 155.
21. Marcus, R.A., J. Chem. Phys., 1965, 43, 679.
22. Janz, D. J.; Tomkins, R.P.T., "Nonaqueous Electrolytes
     Handbook", Vol. 2, Academic Press: New York, 1972.
23. Piebarski, S.; Adams, R.N., in "Physical Methods of
     Chemistry", Part IIA, Weissberger, A. and Rossiter, B.,
     eds., Wiley-Intersciences: New York, 1971, Chapter 7.
24. Galus, Z., Adams, R.N., J. Phys. Chem., 1963, 67, 866.
25. Levich, V.G., "Physicochemical Hydrodynamics", Prentice-Hall:
     Englewood Cliffs, New Jersey, 1962.
26. Yeh, P.; Kuwana, T., J. Electrochem. Soc., 1976, 123, 1334.
27. Lenhard, J.R.; Rocklin, R.; Abruna, H.; William, K.; Kuo,
     K.; Nowak, R.; Murray, R.W., J. Am. Chem. Soc., 1978, 100,
     5213.
28. Oyama, N.; Anson, F.C., J. Electrochem. Soc., 1980, 127,
     247.
29. Yang, E.S.; Chan, M.; Wahl, A.C., J. Phys. Chem., 1975, 79,
     2049.
30. Pladziewicz, J.R.; Espenson, J.H., J. Am. Chem. Soc., 1973,
     95, 56.
31. Farina, R.; Wilkins, R.G., Inorg. Chem., 1968, 7, 514.
32. Armstrong, N.R.; Quinn, R.K.; Vanderborgh, N.E., J. Electro-
     chem. Soc., 1976, 123, 646.
33. Balzani, V.; Carassiti, V., "Photochemistry of Coordination
     Compounds", Academic Press: New York, 1970.
34. King, R.B., "Organometallic Synthesis", Academic Press:
     New York, 1965, p. 73.
35. Broadhead, G.; Pauson, P.L., J. Chem. Soc., 1955, 367.

36. King, R.B., Inorg. Chem., 1966, 5, 2227.
37. Burstall, F.H.; Nyholm, R.S., J. Chem. Soc., 1952, 3570.
38. Delepine, M.; Haller, A., Comptes Rendus de L'Academie
     des Sciences, 1907, 144, 1126.
39. Adams, R.N., "Electrochemistry at Solid Electrodes",
     Marcel Dekker: New York, 1969.

RECEIVED October 3, 1980.

Chemical Control of Surface and Grain Boundary Recombination in Semiconductors

ADAM HELLER

Bell Laboratories, Murray Hill, NJ 07974

The density and distribution of surface and grain boundary states between the edges of valence and conduction bands of a semiconductor, which affect the electron hole recombination velocities, can be controlled by chemisorption reactions. Strongly exoergic chemisorption reactions reduce the density of surface states in the band gap, leading to substantial decrease in the recombination velocity at grain boundaries and surfaces and thus to an improvement in the performance of solar cells. Chemisorption of oxygen reduces the surface recombination velocities in Si, Ge and InP. With single crystals of n-GaAs, reduction of the surface recombination velocity from 10^6 cm/sec to 3×10^4 cm/sec, doubling of the band gap luminescence intensity, and increase in the solar-to-electrical conversion efficiency of the cell n$-$GaAs|0.8M K_2Se-0.1M K_2Se_2-1M KOH|C from 8.8% to 12% are observed upon the chemisorption of a submonolayer quantity of Ru^{3+}. With thin, chemically vapor deposited films of polycrystalline (9μm average grain size) n-GaAs diffusion of Ru^{3+} ions into the grain boundaries increases the current collection efficiency in electron beam induced current measurements by 50% and the solar energy conversion efficiency of the n$-$GaAs|0.8M K_2Se-0.1M K_2Se_2-1M KOH|C cell fourfold to 7.3%. Co-absorbing Ru^{3+} and Pb^{2+} raises the efficiency of the thin film cell to 7.8%.

Recombination of electrons and holes at grain boundaries in thin, polycrystalline films of semiconductors is a key problem that requires solution if efficient and inexpensive solar cells are to be developed. This problem is quite similar to the extensively studied phenomenon of recombination at semiconductor surfaces. Both involve the electronic states associated with the abrupt discontinuity in the chemical bonding at an interface.

Semiconductor surfaces have been the subject of numerous studies during the past three decades. The extent and number of past studies can be appreciated by a review of the references of the books of Many, Goldstein, Grover[1] and of Morrison.[2] Chapters 3, 5, 7 and 9 of the first book and chapter 9 of the second show that workers in the field have been quite aware of chemical effects on surface recombination velocities. Indeed, in their 1952 paper on "Surface Recombination of Germanium" Brattain and Bardeen pointed out that there is relatively little change in the surface recombination velocity (v_s) when n or p crystals are exposed to O_3/O_2, humid N_2 or O_2 and dry O_2. However, following a suggestion of C. S. Fuller, they did observe, "that v_s could be changed from the order of 10^2 cm/sec. to greater than 10^5 cm/sec and back again by exposure . . . to NH_4OH fumes and then to HCl fumes respectively."[3] During the 30 years that have elapsed since the pioneering work of Brattain and Bardeen, information has been gathered on the effect of gases, etchants, ions, adsorbed organic molecules and physical damage at or near the surface. We have analyzed a small part of the accumulated information and will show that a simple chemical model can account for many of these effects. This model allows the application of chemical concepts and techniques to the reduction of the recombination velocity at surfaces and grain boundaries. We have applied these concepts to the case of n-GaAs, and have reduced the surface and grain boundary recombination velocity by chemisorbing Ru^{3+} and other ions.

Chemical Model

Electron-hole recombination velocities at semiconductor interfaces vary from 10^2 cm/sec for Ge[3] to 10^6 cm/sec for GaAs.[4] Our first purpose is to explain this variation in chemical terms. In physical terms, the velocities are determined by the surface (or grain boundary) density of trapped electrons and holes and by the cross section of their recombination reaction. The surface density of the carriers depends on the density of surface donor and acceptor states and the (potential dependent) population of these. If the states are outside the band gap of the semiconductor, or are not populated because of their location or because they are inaccessible by either thermal or tunneling processes, they do not contribute to the recombination process. Thus, chemical processes that substantially reduce the number of states within the band gap, or shift these, so that they are less populated or make these inaccessible, reduce recombination velocities. Processes which increase the surface state density or their population or make these states accessible, increase the recombination velocity.

A hypothetical semiconductor with dangling or highly strained bonds at its surface, such as would be formed if the crystal were cleaved under perfect vacuum at a temperature low enough to prevent surface relaxation and reconstruction will have a high density of surface states and a high recombination velocity. This can be derived from elementary chemical bonding theory. When a chemical bond is formed between two atoms the states occupied by the bonding electrons are split to form a lower energy, occupied or bonding state and a higher unoccupied or antibonding state. States due to weakly or partially bound atoms or to atoms with strained bonds (for example, double carbon carbon bonds on diamond) are therefore always between the product states of strongly bound atoms. As valence and conduction bands evolve from the initial diatomic states, upon the extension of the bonding in the lattice, unbound and partially bound surface atoms, or surface atoms with strained bonds, introduce states within the band gap. (Figure 1)

If such a hypothetical surface with strained bonds or with weakly or partially bound surface atoms is allowed to relax and reconstruct, the bonding at the surface will still be weaker than in the bulk and the splitting usually less than the band gap. Thus, clean, relaxed semiconductor surfaces have high densities of surface states and, unless the cross sections for recombination are unusually low, also exhibit high recombination velocities. This need not be the case, however, if another element, ion or molecule is chemisorbed on the surface. In this case, the atoms with strained bonds and the partially or weakly bound atoms responsible for recombination may react with the chemisorbed species. If the amount of free energy released is of the order of magnitude (or larger than) the band gap, i.e., the chemisorption is strong, there will be adequate displacement of the surface or grain boundary states (Fig. 2) to sweep clean part or all of the region between the edges of the valence and conduction bands.

If the splitting is now adequate to displace the surface states to positions either outside the band gap or to positions where they are less populated or are inaccessible to the majority carrier by both thermal scaling of the surface barrier or by tunneling through the barrier (Fig. 2c), the recombination velocity will be reduced. If the chemisorption is weak, the splitting of the surface states will not be adequate to displace the surface states from within the band gap, or to shift these to positions where they cannot be populated by thermal or tunneling processes. (Fig. 2b) Thus, weak chemisorption merely redistributes or even increases the density of surface states within the band gap, and with it the surface recombination velocity. The concepts of strong and weak splitting

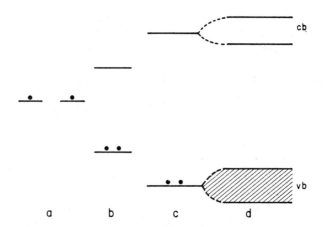

Figure 1. Simple chemical bonding model showing that unbound or partially bound atoms on a semiconductor surface contribute states within the band gap.

The states of unbound atoms (a) are split upon partial bonding (b), then further split when the fully bound species (c) is formed. Evolution of the periodic lattice broadens the bonding states to form the valence band (vb) and the antibonding orbitals to form the conduction band (cb). In the process of band formation, the unbound and partially bound states (a and b) remain between vb and cb.

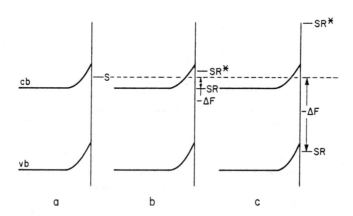

Figure 2. A surface state (S) located near the conduction band (cb) (left) is split upon the chemisorption of a reagent (R).

If the release of energy (ΔF) is small, the splitting is small (center) and the number of product states (SR and SR*) within the band gap, near the conduction band, increases. If the chemisorption is strong, a substantial amount of free energy is released (right) and the number of surface states within the band gap (or near the conduction band) decreases.

are presented in Fig. 2. Whether states between the band edges are added or substantially removed upon "contamination" of the surface by a chemisorbed reagent depends on the initial position of these states and on $-\Delta F$, the change in free energy which produces the splitting. When $-\Delta F$ is large, states are likely to be removed. When $-\Delta F$ is small, states are usually added.

In summary, a simple chemical picture of surface recombination is presented. The surface or grain boundary recombination velocity decreases when the appropriate surface species is reacted with a strongly chemisorbed species. It increases when a species is weakly chemisorbed. We shall now illustrate this concept for six extensively studied semiconductors, Ge, Si, GaP, GaAs, InP and InSb.

Ge

Brattain and Bardeen[3] measured the surface recombination velocity (v_s) of oxidized (ozone and peroxide exposed) germanium and found it to be slow, about 170 cm/sec for p-type and 50 cm/sec for n-type/Ge. Following complete oxidation, these values change only slightly in a variety of ambients. The recombination velocity increases by three orders of magnitude, to $>10^6$ cm/sec, when the oxidized surface is exposed to ammonia, and reverts back to $\sim 10^2$ cm/sec when the surface is exposed to HCl. We interpret these observations as follows. Chemisorption of reactive oxygen on water-free Ge produces a GeO_2 film. The standard-free energy of formation for bulk GeO_2 is -115 Kcal/mole. For the reaction of oxygen at a surface with unsaturated Ge bonds, ΔF is likely to be more negative. Thus, the splitting of the surface states upon oxidation displaces these states to domains above the edge of the conduction band and below the edge of the valence band. As the density of states within the band gap is reduced, the surface recombination velocity is also reduced. Humid ammonia vapor attacks the GeO_2 film. Reoxidation by air in the presence of an acid (humid HCl) restores the film.

Si

The existence of today's silicon based microelectronics technology is evidence for the low surface recombination velocity of oxidized Si. The velocity is less than 10^3 cm/sec.[5,6,7]

We explain the low surface recombination velocity by the sweeping of surface states from the region between the edges of the conduction and valence bands upon oxidation of the surface. The standard free energies of formation of crystalline and fused quartz from bulk Si are -192 and -191 Kcal/mole respectively. The standard free energy change for a Si surface is likely to be more negative. Thus the edges of the conduction and valence bands of Si are well within those of SiO_2.

We note in this context that in Si based MIS (metal-insulator-semiconductor) solar cells one of the roles of the 20-60Å thick SiO_2 layer may well be reduction of the recombination velocity at the Si surface. Chambouleyron and Soucedo noted a decrease in the recombination velocity at the conductive SnO_2/Si interface[8] relative to that at the Si surface and Michel and Lasnier find a recombination velocity of less than $2x10^4$ cm/sec at the conductive indium tin oxide/Si interface.[9] In both cases heating the metal oxides present on the elemental Si produces an intermediate SiO_2 layer.

The grain boundary recombination velocity in polycrystalline Si, which reaches 10^6cm/sec[10], is well above v_s at the air (oxygen)/Si interface.[10] We attribute the difference to the fact that Si grain boundaries are either not oxidized or are oxidized only to SiO, a "black", small band gap material. Were it possible to grow 20-50Å thick dioxide layers at the boundaries, the density of grain boundary states could be decreased yet carriers could tunnel through. This would increase the efficiency of small grained Si based solar cells. Presently available oxidation methods lead to excessively thick oxide near the O_2 exposed surface and to little or no oxidation deeper in the polycrystalline material.

Exposure of silicon to atomic hydrogen increases the surface recombination velocity.[11,12,13] The free energy of formation of SiH_4, the most stable of the hydrides of silicon, is only $\sim$-10Kcal/mole. Since four electron pairs are shared in the formation of the molecule, the free energy of formations per Si-H bond is only $\sim$-2.5 Kcal or about 0.1eV. Because of the weak chemisorption, heating of the silicon to temperatures above 500°C is adequate to release the hydrogen. Our model explains the increase in surface recombination velocity by the weak chemisorption of hydrogen, which may increase the density of surface states within the band gap (see Fig. 2b).

The observation of Seager and Ginley that grain boundaries of polycrystalline silicon are passivated by atomic hydrogen is explained by the filling of empty or part empty states upon reaction with hydrogen and the consequent reduction in their population. In p-type (but not n-type) silicon, the effect may also be due to penetration of hydrogen into the Si surface, which leads to heavy doping of the material near the boundary. Such doping makes the grain boundary states less accessible under circumstances discussed at the end of the paper. We note that penetration of hydrogen into Si and doping were suggested by Law[11,12], and that

Iles[15], Soclof and Iles[16] and Sah and Lindholm[10] have shown that heavy doping near boundaries is a means to reduce the grain boundary recombination velocity.

InP and GaP

Casey and Buehler have shown that the surface recombination velocity of n-InP ($\sim 5 \times 10^{17}$ carriers/cm^3) is low, $\sim 10^3$cm/sec.[17] Suzuki and Ogawa have recently reported a sequence of surface treatments that cause substantial changes in the surface recombination velocity of InP.[18] They found that in freshly vacuum cleaved (110) faces v_s is much greater than at air exposed faces and that the quantum efficiency of band gap luminescence increases by an order of magnitude when the freshly cleaved face is exposed to air. This suggests that the surface recombination velocity is reduced when O_2 is chemisorbed.

The changes are explained as follows: The density of surface states within the band gap on freshly cleaved InP is high. As a result, the surface recombination velocity is high and the luminescence efficiency is low. Chemisorption of oxygen splits the surface states, as large band gap, colorless $InPO_4$ is formed.[19]

Reduction in the surface recombination velocity of GaP, from 1.7×10^5 cm/sec to 5×10^3 cm/sec, is observed upon exposure to a CF_4 plasma in which fluorine is known to be present.[20] Again, the product of chemisorption of fluorine on the surface is likely to be a large band gap material such as GaF_3, which straddles the edges of the conduction and valence bands of GaP.

GaAs and InSb

For both GaAs and InSb, even when exposed to air or oxygen, v_s is high, typically 10^6 cm/sec.[17,21,22,23,24] Thurmond et al[25] have shown that no arsenic containing oxide (i.e. As_2O_3, As_2O_5 $GaAsO_3$ or $GaAsO_4$) will co-exist in thermodynamic equilibrium with GaAs. For example, the standard free energy change for the reaction

$$As_2O_3 + 2GaAs \rightarrow Ga_2O_3 + 4As$$

is -62Kcal/mole. Consequently, elemental As is a constituent of the GaAs/oxide interface in thermal, plasma and anodically produced oxides.[26,27,28] Since surface GaAs is more reactive than bulk GaAs, elemental As must always be a constituent of the first monolayer on GaAs in air.

The presence of arsenic at the interface implies that surface states within the band gap will be introduced (see Fig. 1). We associate the high surface recombination velocity with the presence of arsenic. The formation of elemental As on the GaAs surface explains the difference in behavior of InP and of GaAs. In InP the thermodynamically stable phase that results from oxidation of the surface is colorless $InPO_4$ which straddles the band gap. In GaAs it is Ga_2O_3 and small band gap As.

The standard free energy change in the reaction

$$2InSb + Sb_2O_3 \rightarrow In_2O_3 + 4Sb$$

is -33 Kcal/mole for bulk InSb. For an InSb surface, it is probably more negative. Consequently, Sb will be present at the interface, and will increase v_s. Indeed, Skountzos and Euthymious report $v_s \sim 10^6 cm/sec$ for InSb in air.[29]

Effect of Chemisorbed Ru^{3+} on n-GaAs Surfaces and Grain Boundaries

It is evident that in order to reduce the high surface recombination velocity of GaAs, it is necessary to remove the elemental arsenic at the air interface and to stabilize the surface with a strongly chemisorbed species. This can be accomplished by a sequence of surface treatments consisting of oxidation, dipping into either a base or into a basic selenide-diselenide solution and dipping into an acidic Ru^{3+} solution.[30,31,32] The function of the base or the basic selenide diselenide solution is to clean the GaAs surface. The base dissolves the amphoteric gallium oxide and the acidic arsenic oxides. The selenide-diselenide solution dissolves residual arsenic, when present, by the reaction

$$2As + 3Se_2^{2-} \rightarrow 2AsSe_3^{3-}$$

which will also leave a monolayer of selenide on the surface. The acid Ru^{3+} solution either removes this layer or the selenide layer allows the Ru^{3+} to diffuse through it. In either case, the Ru^{3+} is strongly chemisorbed on the n-GaAs surface. Bulk doping by Ru^{3+} is ruled out by Rutherford backscattering studies.[33] The latter show that the Ru^{3+} remains at the surface and does not diffuse into the bulk even at 300°C. The number of Ru atoms on the surface corresponds to substantially less than a monolayer.[33] Measurement of the decay time of the band gap luminescence following Ru^{3+} treatment reveals a dramatic increase in the carrier lifetime. Figure 3 shows the luminescence decay following excitation by light pulses of several nanosecond duration of a GaAs crystal that had received three surface treatments.[33] The slowest decay is observed for the crystal with an epitaxial GaAlAs layer grown on it. The carrier recombination velocity at the GaAlAs-GaAs interface is less than 500 cm/sec.[34] Here, the observed decay rate represents the sum of the nonradiative bulk recombination and radiative bulk recombination processes. When the GaAlAs layer is removed by HCl to expose the GaAs surface, v_s increases to 10^6 cm/sec.[21-23] The luminescence decay is now fast and is dominated by the surface recombination process. Chemisorption of Ru^{3+} increases the carrier lifetime by reducing the surface recombination velocity. Analysis of the luminescence decay time as a function of the thickness[33] of the GaAs sample shows a decrease in recombination velocity from 10^6 cm/sec to 3×10^4 cm/sec[33] following Ru^{3+} treatment. It is notable that the effect of Ru^{3+} persists in spite of oxidation of the surface in air, suggesting that the ruthenium stays at the interface between the oxide layer and n-GaAs. Recent results of Rowe, who measured the depth profile of the Ru concentration at the air-GaAs interface by Auger spectroscopy, prove this point.[35]

Woodall et al.[36] have analyzed the relationship between surface recombination velocity and the steady state band gap luminescence in GaAs. They calculate for 534nm excitation that a decrease in v_s from 10^6 cm/sec to 10^4 cm/sec will triple the quantum efficiency at a 2.5μm deep p-n junction if the hole diffusion length, L_p, is 3μm, and the electron diffusion length, L_n is 4μm.

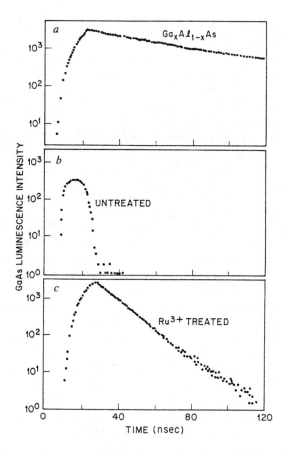

Figure 3.   Decay of the band gap luminescence in the same n-GaAs crystal (a) with GaAlAs windows on both sides; (b) with a GaAlAs window on one side and an air-exposed GaAs surface on the other; (c) same as (b) after chemisorption of Ru^{3+} on the GaAs surface. For details, see Ref. 33.

They also show that as L_p and L_n decrease or as carrier concentration increases there is less improvement in the quantum yield. In agreement with these calculations, we find for samples with $L_p \sim 2\mu$m a doubling of the luminescence intensity.[33]

Chemisorption of a fraction of a monolayer of Ru^{3+} also improves the performance of n-GaAs based solar cells. The solar-to-electrical conversion efficiency of a n–GaAs|0.8M K_2Se–0.1M K_2Se_2–1M KOH|C semiconductor-liquid junction solar cell increases from 8.8%[35] to 12%.[30,31] Current voltage characteristics for the cell, before and after Ru^{3+} chemisorption, are shown in Fig. 4. The improvement is in the fill factor. At potentials approaching open circuit, fewer electrons recombine at the semiconductor liquid interface with holes after Ru^{3+} is chemisorbed. The exceptionally strong bond between Ru^{3+} and the n-GaAs surface is evidenced by the fact that the improvement in cell performance persists for weeks.

Ru^{3+} is not the only ion capable of improving the efficiency of n-GaAs based solar cells. Lead chemisorbed from a basic plumbite solution,[36] as well as Ir^{4+} and Rh^{3+} chemisorbed from acids[32] are also effective. These ions are, however, not as strongly chemisorbed as Ru^{3+}. Consequently, they are more readily desorbed and produce a lesser improvement. Experiments with combined plumbite and ruthenium chemisorption show a further small improvement.[38]

Ions that are not chemisorbed do not affect the performance of semiconductor liquid junction solar cells.[32] Weakly chemisorbed ions produce inadequate splitting of surface states between the edges of the conduction and valence band and increase rather than decrease the density of the surface states in the band gap and thus the recombination velocity. Bi^{3+} is an example of such an ion. As seen in Figure 5, it decreases the efficiency of the n–GaAs|0.8M K_2Se–0.1M K_2Se_2–1M KOH|C cell.[30] Since the chemisorption of Bi^{3+} is weak, the deterioration in performance is temporary. The ion is desorbed in $\sim$10 min. and the cell recovers.

It is notable that once strongly chemisorbed Ru^{3+} has been adsorbed on an n-GaAs surface, Bi^{3+} causes little deterioration in cell performance. This suggests that the two ions react with the same chemical species or "site" on the GaAs surface.

The basic thesis of this paper, the displacement of interface states by strongly chemisorbed species, is also applicable to grain boundaries.

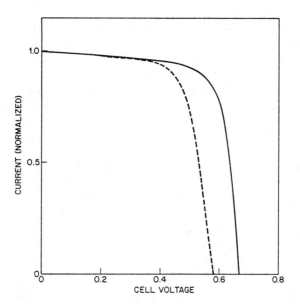

Figure 4.   Effect of chemisorbed Ru³⁺ on the current–voltage characteristics of the n-GaAs/0.8M K₂Se–0.1M K₂Se₂–1M KOH/C solar cell ((– – –) freshly etched; (——) Ru³⁺ chemisorbed)

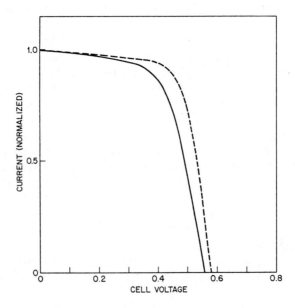

Figure 5.   Effect of chemisorbed Bi³⁺ on the current–voltage characteristics of the n-GaAs/0.8M K₂Se–0.1M K₂Se₂–1M KOH/C solar cell ((– – –) freshly etched; (——) Bi³⁺ chemisorbed)

Our model predicts that a strong chemisorption reaction which is effective in reducing the recombination velocity on the surface of a crystal, will do the same at its grain boundary. Like a surface, a grain boundary is a discontinuity which introduces states in the gap between the edges of the conduction and valence bands. These can be shifted by chemisorption. To accomplish the required grain boundary reaction, advantage is taken of the fact that the rate of diffusion of a reactant in a grain boundary is much faster than the rate of its diffusion in the bulk. Thus, chemisorption reactions can be carried out without substantial doping of the bulk of the semiconductor. It is easier to improve the efficiency of thin film cells of direct gap semiconductors (which have light adsorbtion coefficients of ($\alpha \approx 10^5 cm^{-1}$), than of indirect band gap materials ($\alpha \approx 10^2$-$10^5 cm^{-1}$). The reason is that grain boundary diffusion to depths of $10^{-4} cm$ is adequate in the first, while the second requires diffusion to depths of 10^{-3} - 10^{-2} cm.

GaAs is a direct band gap semiconductor. Films of $1 \mu m$ thickness absorb nearly all the photons with energies exceeding the 1.4 eV bandgap. Upon diffusion of Ru^{3+} into boundaries of 3-$4 \mu m$ grains of n-GaAs (produced by chemical vapor deposition onto a graphite substate coated with a 500Å film of germanium) we observed a fourfold increase in the solar conversion efficiency (from 1.2% to 4.8%) in the n–GaAs|0.8M K$_2$Se–0.1M K$_2$Se$_2$–1M KOH|C cell.[39] Since the relevant area over which electron-hole recombination may take place is much larger than in a single crystal, the improvement is far more dramatic. Fig. 6 shows the current voltage characteristics obtained with a polycrystalline n-GaAs photoanode before and after chemisorption of Ru^{3+} at the grain boundaries. The deep diffusion of Ru^{3+} into the grain boundaries is evidenced by the fact that etchants, which attack the surface of n-GaAs and completely reverse the improvement in solar cells following Ru^{3+} chemisorption in single crystals, reverse the improvement in polycrystalline films only in part, unless substantial (>1000Å) film thickness is removed.

Using an improved chemically vapor deposited film of n-GaAs on graphite, with a $9 \mu m$ average grain size, a solar-to-electrical conversion efficiency of 7.3% is reached after Ru^{3+} is chemisorbed at the grain boundaries.[40] By chemisorbing first Ru^{3+} from an acid, then Pb^{2+} from a basic aqueous solution, the efficiency of the polycrystalline thin film n–GaAs|0.8M K$_2$Se–0.1M K$_2$Se$_2$–1M KOH|C cell

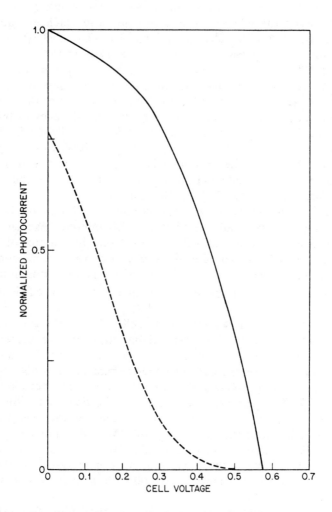

Figure 6.   Effect of chemisorption of Ru^{3+} on the grain boundaries of a chemically vapor-deposited, thin-film n-GaAs photoanode. The grains are of 3–5 μm in size. The current–voltage characteristics shown are for the n-GaAs/0.8M K_2Se–0.1M K_2Se_2–1M KOH/C cell. ((– – –) freshly etched; (———) Ru^{3+} chemisorbed)

is increased to 7.8%, the highest value for any GaAs cell of such small grain size.[38] The current voltage characteristics of the 7.8% efficient cell following Ru^{3+} and Pb^{2+} treatments are shown in Fig. 7.

The improvement in performance following chemisorption of Ru^{3+} at n-GaAs grain boundaries is not limited to semiconductor liquid junction solar cells. Charge collection scanning electron microscopy at a gold-n-GaAs Schottky junction shows a drastic reduction in grain boundary recombination following the Ru^{3+} and Ru^{3+} plus Pb^{2+} treatments. Figure 8 shows charge collection scanning electron micrographs for the same polycrystalline film before and after Ru^{3+} treatment.

While chemisorbed Ru^{3+} reduces the surface and grain boundary recombination on n-GaAs, it has no effect on p-GaAs. This can be accounted for if the chemisorption reaction decreases only the density of electron trapping states near the conduction band edge but not of hole trapping states near the valence band edge. It appears that in GaAs chemisorption of Ru^{3+} splits electron trapping surface states near the edge of the conduction band to form states above the edge of this band and states near the valence band (Fig. 9). While such a reaction prevents electrons from reaching the surface in n-type materials (Fig. 9, left) it does not prevent holes from doing the same in p-type materials (Fig. 9, right). Since holes and electrons are, respectively, abundant at surfaces of illuminated n and p type GaAs, only recombination at surfaces of n-type materials is reduced.

One may speculate about the causes of strong chemisorption of some ions and the weak chemisorption of others. The four metals, with strongly chemisorbed ions on GaAs, (Ru, Pb, Ir, Rh) have several stable oxidation states and thus radii, some of which approach those of the lattice components. The orbitals of these metals and ions also have substantial mixed sp (~60%) and d(~40%) character, which makes varying orbital hybridizations and thus a range of orbitals of different directionality possible.[41] In some cases, orbitals binding at two or more surface sites can be envisaged.

The recombination velocity at a grain boundary can be reduced not only by reducing the density of grain boundary states, but also by diffusing a dopant into the boundary and heavily doping the nearby region of the grain. If n^+-n or p^+-p junctions are formed, the space charge region associated with the boundary is shrunk and minority carriers are no longer pulled to the grain boundary (by the field associated with the space charge region)

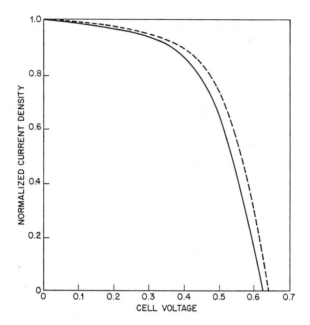

Figure 7.   Current–voltage characteristics of an n-GaAs/0.8M K$_2$Se–0.1M K$_2$Se$_2$– 1M KOH/C cell made with a chemically vapor-deposited, thin film of n-GaAs on W-coated graphite with Ru^{3+}(——) and with Ru^{3+} and Pb^{2+} (– – –) chemisorbed on the grain boundaries. The grains are of 9-μm average size. The GaAs layer is 20 μm thick.

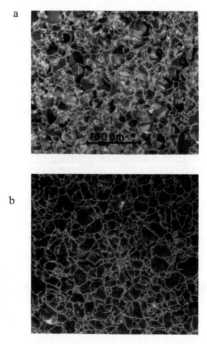

Figure 8.   Charge collection scanning electron micrographs for a gold-polycrystalline n-GaAs junction (a) before and (b) after Ru^{3+} treatment. The darker the area the less the recombination.

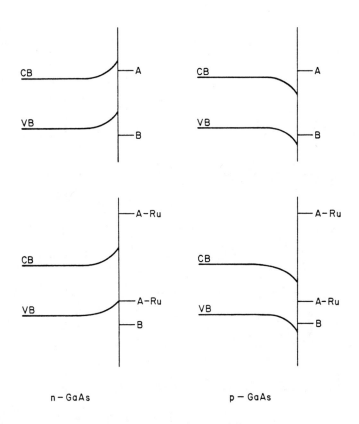

*Figure 9.   Model for splitting of surface states upon chemisorption of Ru³⁺ on n-
and p-GaAs.*

*Surface states before Ru³⁺ chemisorption (top) and after (bottom) are shown for n-GaAs
(left) and p-GaAs (right). Note that the splitting of the electron trapping surface states
decreases the density of State A, potentially accessible to the majority carrier in n-GaAs,
but not in p-GaAs, where the recombination controlling State B is a hole trap.*

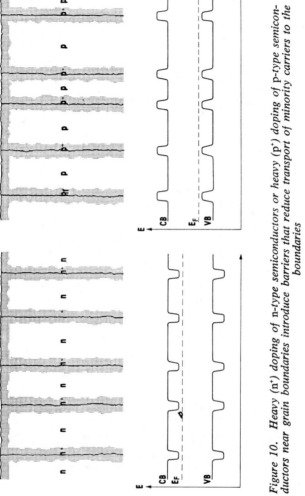

Figure 10.   Heavy (n⁺) doping of n-type semiconductors or heavy (p⁺) doping of p-type semiconductors near grain boundaries introduce barriers that reduce transport of minority carriers to the boundaries

from the bulk of the grain. The high-low doping also produces a reflecting barrier to minority carrier diffusion (Fig. 10). Heavy doping has been proposed earlier[10,15,16] as a means to reduce the recombination velocity at grain boundaries and Di Stefano and Cuomo[42], who phosphorus doped p-Si grain boundaries, were successful in reducing recombination.

In retrospect, we also explain as being due to the formation of n-n$^+$ junctions the improvement in the conversion efficiency of solar cells made with hot-pressed, polycrystalline n-CdSe upon diffusion of cadmium metal, and n-type dopant, into the boundaries.[43]

The effectiveness of the doping approach is limited by the residual thermal diffusion of minority carriers to grain boundaries if n$^+$ -n or p$^+$ -p junctions are formed. Such a loss does not exist if the grain boundary recombination velocity is reduced by the strong chemisorption process proposed in this paper. With this method single crystal efficiency is approached in polycrystalline films made of small grained semiconductors. Indeed, we reach two thirds of the efficiency of single crystal n-GaAs cells with 20μm thick chemically deposited films of n-GaAs on graphite.[38,40] We would have reached a still higher fraction of the single crystal efficiency had we been able to achieve a non-reflective surface by etching wavelength sized hillocks into the film surface, as we did in single crystal GaAs.[30,31]

Based on our observations, we have reason to believe that single crystal performance will be approached in future thin film, polycrystalline semiconductor based solar cells with grain boundary recombination velocities reduced by strongly chemisorbed species.

ACKNOWLEDGMENTS

Extensive and enjoyable discussions with Harry J. Leamy and Barry Miller are acknowledged. The author also thanks H. D. Hagstrum and J. C. Philips for reviewing the manuscript.

REFERENCES

(1) Many, A., Goldstein, Y. and Grover, N. B., "Semiconductor Surfaces", 1965, North Holland Publishing Co., Amsterdam.

(2) S. R. Morrison, "The Chemical Physics of Surfaces", 1977, Plenum Press, N. Y.

(3) Brattain, W. H., and Bardeen, J., *Bell System Journal*, 1952, *32*, 1.

(4) Pastrzebski, L., Gatos, H. C., and Lagowski, J., *J. Appl. Phys.* 1977, *48*, 1730.

(5) Moore, A. R., and Nelson, H., *RCA Rev.*, 1956, *17*, 5.

(6) Hogarth, C. A., *Proc. Phys. Soc. (London)*, 1956, *B69*, 791.

(7) Petrusevich, V. A., *Fiz. tver. Tela*, 1959, *1*, 1695.

(8) Chambouleyron, I. and Saucedo, E., *Sol. Energy Mater.* 1979, *1*, 299.

(9) Michel, J. and Lasnier, M. F., Proc. Intnl. Conf. on Photovoltaic Solar Energy, Luxembourg, 1977, p. 125.

(10) Sah, C. T., and Lindholm, F. A., IEEE 12th Photovoltaic Specialists Conf., Baton Rouge, LA, 1976, p. 93.

(11) Law, J. T., *J. Appl. Phys.*, 1961, *32*, 600.

(12) Law, J. T., *J. Appl. Phys.*, 1961, *32*, 848.

(13) Law, J. T., "Semiconductor Surfaces", in Zemel, J. N., Editor, Pergamon Press, Oxford, 1960, p. 9.

(14) Seager, C. H., and Ginley, D. S., *Appl. Phys. Lett.* 1979, *34*, 337.

(15) P. A. Iles, Workshop Proceedings on Photovoltaic Conversion of Solar Energy and Terrestrial Application, Cherry Hill, October 1973, p. 71.

(16) S. I. Soclof and P. A. Iles, Proc. 11th IEEE Photovoltaic Specialists Conference, May 6-8, 1975, p. 56.

(17) Casey, H. C., Jr. and Buehler, E., *Appl. Phys. Lett.* 1977, *30*, 247.

(18) Suzuki, T., and Ogawa, M., *Appl. Phys. Lett.* 1979, *34*, 447.

(19) L. L. Vereikina, Khim. Svoistva Metody. Anal. Tugoplavkikh Sodein, Edited by G. V. Samsonov, Naukova Dumka, Kiev, 1969, p. 37.

(20) Stringfellow, G. B., *J. Vac. Sci. and Technol.* 1976, *13*, 908.

(21) Jastrzebski, L., Lagowski, J., and Gatos, H. C., *Appl. Phys. Lett.* 1975, *27*, 537.

(22) Jastrzebski, L., Gatos, H. C. and Lagowski, J. *J. Appl. Phys.*, 1977, *48*, 1730.

(23) Hoffman, C. A., Jarasiunas, K., Gerritsen, H. J., and Nurmikko, A. V., *Appl. Phys. Lett.*, 1978, *33*, 536.

(24) Skountzos, P. A., and Euthyimiou, P. C., *J. Appl. Phys.*, 1977, *48*, 430.

(25) Thurmond, C. D., Schwartz, G. P., Kamlott, G. W., and Schwartz, B., *J. Electrochem. Soc.*, 1980., *127*, 1366.

(26) Schwartz, G. P., Griffiths, J. E., DiStefano, D., Gualteri, G. J., and Schwartz, B., *Appl. Phys. Lett.,* 1979, *34,* 742.

(27) Watanabe, K., Hashiba, M., Hirohata, Y., Nishino, M., and Yamashina, T., *Thin Solid Films,* 1979, *56,* 63.

(28) Schwartz, G. P., Griffiths, J. E., and Schwartz, B., *J. Vac. Sci. Technol.,* 1979, *16,* 1383.

(29) Skountzos, P. A., and Euthymiou, P. C., *J. Appl. Phys.,* 1977, *48,* 430.

(30) Parkinson, B. A., Heller, A. and Miller, B., *Appl. Phys. Lett.,* 1978, *33,* 521.

(31) Heller, A., Parkinson, B. A., and Miller, B., Proc. 12th IEEE Photovoltaic Specialists Conf., Washington, June 5-8, 1978, p. 1253.

(32) Parkinson, B. A., Heller, A., and Miller, B., *J. Electrochem. Soc.,* 979, *126,* 954.

(33) Nelson, R. J., Williams, J. S., Leamy, H. J., Miller, B., Parkinson, B. A., and Heller, A., *Appl. Phys. Lett.,* 1980, *36,* 76.

(34) Thrush, E. J., Selway, P. R., and Henshall, G. D., *Electron. Lett. (GB),* 1979, *15,* 156.

(35) Rowe, J. E., private communication.

(36) Woodall, J., Pettit, G. D., Chappell, T., and Hovel, H. J., *J. Vac. Sci. Tech.,* 1979, *16,* 1389.

(37) Chang, K. C., Heller, A. and Miller, B., *Science,* 1977, *196,* 1097.

(38) Heller, A., Lewerenz, H. J., and Miller, B., *Ber. Bunsengesellschaft, Phys. Chem.,* 1980, *00,* 0000.

(39) Johnston, W. D., Jr., Leamy, H. J., Parkinson, B. A., Heller, A., and Miller, B., *J. Electrochem. Soc.,* 1980, *127,* 90.

(40) Heller, A., Miller, B., Chu, S. S., and Lee, Y. T., *J. Am. Chem. Soc.,* 1979, *101,* 7633.

(41) L. Pauling, "The Nature of the Chemical Bond", 3rd Edition, Cornell University Press, Ithaca, 1960, pp. 417-9.

(42) DiStefano, T. H., and Cuomo, J. J., *Appl. Phys. Lett.,* 1977, *30,* 351.

(43) Miller, B., Heller, A., Robbins, M., Menezes, S., Chang, K. C., and Thomson, J. Jr., *J. Electrochem. Soc.,* 1977, *124,* 1019.

RECEIVED October 3, 1980.

Charge-Transfer Processes in Photoelectrochemical Cells

D. S. GINLEY and M. A. BUTLER

Sandia National Laboratories, Albuquerque, NM 87185

The promise of photoelectrochemical devices of both the photovoltaic and chemical producing variety has been discussed and reviewed extensively.(1,2,3,4) The criteria that these cells must meet with respect to stability, band gap and flatband potential have been modeled effectively and in a systematic fashion. However, it is becoming clear that though such models accurately describe the general features of the device, as in the case of solid state Schottky barrier solar cells, the detailed nature of the interfacial properties can play an overriding role in determining the device properties. Some of these interface properties and processes and their potential deleterious or beneficial effects on electrode performance will be discussed.

Due to the chemical potential difference for species in the electrolyte and the photoelectrode, and by virtue of the fact that the electrode can be run in forward and reverse bias configurations, a number of important processes at the interface can be discerned. In each case, we will be concerned with the energy required for the process under consideration to occur and its resulting effects on photoelectrode performance. We can think of these processes as being of four basic types: chemisorption, the desired electron or hole charge transfer, surface decomposition and electrochemical ion injection. In the rest of the paper we will briefly summarize our present understanding of each.

Chemisorption

Chemisorption is the process by which various ions are adsorbed on the semiconductor surface with the formation of a chemical bond and can affect a number of important cell parameters.

As has been previously discussed,(5,6) the amount of band bending in the depletion region determines the amount of external bias (V_{bais}) needed for chemically producing PECs (photoelectrochemical cells) or the open-circuit potential (V_{oc}) for wet photovoltaic cells. Figure 1 illustrates the energy level diagram for a typical PEC. Under equilibrium conditions and no illumination, the Fermi level in the semiconductor should equilibrate at the redox potential in the electrolyte. Therefore, the amount of band bending is determined by the difference between the redox level in the electrolyte, E_{redox}, and the flatband potential of the semiconductor, V_{fb}. The flatband potential is a measure of the semiconductor electron affinity relative to a reference electrode in the electrolyte. It includes not only the intrinsic electron affinity which can be calculated using electronegativity arguments(3) but also the potential drop across any dipole layer that may exist at the interface. Since a dipole layer can arise due to chemisorption of ions, the flatband potential and therefore V_{oc} or V_{bias} will be determined to some extent by the chemisorption process.

A cut and polished semiconductor surface generally contains a large amount of disorder and a significant density of defect states. The electrochemical potential for an ion adsorbed on this surface is generally quite different from that of the same ion in the electrolyte. If the free energy of the adsorbed ion is significantly more negative than the hydration energy of the ions, they will be adsorbed. Consequently, there is a strong interaction of those ions with the surface. If the chemisorption energy is different for the anion and the cation, one will be preferentially adsorbed giving rise to a net surface charge. A compensating charge layer will exist in the electrolyte resulting in a capacitor like structure. The potential drop across this dipole layer is given by $\Delta V = Q/C$ where Q is the net adsorbed charge and C the capacitance of the dipole layer. Since at equilibrium the electrochemical potential of the adsorbed ions must equal the electrochemical potential of the same ions in solution and since the electrochemical potential of a species in a given phase depends on its activity, we can change the number of adsorbed anions and cations by altering their concentration in solution. This will result in a change in the net adsorbed charge and give rise to the Nernstian dependence of V_{fb} on ion concentration:

$$V_{fb} = V_{fb}(PZZP) - \frac{RT}{zF} \ln \frac{a}{a_{PZZP}} \qquad (1)$$

where V_{fb} (PZZP) is the intrinsic flatband potential determined by the semiconductor electron affinity, "z" is the ion charge F is Faraday's constant and "a" the activity. Clearly, at some point the number of adsorbed positive and negative charges will be equal and the potential drop across the Helmholtz layer will be effectively zero. Only the small dipolar contribution of adsorbed water or other neutral molecules will provide a potential gradient. At this point, the Point of Zero Zeta Potential

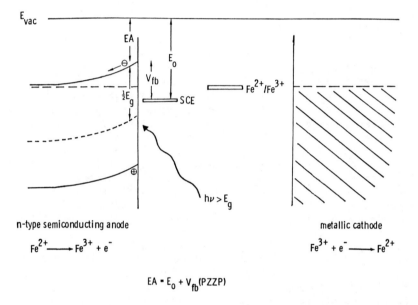

Figure 1. Energy level diagram for a photoelectrochemical cell illustrating the relationship between the electron affinity (EA) and the flatband potential (V_{fb}). Energy levels are shown for zero external bias ($E_o = 4.75$ eV).

(PZZP), the flatband potential is a direct measure of the semi-conductor electron affinity (EA).

This point may be located analytically by virtue of the fact that a semiconductor powder in a solution of adsorbing ions acts as a buffer for those ions everywhere but at the PZZP. Thus, potential drift or differential potentiometric titrations(7,8) can be employed to determine the PZZP as illustrated in Figure 2 for CdS. Once the PZZP is determined in this fashion, a direct comparison of EA and V_{fb} is possible and has been done for a variety of semiconductors.(9) Figure 3 illustrates the V_{fb} vs. pH data for p-GaP and shows good agreement between the predicted V_{fb} at the PZZP from electronegativity calculations and the observed value.(9)

The potential drop across the Helmholtz layer is thus impor-tant to the biasing requirements of a PEC and is directly affected by the chemisorption of ions at the semiconductor/electrolyte interface.

The chemisorption of ions plays another role in determining the properties of PEC devices. The adsorbed ions may create the chemical intermediates or specific reaction sites necessary for charge transfer and chemical product formation. The rapid kinet-ics in photoelectrolysis and wet photovoltaic cells are in no small part due to the fact that in most of these systems the redox species are strongly adsorbed. Knotek(10,11) and others (12,13) have shown that the nature of the TiO_2 surface and the species adsorbed on it greatly affect its catalytic properties. We have observed that p-GaP seems to be an effective hydrogen electrode because of the adsorbed aqueous species on the surface as reflected by the Nernstian dependence of V_{fb} on pH in Figure 3. Thus, it appears in many cases the chemisorption of ions at the interface is a necessary step for rapid charge transfer to occur. This clearly has a strong bearing on recent observations of Fermi level pinning in various PEC devices where strong chemi-sorption does not occur.(14,15)

The influence of chemisorbed ions is thus seen in the poten-tial drop across the Helmholtz layer and in the catalytic ability of the surface. Considerable work remains to be done on the chemical pretreatment of surfaces to maximize the catalytic nature of the surface and enhance the adsorption of appropriate ionic species.

Surface Decomposition

While it is related to and sometimes dependent upon chemi-sorption effects, one pathway for electrode decomposition can be looked upon as a separate type of charge transfer phenomenon. Here, charge transfer generally occurs between species in the semiconductor near the surface and not across the interface to species in the electrolyte. This charge transfer can be highly

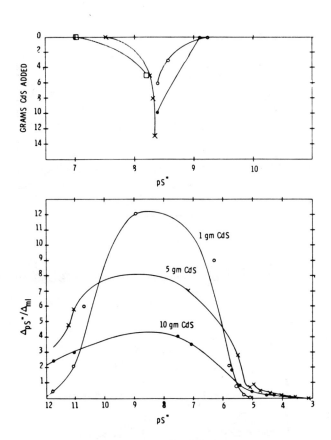

Figure 2.   Determination of the $S^=$ concentration at the PZZP of the CdS anode.

(top) Data taken by the pX drift technique. In particular, we plot grams of CdS added to a 600 mL-sulfide solution of known initial $pS^=$ and 0.1M KOH vs. measured $pS^=$. Points on continuous lines represent successive additions to the same solution. (bottom) Three differential potentiometric titrations of a suspended amount of CdS, 1, 5, 10 gm in 600 mL 0.1M KOH with Na_2S solution. The range of change of $pS^=$ with added titrant, $\partial pS^=/\partial S$, is plotted vs. $pS^=$. It is of note that the same peak for the PZZP is observed in both curves and that $\partial pS^=/\partial x$ scales inversely with the amount of added CdS.

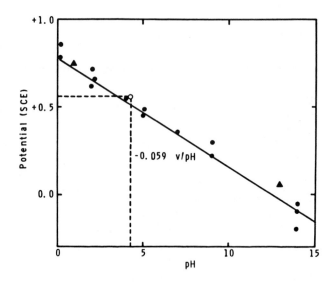

Figure 3.   Experimental flatband potential for p-GaP vs. pH (———) has a slope of 59 mV/pH unit; (○) calculated V_{fb} at the measured PZZP; (▲) this work; (●) G. Horowitz (26))

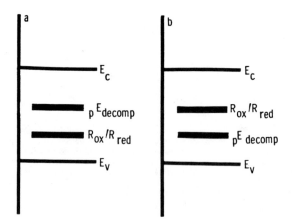

Figure 4.   Relative positions of hole decomposition potentials for the semiconductor and desired redox potentials for a photoelectrochemical cell using n-type semiconducting electrodes.   All examples are thermodynamically unstable, but (b) is kinetically more stable than (a).

activated by the adsorption of ions on the surface since hydration energies play an important role in electrode dissolution. However, the fundamental decomposition pathway is bond breaking in the semiconductor surface.

Gerischer($\underline{16}$), Bard and Wrighton($\underline{17}$) have recently discussed a simple model for the thermodynamic stability of a range of photoelectrodes. As has been discussed previously, except for the rare case where the anodic and cathodic decomposition potentials lie outside the band gap, the electrode will be intrinsically unstable anodically, cathodically, or both.($\underline{16}$) It is the relative overpotential of the redox reaction of interest compared to that of the appropriate decomposition potential which determines the relative kinetics and thus stability of the electrode as illustrated in Figure 4. The cathodic and anodic decomposition potentials may be roughly estimated by thermodynamic free-energy calculations but these numbers may not be truly representative due to the mediation of surface effects.

Once again the precise nature of the interface plays a crucial role in determining the decomposition products observed. This is clearly illustrated in the case of p-GaP as shown in Figure 5. In a 0.1 M NaOH solution decomposition is by Ga loss from the surface as the hydroxide, and photocurrent decay is rapid as a phosphate layer builds on the surface. In 0.1 M H_3PO_4 solution decay is again rapid but in this case the Ga is not solubilized but deposits on the surface as the metal. In the more oxidizing acids such as H_2SO_4 both Ga and P are removed from the surface and the photocurrent remains high as the surface is essentially photoetched.

While desorption is operative to some extent at most semiconductor/electrolyte interfaces, in some cases surface passivation can occur. Here charge transfer across the interface is used to establish covalent bonds with electrolyte species, which results in changes in surface composition. This is typified by the well-known oxide film growth on n-GaP and n-GaAs surfaces in aqueous solutions. In these cases, however, the passivation process can be competed with effectively by the use of high concentrations of other redox species such as the polychalcogenides.

Charge Transfer

We have thus far talked about the chemisorption of ions at the semiconductor/electrolyte interface and charge transfer in the semiconductor surface layer. The main charge transfer process of interest is the transfer of electrons and holes across the semiconductor/electrolyte interface to the desired electrolyte species resulting in their oxidation or reduction. For any semiconductor, electrode charge transfer can occur with or without illumination and with the junction biased in the forward or reverse direction.

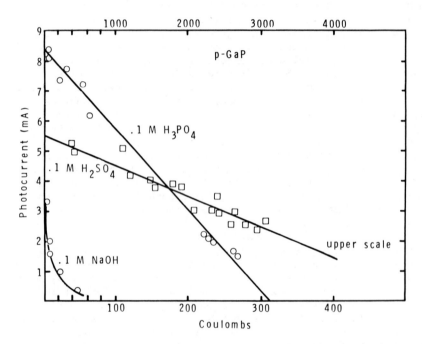

Figure 5. Photocurrent for p-GaP illuminated with white light as a function of total charge passed through the interface. Note the different scales for different electrolytes.

One of the major questions that remains to be answered is
the detailed mechanism of charge transfer. For redox couples
which lie in the gap of the semiconductor, isoenergetic electron
transfer would require the existence of an appropriate surface
state. While such states have been postulated, little direct
evidence of their existence is available. An alternate possibility is an inelastic (non-isoenergetic) electron transfer
process such as is commonly observed in solid state devices.(18)
    As indicated in Figure 1, if a semiconductor is biased to
depletion in contact with an electrolyte, a photocurrent can be
generated upon illumination. This occurs because the photoexcited majority carriers are driven by the electric field in
the depletion layer to the counter electrode and minority carriers migrate to the interface where they are trapped at the band
edge. Nozik has recently speculated that hot minority carrier
injection may play a role in supra-band edge reactions.(19)
Here the transfer time of the hot carriers is balanced against
their thermalization time constant.
    In many PEC systems the chemical kinetics for the primary
charge transfer process at the interface are not observed at the
light intensities of interest for practical devices and the
interface can be modeled as a Schottky barrier. This is true
because the inherent overpotential, the energy difference
between where minority carriers are trapped at the band edge and
the location of the appropriate redox potential in the electrolyte, drives the reaction of interest. The Schottky barrier
assumption breaks down near zero bias where the effects of interface states or surface recombination become more important.(13)
    The precise nature of the interaction of the potential deter-
mining species with the semiconductor surface will determine the
apparent redox potentials in the electrolyte. Kinetics would be
anticipated to be of importance in those cases where there is
not strong chemisorption of the potential determining species.
This is the case for CdS in acidic or basic aqueous solution
where photocurrents are nonlinear at low-light intensities and
the dependence of V_{fb} on pH is non-Nernstian.(20) Recent observations by Bard and Wrighton(14,15) indicate that Fermi level
pinning and therefore supra-band edge charge transfer can occur
in Si and GaAs in those systems (i.e., $CH_3CN/[n-Bu_4N]ClO_4$) with
various redox couples where little electrolyte interaction is
anticipated.
    That surface interactions play such a role clearly demands
that some sort of surface state concept be invoked. However, no
simple techniques have yielded direct information about the
nature of such states. To explain charge transfer, isoenergetic
electron or hole processes are normally invoked with a subsequent thermalization of the electron or hole in the semiconductor.
This unfortunately necessitates the existence of a surface state
at the level of the redox potential. This may of necessity occur
when strong chemisorption is present. However, in those cases,

where this is not the case, inelastic electron transfer may
occur. Such charge transfer processes at solid state interfaces
are well known(18) and involve the excitation of vibrational
modes, discrete electronic transitions and collective excitations
such as surface plasmons. It seems very plausible that similar
processes should occur at the semiconductor/electrolyte interface.

One way to probe the existence of such surface states is to
utilize another charge transfer phenomenon, that of photoemission.(21) Here electrons or holes may be injected directly
into the electrolyte from an illuminated, biased electrode as
illustrated for electron photoemission in Figure 6. Though
photoemission from a semiconductor directly into an aqueous
electrolyte has yet to be observed conclusively(22), the technique shows considerable promise as a surfce sensitive probe for
PECs. Figure 7 illustrates the yield curves for a WO_3 electrode
in 1 N - H_2SO_4.(23) A positive photocurrent is observed until a
sufficiently negative potential is achieved so that hydrogen
bronze formation occurs at the electrode surface. At this point
electron photoemission occurs. The nature of the positive photocurrent is not known. Threshold data, as shown in the insertion
Figure 7, indicate that it is true photoemission being observed
since the thresholds shift linearly with potential and agree
with those observed by others for photoemission from metal
electrodes.(23) The peak in the yield curve at 3.5 eV is
interesting and unexpected and may be due to reflectivity changes
or modifications in the final density of states.

Hole photoemission may also occur in appropriately biased
PECs, although this process has not yet been observed. The
major problem with the observation of photoemission from semiconducting electrodes is interference from the much larger photocurrents produced by the existence of the Schottky barrier at
the interface.

Thus hole or electron transfer can follow a number of path-
ways across the semiconductor/electrolyte interface. First, one
can have direct oxidative or reductive charge transfer to solution species resulting in desired product formation. Second,
one can have direct charge transfer resulting in surface modification, such as oxide film growth on GaP or CdS in aqueous PECs.
Finally, one can have photoemission of electrons or holes
directly into the electrolyte. All of these processes provide
some information about the electronic structure of the interface.

Electromigation of Ions

We have discussed the effects of chemisorption and of electron and hole transport across the semiconductor/electrolyte
interface. These have been shown to play a large role in determining electrode properties. Recently another mechanism of

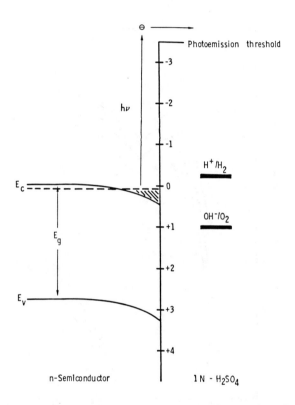

Figure 6.   The energy level structure for an n-*semiconductor–electrolyte interface as is appropriate for electron photoemission.*

The vertical axis is in volts relative to the saturated calomel reference electrode (SCE). Photoemission in aqueous media is best performed in the electrochemical window, between the hydrogen and oxygen evolution redox potentials, where small dc currents allow easier detection of the photoemission current. Protons are used for scavenging the photoexcited electrons.

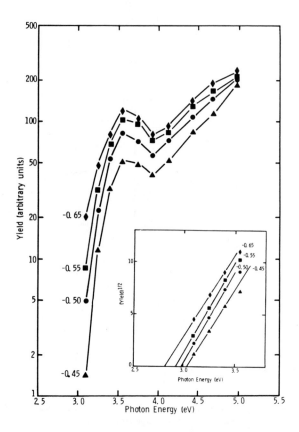

Figure 7. Normalized photoresponses yield spectra for WO$_3$ in 1N H$_2$SO$_4$. The data are taken in order of decreasing negative potential at the potentials indicated. All of the currents are cathodic. The insert shows Fowler plots used to determine the photoemission threshold at each potential.

charge transport across the interface has been recognized and
demonstrated to play a role in determining PEC characteristics.
(11,24) The field in the depletion region is only ~1 volt but
this potential is normally dropped across a depletion layer of
<1 μm depth. This gives rise to electric fields of 10^4
volts/cm and larger. These fields, which are large enough to
promote the movement of small charged ions, can easily be
manipulated by the external bias. This movement of ions can
effect the stability, flatband and spectral response of a semiconducting electrode.

Figure 8 illustrates the crystal structures for TiO_2 and
GaP, which are typical examples of a n-type semiconducting oxide
and a p-type main group semiconductor which show promise as photoelectrodes. Both materials have channels through the lattice
perpendicular to the exposed face. From a simple geometric
argument a critical radius for ions that can fit into the channels has been determined for each material. For rutile this is
R_c = 0.94 Å and for GaP it is R_c = 0.85 Å. These numbers
tell us immediately that we will only have to worry about the
migration of positive ions in these and most other semiconductors
since negative ions have too large a radius, the smallest being
F^- at 1.33 Å and $O^=$ at 1.32 Å. There do exist, however, a
large number of positively charged ions with an appropriate
radius to fit into the channels. The influence on anions at the
semiconductor surface can be large however and this will be
illustrated later for F^- on $SrTiO_3$.

Under normal operating conditions, the fields in the depletion
region would be expected to cause the migration of positive ions
for n- and p-type semiconductors as shown in Figure 9. This
clearly has large consequences for the defect doped metal oxides
where interstitial metal atoms would be expected to be leached
slowly from the surface changing the doping profile. For TiO_2
and $SrTiO_3$ this is in fact exactly what is observed. Figure 10
shows the results of the anodic aging of a lightlydoped $SrTiO_3$
wafer on its spectral response. As expected, the removal of
interstitial Ti^{3+}(R_c = .76 Å) alters the doping profile and
this widens the depletion region width allowing for the collection of electron-hole pairs from photons with a longer wavelength
and a deeper penetration depth. This, in effect, red shifts the
photoresponse.

The doping profile after anodic aging has been modeled by
Butler(5) and a typical example is shown in Figure 11. After
aging the TiO_2 the resultant concentration gradient causes the
room temperature migration of Ti^{3+} interstials and over a period
of one week the doping profile is restored to near that of the
virgin sample. For $SrTiO_3$ the activation energy for movement of
the interstitials is considerably larger and the anodically aged
profile is relatively stable at room temperature. At ~300°C
and above the interstitials can move more freely. Since only
small changes in sample stoichiometry <1 ppm occur no changes in

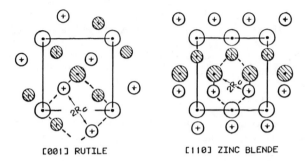

Figure 8.   *Crystal structures for TiO$_2$ (rutile) (left) and GaP (zinc blende) (right). The critical radii for these two are 0.94 Å and 0.85 Å, respectively.*

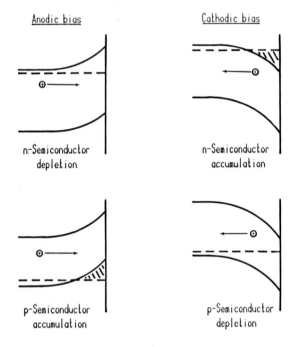

Figure 9.   *The electric fields and directions of positive ion electromigration in the near-surface region of both n- and p-type semiconductors under anodic and cathodic bias*

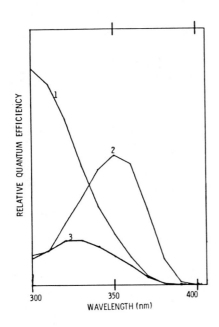

Figure 10.   Electrochemical anodic aging experiments on a lightly doped SrTiO₃ wafer. All experiments were performed in 1M NaOH. ((1) virgin SrTiO₃; (2) anodically aged 21 h at 5 V vs. SCE; (3) annealed in air 16 h at 300°C)

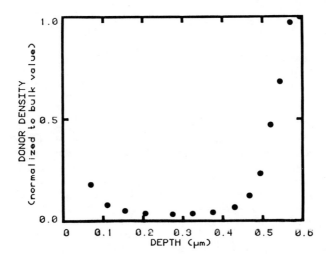

Figure 11.   The measured defect density profile in 3 Ω-cm TiO₂ sample aged under constant voltage (+5 V) for 30 h. The data is normalized to the bulk density.

V_{fb} are expected and none are observed as shown in the top half
of Table I.   While enhanced red response is clearly useful in
these photoelectrodes, the consequences in terms of overall
electrode stability may outweigh any advantage gained.   Clearly
substitutional doping would avert these problems and may be a
preferable technique.

Table 1

DEPENDENCE OF V_{fb} ON AGING CONDITIONS FOR $SrTiO_3$

Bias	Illumination	Conditions[†]	V_{fb} vs SCE
		Virgin	−1.11
Anodic	On	+5 V vs SCE PC* 21 hrs.	−1.15
Anodic	On	+1 V vs SCE PC 4 days	−1.15
		Virgin	−1.17
Cathodic	Off	.2 mA-CC** 25 hrs.	− .95
Cathodic	Off	.2 mA-CC 21 hrs.	− .96

*PC — Potential Control
**CC — Current Control
[†]The anodic and cathodic aging experiments were done
consecutively.

    For a p-type semiconductor operated in the normal fashion or
for an n-type material under cathodic bias (accumulation) it is
possible to get positive ion injection into the semiconductor
surface.   As we have previously discussed, the effects of such
injection can be quite dramatic including fracture of the electrode surface as illustrated in Figure 12.   In aqueous media
proton injection is most likely to occur and can have pronounced
effects on electrode performance.   Figure 13 illustrates the
changes in the spectral response of a $SrTiO_3$ electrode that has
had H^+ injected into the near—surface region by cathodic aging.
The shape of the spectral response curve remains basically
unchanged as a consequence of the fact that the depletion layer
width is relatively unchanged.   The overall quantum efficiency
is significantly reduced, which suggests that the injected H^+
significantly increases recombination center densities.   Since

Figure 12. Photograph (4× magnification) of an electrochemically deuterium-doped rutile wafer. Doping was accomplished by current-controlled cathodic aging an undoped TiO_2 wafer in LiOD, D_2O for 3 days at 10 mA. This resulted in a shattering of the sample in the region exposed to the electrolyte.

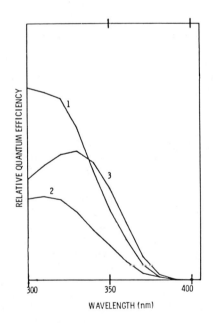

Figure 13. Electrochemical cathodic aging experiments under current-controlled conditions (cc) on a lightly doped $SrTiO_3$ wafer. All experiments were performed in 1M NaOH. ((1) virgin $SrTiO_3$; (2) cathodically aged 0.2 mA-cc for 25 h; (3) 0.0 V vs. SCE for 27 h)

the amount of hydrogen injected is large, the electron affinity
of the surface region should change and therefore changes would
be expected in V_{fb}. The lower portion of Table I illustrates
V_{fb} as a function of cathodic aging. The large change initially
represents a saturation of available sites. The subsequent lack
of change is an indication that additional hydrogen penetrates
further and that the concentration of hydrogen cannot go beyond
the original saturated level. We can use the electronegativity
model(7,8) to quantify the amount of hydogen incorporated. The
bulk electronegativity for hydrogenated $SrTiO_3$ will be:

$$X(SrTiO_3H^x) = [X(Sr)X(Ti)X^3(O)X^x(H)]^{1/(5+x)} \qquad (2)$$

From the measured flatband potential and the relationship(24)

$$EA = V_{fb} + E_o + \Delta_{fc} + \Delta_{px} \qquad (3)$$

where E_o is the difference between the vacuum level and the SCE,
Δ_{fc} is a small correction term for the difference between the
Fermi level and the conduction band and Δ_{px} is the potential
drop across the Helmholtz layer due to adsorbed charge, we can
calculate the materials apparent electron affinity. We know as
well that

$$EA(mat) = X(mat_{bulk}) - 1/2\ E_g(mat) \qquad (4)$$

thus

$$EA(observed) = [X(Sr)X(Ti)X^3(O)X^x(H)]^{1/(5+x)}$$
$$- 1/2\ E_g(SrTiO_3) \qquad (5)$$

Since E_g remains unchanged from the virgin value, (Fig. 13) substituting we find X equals 0.2 or there is one hydrogen for every
five titaniums. This ratio is also found for TiO_2 where it is
substantiated by stimulated desorption experiments.(11) The
hydrogen occupies a number of different sites both hydroxyl and
hydride as observed by infrared spectroscopy.(11) As curve three
in Figure 13 shows, anodic aging of the sample removes a large
portion of the hydrogen associated with recombination centers.
But as the flatband indicates and surface measurements show, a
significant portion of the hydrogen cannot be removed in this
fashion.
       An interesting effect has been observed in p-GaP.(25) If the
cathodic breakdown regime is attained (10 V or greater vs SCE) in
the dark at cathodic currents of 10 mA or greater, luminescence
is observed from the electrode. This luminescence is associated
with the injection of ions into the semiconductor. The luminescence is broad band and occurs both above and below the band gap
as illustrated in Figure 14a. Table II illustrates the depen-

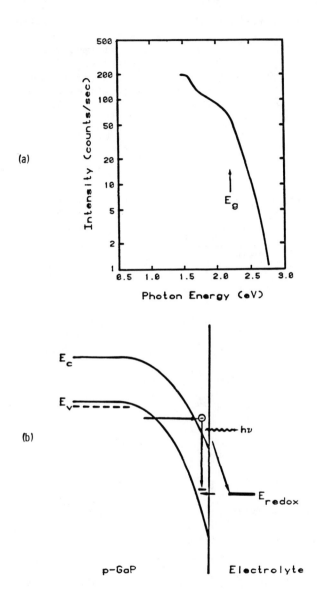

Figure 14. (a) Spectral distribution of the luminescence at a reverse-biased p-GaP/electrolyte interface. The electrolyte is 0.15M HNO₃ and the current density flowing through the interface is ~ 20 mA/cm². The low-energy limit of the spectrum is determined by the photomultiplier sensitivity. (b) Strongly cathodically biased p-GaP/electrolyte interface. Hot electrons are created by tunneling from valence to conduction bands. These may decay radioactively to fill empty states created by cation injection or drive other redox reactions.

dence of the luminescence on the ion size. Clearly ions of radius greater than R_c cannot be injected and thus do not cause luminescence. Ion injection has been confirmed with forward scattering experiments which have detected Li injection to depths of greater than 0.5 microns.(25) A possible luminescence mechanism is illustrated in Figure 14b. Injected cations create vacant electronic states below the conduction band of the semiconductor which may be filled by radiative decay of hot electrons injected into the conduction band.

Table 2

EFFECTS OF VARIOUS IONS ON THE LUMINESCENCE IN p–GaP

Ion	Crystal Ionic Radii (Å)	Steady State Luminescence
H^+	--	Yes**
Mg^{++}	0.66	Yes
Li^+	0.68	Yes
Zn^{++}	0.74	Yes
-------------- R_c(GaP) = 0.85 Å --------------		
Na^+	0.97	No
Cd^{++}	0.97	No
K^+	1.33	No
NH_4^+	1.43	Yes

*Handbook of Chemistry and Physics, ed. by C. D. Hodgman
  (Chem. Rubber. Co., Cleveland, 1962) pg. 3507.
**Only at concentrations less than 0.5 M.

Anions are nominally too large to be injected down channels in the semiconductor. However, under strong anodic bias there may be electrochemical interactions of the anions with the semiconductor surface (possibly by an ion exchange mechanism). Figure 15 shows the spectral response and I–V characteristics for a $SrTiO_3$ electrode in 1 M KOH after being aged for two days at +5 V vs SCE in 1 M KF. These electrode properties appear constant with respect to operation in a PEC. The quantum efficiency at 0 V vs SCE is substantially improved as is the collec-

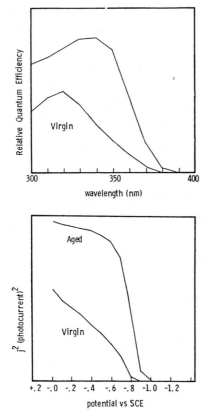

Figure 15. (top) Spectral response curves for a reduced SrTiO₃ electrode before and after aging in F⁻ solution. (bottom) I–V data for the same aging conditions. The I–V and spectral response curves were obtained in 0.1M NaOH, and the electrode was aged in 1M KF for 2 days at +5 V vs. SCE.

tion efficiency near the flatband potential. This increase in the fill factor and enhancement of surface charge transfer may be indicative of improved electrode kinetics or the elimination of surface recombination. Work is currently in progress to determine which process is important.

Summary

Four main charge transfer phenomena appear important in PECs. Chemisorption of ions creates a potential drop across the Helmholtz double layer and provides chemical intermediates which can significantly alter the kinetics of the reaction. Surface bond breaking is a potent means of electrode degradation. Electron and hole transport across the semiconductor/electrolyte interface may be elastic or inelastic and give rise to classical electrochemistry. Photoemission may also occur giving rise to nonclassical products and can be used as a useful probe of the interfacial electronic structure. Electromigration of ions can change doping profiles and be used by way of electrochemical ion injection to modify electrode surface and near-surface regions. Much remains to be understood concerning the details of these processes. Such an understanding is vital to the successful production of an optimum photoelectrode.

*This work was supported by the Materials Sciences Division, Office of Basic Energy Sciences, U. S. Department of Energy under Contract DE-AC04-76-DP00789.

Literature Cited

1. Gerischer, H. in "Physical Chemistry: An Advanced Treatise," Eds. Eyring, H., Henderson, D., Jost, W. (Academic Press, New York, 1970) Vol. 9A.
2. Nozik, A. J.; Ann. Rev. Phys. Chem., 1978, 29, 189.
3. Butler, M. A. and Ginley, D. S.; J. Mat. Sci., 1980, 15, 1.
4. Gerrard, W. A. and Rouse, L. M.; J. Vac. Sci. Technol., 1978, 15, 1155.
5. Butler, M. A.; J. Electrochem. Soc., 1979, 126, 338.
6. Schwerzel, R. E.; Brooman, E. W.; Craig, R. A.; V. E. Wood in "Semiconductor Liquid-Junction Sola Cells," edited by Heller, A. (Electrochemical Society, Princeton, 1977) p. 293.
7. Butler, M. A. and Ginley, D. S.; J. Electrochem. Soc., 1978, 125, 228.
8. Ginley, D. S. and Butler, M. A.; J. Electrochem. Soc., 1978, 125, 1968.
9. Butler, M. A. and Ginley, D. S.; J. Electrochem. Soc., (June 1980).

10. Knotek, M. L.; Abstract 339, p. 869, The Electrochemical Society Extended Astracts, Philadelphia, PA, May 8–13, 1977.
11. Ginley, D. S. and Knotek, M. L.; J. Electrochem. Soc., 1979, 126, 2163.
12. Wilson, R. H.; Abstract 415, p. 1038, The Electrochemical Society Extended Abstracts, Seattle, WA, May 21–26, 1978.
13. Somorjai, G. A.; Abstract, Solar Energy and Photoelectronic Processes Symposium, p. 420, 109th AIME Annual meeting, Las Vegas, NV, February 24–28, 1980.
14. Bocarsly, A. B.; Bookbinder, D. C.; Dominey, R. N.; Lewis, N. S.; Wrighton, M. S.; J. Amer. Chem. Soc., 1980, 102, 0000.
15. Bard, A. J.; Bocarsly, A. B.; Fan, F. F.; Walton, E. G.; Wrighton, M. S.; J. Amer. Chem. Soc., 1980, 102, 0000.
16. Gerischer, H.; J. Electroanal. Chem., 1977, 82, 133.
17. A. J. Bard and M. S. Wrighton; J. Electrochem. Soc., 1977, 124, 1706.
18. Duke, C. B.; "Tunneling in Solids," Academic Press, New York 1969.
19. Turner, J. and Nozik, A. J.; Abstract 84, Coll. and Surface Science Division, American Chemical Society Abstracts, Houston, TX, March 23–28, 1980.
20. Watanabe, T.; Fujishima, A.; Honda, A.; Chem. Lett., 1975, 897.
21. Gerischer, H.; Kolb, D. M.; Sass, J. K.; Adv. in Phys., 1978, 27, 437.
22. Gurevich, Y. Y.; Krotova, M. D.; Pleskov, Y. V.; J. Electroanal. Chem., 1977, 75, 339.
23. Butler, M. A.; Surface Science, in press.
24. Butler, M. A. and Ginley, D. S.; Chem. Phys. Lett., 1977, 47, 319.
25. Butler, M. A. and Ginley, D. S.; Appl. Phys. Lett., 1980, 36, 845.
26. Horowitz, G.; J. Appl. Phys., 1978, 49, 3571.

RECEIVED October 15, 1980.

The Role of Interface States in Electron-Transfer Processes at Photoexcited Semiconductor Electrodes

R. H. WILSON

Corporate Research and Development, General Electric Company, P.O. Box 8, Schenectady, NY 12301

Electronic energy levels localized at the surface of a semiconductor have frequently been used to explain experimentally observed currents at semiconductor-electrolyte junctions.[1-9] These surface or interface states are invoked when the observations are inconsistent with direct electron transfer between the conduction or valence band of the semiconductor and electronic states of an electrolyte in contact with the semiconductor. Charge carriers in the semiconductor bands are transferred to surface states that have energies within the bandgap of the semiconductor. From these states electrons can move isoenergetically across the interface to or from electrolyte states in accordance with the widely accepted view of electron transfer.

The process by which the semiconductor carriers reach the surface to react with surface states must be considered. The case of greatest importance under photoexcitation is with the semiconductor biased to depletion as shown in Figure 1. While it is possible for semiconductor carriers to reach the surface of the semiconductor through tunneling, or impurity conduction processes, these processes have not been shown to be important in most examples of photoexcited semiconductor electrodes. Consequently, these processes will be ignored here in favor of the normal transport of carriers in the semiconductor bands. Furthermore, only carriers within a few kT of the band edges will be considered, i.e., "hot" carriers will be ignored.

Figure 1 illustrates different modes of electron transfer between electrolyte states and carriers in the bands at the semiconductor surface. If the overlap between the electrolyte levels and the semiconductor bands is insufficient to allow direct, isoenergetic electron transfer, then an inelastic, energy-dissipating process must be used to explain experimentally observed electron transfer. Duke[10] has argued that a complete theory for electron transfer includes terms that allow direct, inelastic processes. The probability of such processes, however, has not been treated quantitatively.

On the other hand, inelastic transfer of carriers in the bands to surface states is well known[11,12,13] and reaction rates sufficient to

0097-6156/81/0146-0103$05.00/0

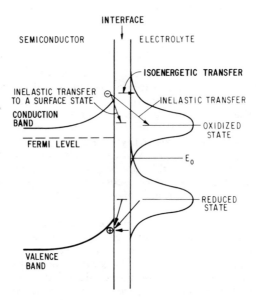

*Figure 1.   Schematic of various electron transfer processes between semiconductor
carriers at the surface and electrolyte and surface states*

explain many of the experimental observations can be readily estimated. The objective of this paper is to focus on this reaction between carriers and surface states and to emphasize the reaction cross section as an important parameter in charge transfer as well as in surface recombination. The role of surface states in charge transfer at the semiconductor-electrolyte interface is contrasted with effects at semiconductor-vacuum, semiconductor-insulator and semiconductor-metal interfaces. Factors affecting the magnitude of the reaction cross section are discussed and the theoretical understanding of the capture process is briefly reviewed.

Finally, some experimental observations are discussed in which charge transfer to surface states is important. The emphasis is on methods to be quantitative in describing the role of surface states by determining their density and reaction cross sections. Some previously published observations as well as preliminary new results are used to illustrate the role of surface bound species as charge transfer surface states.

Quantitative Reaction Rates

Localized states in the bulk of a semiconductor that have energies within the bandgap are known[10] to capture mobile carriers from the conduction and valence bands. The bulk reaction rate is determined by the product of the carrier density, density of empty states, the thermal velocity of the carriers and the cross-section for carrier capture. These same concepts are applied to reactions at semiconductor surfaces that have localized energy levels within the bandgap.[11,12] In that case the electron flux to the surface, F_n, reacting with a surface state is given by

$$F_n = n_s N_e \sigma_n v_n \tag{1}$$

where N_s is the density of electrons in the conduction band at the surface, N_e is the area density of empty surface states, σ_n is the electron capture cross section of the surface states and v_n is the thermal velocity of electrons. This is clearly an oversimplified description of the capture process which hides many of the complexities in the phenomenological parameter, σ. Nevertheless, this approach has been usefully applied in describing the kinetics of semiconductor surface states in a variety of circumstances.

For holes near the edge of the valence band with density, p_s, at the surface the analogous expression for the hole flux to the surface, F_p, reacting with a filled surface state of area density, N_f, is

$$F_p - p_s N_f \sigma_p v_p \tag{2}$$

By including terms to describe the rate of emission of electrons and holes from filled and empty surface states respectively, constraining the total number of surface states to N_t by $N_f + N_e = N_t$ and assuming that the states do not interact, the electron-hole recombination at the surface has been analyzed[11,12] in analogy with Hall-Shockley-Read[15,16] recombination. These methods are important in the study of semiconductor-

insulator interfaces.[17]   Similar methods are applied to the kinetics at
semiconductor-gas interfaces.[18,19]   When the semiconductor surface is
coupled through a capacitance small compared with the semiconductor
space charge capacitance (semiconductor-gas, semiconductor-vacuum,
metal-thick insulator-semiconductor) the surface potential is very sensitive to the net change in surface states. (In the standardized terminology
for metal-insulator-semiconductor structures this charge is referred to as
interface trapped charge).[20]

In the case of a metal-semiconductor junction the semiconductor
surface is closely coupled to the metal. As a result electrons in the
conduction band at the surface see a high density of empty metal states
into which they can cross the interface isoenergetically.   Similarly,
valence band holes see a high density of filled electron states in the
metal. As a result, electron transfer is usually treated as direct transfer
in a thermionic process and surface states are relegated to a minor role in
surface generation and recombination effects.[21,22]

By contrast, electrolyte states are much more limited in their
distribution than metal conduction band states so that in many cases
electron transfer through surface states may be the dominant process in
semiconductor-electrolyte junctions. On the other hand, in contrast to
vacuum and insulators, liquid electrolytes allow substantial interaction at
the interface. Ionic currents flow, adsorption and desorption take place,
solvent molecules fluctuate around ions and reactants and products
diffuse to and from the surface. The reactions and kinetics of these
processes must be considered in analyzing the behavior of surface states
at the semiconductor-electrolyte junction. Thus, at the semiconductor-
electrolyte junction, surface states can interact strongly with the electrolyte but from the point of view of the semiconductor the reaction of
surface states with the semiconductor carriers should still be describable
by equations 1 and 2.

In the next section factors that affect the reaction cross section are
discussed. It is argued that electrolyte species on the surface of the
semiconductor can qualify as surface states. In the subsequent section
several examples of such surface states will be discussed.

Factors Affecting Cross Section

As a hole or electron moves through a semiconductor its cross
section for encountering a neutral impurity atom should be an atomic
dimension, about 10^{-15} cm^2. The probability that the carrier be captured
during that encounter, however, may be much less than one, depending on
the energy of the available impurity level and the quantum state of the
carrier and impurity levels. When the capture occurs an amount of energy
equal to the difference between the band edge and the impurity level is
lost by the electron. This energy must be carried away by photon and/or
phonon emission. The maximum phonon energy in the available phonon
spectrum of the semiconductor crystal is generally less than one-tenth of
an electron-volt. Consequently, for radiationless capture (no photon
emission) of a carrier by a deep level more than a tenth of an electron-
volt from the band, emission of many phonons is required. In that case

the capture probability per encounter might be quite low so that cross sections on the order of 10^{-22} cm^2 are expected. Observed cross sections, however, are frequently on the order of 10^{-15} cm^2. (For a more complete discussion see Ref. 14, p. 91.) Theory to explain such large cross sections is not well established. The cascade capture model of Lax[23,24] has had some success but a multiphonon emission process is more appropriate for deep levels.[25]

A similar situation exists for carrier capture by surface states. Relatively large capture cross sections are observed but no adequate theoretical treatment exists. Theory to describe the capture process is greatly complicated by presence of the surface. The carrier motion as well as the vibrational behavior of the crystal is perturbed by the surface. What does seem clear, however, is that the surface state should be tightly enough bound to the crystal lattice so that phonon emission is possible. In addition, the state should be close enough to the semiconductor to overlap the wave function of the semiconductor carrier.

As a working definition, a surface state can be any electron energy level within the bandgap of the semiconductor located at its surface that is coupled to the semiconductor lattice strongly enough to allow inelastic capture of carriers from the semiconductor bands. Several examples of possible surface states are illustrated in Figure 2. In the next section experimental manifestations of some of these are described.

Experimental Examples

Sulfide ions on CdS. When Na_2S is added to an aqueous electrolyte the flatband voltage of a CdS electrodes shifts to more negative potentials by an amount linear in the log of the Na_2S concentration.[26] The change in flatband voltage is due to a change in voltage across the Helmholtz layer. The Nernstian dependence on concentration suggests that some form of negative sulfide ion is bound to the semiconductor surface by a chemical reaction. An estimate of the sulfide ion concentration on the surface has been made using a simple capacitive model of the Helmholtz layer.[27] The result is shown in Figure 3 based on the flatband voltage measurements of Inoue et al.[26] In addition, when their current-voltage data are replotted relative to the flatband voltage as shown in Figure 4, significant effects in addition to the flatband voltage changes are evident. The photocurrent onset is shifted to more negative potentials than observed in the absence of sulfide. The role of the sulfide ions can be interpreted in several ways.

One interpretation presumes that the photocurrent onset in the absence of sulfide is determined by electron-hole recombination. The sulfide ions on the surface are then supposed to be bound to these surface recombination levels rendering them unavilable for recombination reactions. The charge transfer reactions could then proceed at lower voltages. In this case the corrosion suppression role of the sulfide ions would be to reduce the oxidized corrosion site before a cadmium ion could go into solution. A variation on this theme is to consider the corrosion site to be the recombination state, i.e., the site on the surface that normally leads to corrosion when oxidized by a photoexcited hole can be

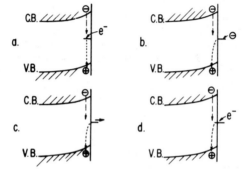

Figure 2.   Illustration of various kinds of surface states that can react with carriers at the surface of the semiconductors.

(a) an intrinsic surface state of the semiconductor that may be due to lattice termination or a dangling bond. Subsequent reactions of such a state may lead to corrosion. (b) A specfically adsorbed electrolyte species. After reacting with a carrier it may desorb. (c) An intermediate in a multielectron reaction. After first hole capture, for example, the intermediate product may form a surface state capable of a second hole capture with subsequent release of the final product to solution. (d) A molcule attached to the surface by chemical treatment prior to immersing in the electrolyte. Such a surface state could, for example, be oxidized by a hole and subsequently be reduced from the electrolyte.

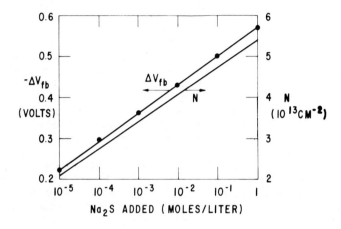

Journal of the Electrochemical Society

Figure 3.   Surface density N of adsorbed sulfide ions calculated from the change in flatband potential V_{fb} reported in Ref. 26 (27).

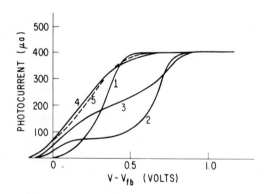

Journal of the Electrochemical Society

Figure 4.   Current–voltage curves of a CdS electrode in 0.2M Na_2SO_4 from Ref. 26. Figure 10 replotted relative to flatband voltage. (1) No Na_2S added; (2) 5×10^{-4}M Na_2S; (3) 10^{-3}M Na_2S; (4) 10^{-2}M Na_2S; (5) 10^{-1}M Na_2S (27).

reduced by an electron from the conduction band before the next step in the corrosion process takes place. In the absence of sulfide this recombination would be the dominant process at low voltages when the surface concentration of electrons is high. When sulfide is present in the electrolyte the sulfide ions could deposit on the corrosion sites and prevent their oxidation thereby simultaneously suppressing recombination and allowing current onset at lower voltages.

Another interpretation would be to suppose that the adsorbed sulfide ion forms a surface state that can be directly oxidized by a hole in the valence band. In this case the shift in current onset to lower voltages would be due to an increase in the charge transfer rate rather than the decrease in the recombination rate discussed in the preceeding paragraph. The corrosion suppression associated with the sulfide could then be partially attributed to the rapid kinetics of hole capture by these surface sulfide ions and partially due to reduction of oxidized corrosion sites by sulfide ions in solution.

It is not the objective here to choose between these alternatives. Additional experimental work is needed to do this. The objective of this work is to show how both the corrosion sites and adsorbed sulfide ions can be treated as surface states and to further suggest that quantitative treatment of these surface states can be useful in considering experiments to elucidate the role of the sulfide ions.

First, by referring to Figure 4 it can be seen that change of more than two orders of magnitude in sulfide concentration in solution causes only a small change in the photocurrent onset voltage. This is readily understood from Figure 3 if the adsorbed sulfide ions rather than the solution ions are responsible for the change in current onset. Thus, about 3×10^{13} sulfide ions per square centimeter are responsible for the shift in photocurrent onset. Since this number is only a few percent of the available atomic sites on the surface it would be surprising if that few sulfide ions could passivate all the corrosion sites. On the other hand, if the adsorbed sulfide ions act as surface states for direct, inelastic capture of valence band holes their surface reaction constant for holes is $S_s = N_s \sigma_{sp} v > (3 \times 10^{13}\ \mathrm{cm}^{-2})(10^{-15}\ \mathrm{cm}^2)(10^7\ \mathrm{cm/sec}) = 3 \times 10^5\ \mathrm{cm/sec}$ on the assumption that the coulomb attraction of the negatively charge sulfide ions for the positive holes would make the cross section greater than $10^{-15}\ \mathrm{cm}^2$.

If the corrosion sites are treated as surface states on the CdS (as seems logical since the cadmium atoms that eventually receive the holes are part of the semiconductor surface) the first step corrosion rate constant for holes, S_c, may be analyzed in a similar fashion $S_c = N_c \sigma_c v_p$. To explain the shift in current onset, $S_c \ll S_s$ is required. Assuming $N_c = 10^{15}$ cm^{-2}, $S_c < 10^5$ cm/sec if $\sigma_c < 10^{-17}$ cm^2. This is a reasonable number for a neutral surface state, therefore, this explanation for the observed shift in current onset passes a test of reasonableness. On the other hand, it is hard to be more quantitative without some independent measure of capture cross sections. A method of measuring cross sections applicable in some cases has been reported[28] and will be discussed later in this section.

Some additional experiments relevant to the data in Figure 4 are suggested by the preceding discussion. In particular, if the corrosion sites also act as the recombination centers that control current onset in the absence of sulfide ions (as discussed earlier) then there are no oxidized recombination centers before exposure to light. In that case a CdS electrode biased at a voltage below the saturated portion of curve 1 in Figure 4 would show a higher initial current than indicated in curve 1 and then decay in time to the curve 1 value. This situation can be analyzed by:

$$\frac{dN}{dt}_{ce} = P_s N_{cf} \sigma_{cp} v_p - n_s N_{ce} \sigma_{cn} v_n - P_s N_{ce} \sigma'_{cp} v_p - N_{ce} k \qquad (3)$$

where N_{ce} is the surface density of oxidized corrosion sites that have captured one hole, N_{cf} is the density of corrosion sites that are still filled ($N_{cf} + N_{ce} = N_c$, the total density of corrosion sites) and k is a rate constant for the reaction of the oxidized corrosion site with the electrolyte.

The net photo current, j_c, associated with these processes is

$$\frac{j_c}{q} = P_s N_{cf} \sigma_{cp} v_p + P_s N_{ce} \sigma'_{cp} v_p - n_s N_{ce} \sigma_{cn} v_n \qquad (4)$$

When an n-CdS electrode is suddenly illuminated with light capable of producing holes in the CdS, j_c, would almost immediately reach some large value (equal to or less than the saturation current of curve 1) and then decay to the steady state value of curve 1 as the steady state value of N_{ce} is approached according to equation 3. If such a transient does not occur the oxidized corrosion site acting as a recombination state is not the controlling factor in the photocurrent onset.

Another quantitative consideration involves the rate of replacement of the sulfide ions on the surface during illumination assuming that they are the states with which the holes react. The rate at which sulfide ions reach the CdS surface for 5×10^{-4} M concentration can be estimated to be greater than 10^{21} cm^{-2} sec^{-1} while for curve 2 of Figure 4 the first plateau occurs at $j/q \simeq 10^{15}$ cm^{-2} sec^{-1} suggesting that the value of N_s is not limited by the replacement of the sulfide ions. The plateau in curve 2 has been quantitatively attributed to a change in sulfide ion concentration at the surface due to a diffusion-limited process.[27] If that is the case capacitance measurements made under the operating conditions of the plateau in curve 2 would indicate a reversion of the flatband voltage to that of curve 1 with no sulfide in solution. If such a shift is observed then operation in the plateau area does not help elucidate the mechanism of charge transfer since the same diffusion limitation occurs for surface ion or solution ion transfer.

The objective of the preceding discussion of CdS electrodes has been to point out the possible role of two types of identifiable surface states: one a state intrinsic to the CdS surface, presumably associated with Cd, which is eventually ionized by reaction with a hole and dissolved into solution; the other is a sulfide ion deposited on the CdS surface from solution. By considering these as surface states capable of inelastic capture of carriers from the semiconductor bands quantitative estimates of their role in charge transfer were made to illustrate the usefulness of this approach.

Oxygen evolution on TiO_2. In a previous report[28] the dark reduction peak (curve 3 in Figure 5) produced by photoexciting a TiO_2 electrode was discussed. The extra dark current was analyzed using equation $\pmb{1}$ and

$$n_s = n_b \, exp- \; q(V-V_{fb})/kT$$

where n_b is the bulk electron density and V_{fb} is the flatband voltage. The resulting current during a negative voltage sweep at a rate dV/dt is

$$j = q \, N_i \; \frac{q}{kT} \; \frac{dV}{dt} \; e^{\frac{-q(V-Vp)}{kT}} \quad exp\left[-e \; \frac{-q(V-Vp)}{kT}\right] \qquad (5)$$

where N_i is the area density of oxidized surface states before the sweep was initiated and V_p is the voltage at which the current, j, is maximum. The cross section, σ^p, for electron capture by these states is given by

$$\sigma = -\frac{dV}{dt} \frac{q}{kTn_b v_n} \quad e^{\left[\frac{q(V_p - V_{fb})}{kT}\right]} \qquad (6)$$

All of the parameters except N_i and σ are determined by other observations. N_i is determined by the area under the curve and σ by the position of the current peak. Comparison of experimental and calculated curves are shown in Figure 6. A better fit could be achieved by assuming a distribution of cross sections, but that did not seem justified. More experimental work is needed but the surface state analysis fits the available experimental observations in all essential details with reasonable values for N_i and σ.

As discussed previously, the surface states responsible for the reduction peak could be intrinsic surface states or states associated with a surface-attached intermediate in the series of reactions leading to O_2 evolution. The latter possibility was deemed to be more likely since no change in voltage across the Helmholtz layer (no change in capacitance) was observed when these states are in the oxidized form.

Regardless of the nature of the surface state it is clear that it can capture an electron from the conduction band producing cathodic current. This cathodic current balances the anodic current produced when the photoexcited holes produced the oxidized surface state. The net result of these two processes is electron-hole recombination leading to no net current. This recombination process is what controls the voltage of photocurrent onset as can be seen in curve 2 of Figure 5.

The results summarized here illustrate the important role surface states play in O_2 evolution from photoexcited TiO_2 and provide an example of a quantitative determination of the density and electron capture cross section of these states.

Attached molecules. A molecule attached to a semiconductor surface could behave as a surface state if it has an energy level within the

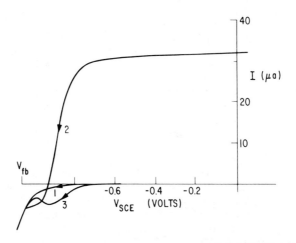

Journal of the Electrochemical Society

Figure 5.   Current–voltage curves of a TiO_2 electrode in 1.0M KOH with 200 mV/s scan from 0 V (SCE). Curve 1, in dark; Curve 2, with 350-nm light on; Curve 3, in dark after 30 s illumination with 350-nm light at 0 V (SCE) (28).

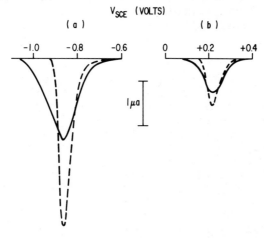

Journal of the Electrochemical Society

Figure 6.   Comparison of extra experimental reduction current ((———); analogous to the difference in Curves 1 and 3 in Figure 5) with calculated ((– – –) using Equation 5 in (a) 1.0M KOH and (b) 0.5M H_2SO_4 using $N_i = 6.7 \times 10^{13}$ cm^{-2} and 1.9×10^{13} cm^{-2}, respectively with peak positions matched to experimental curves (28).

bandgap. The cross section for carrier capture would depend on the nature of the attachment process, the strength of the bond and the distance from the surface.

Figure 7 shows some preliminary results obtained from a TiO_2 electrode to which Ru(4-OH$_{30}$ 3-COOH o-phen)(bipy)$_2$(PF$_6$)$_2$[29] was attached using a silane process.[30] The curves are analogous to Figure 5 except the photoexcitation prior to curve 3 in Figure 7 was with 470nm light. This wavelength light does not produce electron hole pairs in TiO_2 but is at the maximum of absorption of the ruthenium compound. The photocurrent is presumed to be due to transfer of an electron from the excited state of the attached material to the conduction band of the TiO_2 leaving behind the oxidized compound. The current decreases in time indicating that the reduced form is not being regenerated.

The reduction peaks in curve 3 of Figure 7 are presumed to be due to the transfer of an electron from the conduction band to the oxidized compound. After the voltage sweep of curve 3 the current reverts to curve 1 and the 470 nm photocurrent reverts to its original value. The process is repeatable and delays of several minutes between photoexcitation and voltage sweep show little change in curve 3.

Peaks at about -0.25V. and -0.55V can be distinguished. Using -0.2 V/sec scan rate, 10^{18} cm^{-3} for n_b, 10^7 cm/sec for v_n and -0.7V for V_{fb} in equation 6 gives cross sections of 5 x 10^{-17} cm^2 and 2 x 10^{-20} cm^2 respectively for those peaks. Each yields about 3 x 10^{13} cm^{-2} for its surface density.

It should be emphasized that these results are preliminary and subject to modification as further checks and measurements are made. They are included here to illustrate how a surface attached molecule can be treated quantitatively as a surface state.

Conclusions

There is a growing tendency to invoke surface states to explain electron transfer at semiconductor-electrolyte interfaces. Too frequently the discussion of surface states is qualitative with no attempt to make quantitative estimates of the rate of surface state reactions or to measure any of the properties of these surface states. This article summarizes earlier work in which charge transfer at the semiconductor-electrolyte interface is analyzed as inelastic capture by surface states of charge carriers in the semiconductor bands at the surface. This approach is shown to be capable of explaining the experimental results within the context of established semiconductor behavior without tunneling or impurity conduction in the bandgap. Methods for measuring the density and cross section of surface states in different circumstances are discussed.

A working definition of a surface state as any energy level within the bandgap that is bound to the surface sufficiently to allow inelastic electron transfer to or from the semiconductor bonds was introduced. This allows adsorbed electrolyte species, reaction intermediates and attached layers to be considered as surface states. The experimental observations discussed illustrate such states.

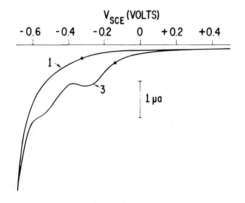

Figure 7.   Current–voltage curves for a Ru(4-OH,3-COOH,o-phen)(bipy)₂(PF₆)₂ coated TiO₂ electrode in pH 7 Na₂SO₄. Current curves analogous to Curves 1 and 3 in Figure 5 except the illumination was at +0.5 V and with 470 nm light.

While the ability to treat capture cross sections theoretically is very primitive and the experimental data on capture cross sections are very limited this phenomenological parameter seems to be an appropriate meeting place for experiment and theory. More work in both of these areas is needed to characterize and understand the important role of surface states in electron transfer at semiconductor-electrolyte interfaces.

Acknowledgements

The author gratefully acknowledges many helpful discussions with LA Harris and the TiO_2 coating work of PA Piacente. This work was supported in part by the Department of Energy Office of Basic Energy Sciences.

LITERATURE CITED

1.   Freund, T. and Morrison, SR, Mechanism of Cathodic Processes on the Semiconductor Zinc Oxide, Surface Science, 9, 119, 1968.

2.   Memming, R. and Schwandt, G., Electrochemical Properties of Gallium Phosphide in Aqueous Solution, Electrochim, Acta, 13, 1299, 1968.

3.   Gerlscher, H., On the Role of Electrons and Holes in Surface Reactions on Semiconductors, Surface Science, 13, 265, 1969.

4.   Tyagai, VA and Kolbasov, GY, The Contribution of Surface States to the Charge Transport Process Across CdS, CdSe-Electrolyte Interface, Surface Science, 28, 423, 1971.

5.   Frank, SN and Bard, AJ, Semiconductor Electrodes. II. Electrochemistry at n-type TiO_2 Electrodes in Acetonitrile Solutions, J. Amer. Chem. Soc., 97, 7427, 1975.

6.   Bolts, JM, and Wrighton, MS, Correlation of Photocurrent-Voltage Curves with Flat-band Potential for Stable Photoelectrodes for Photoelectrolysis of Water, J. Phys. Chem., 80, 2641, 1976.

7.   Wilson, RH, A model for the Current-Voltage Curve of Photoexcited Semiconductor Electrodes, J. Appl. Phys., 48, 4292, 1977.

8.   Nishida, M., Charge Transfer by Surface States in the Photoelectrolysis of Water Using a Semiconductor Electrode, Nature, 277, 202, 1979.

9.   Vandermolen, J., Gomes, WP, Cardon, F., Investigation of the Kinetics of Electroreduction Processes at Dark TiO_2 and $SrTiO_3$ Single Crystal Semiconductor Electrodes, J. Electrochem. Soc., 127, 324, 1980.

10. Duke, CB, Concepts in Quantum Mechanical Tunneling in Systems of Electrochemical Interest, to be published in the proceedings of the Third Symposium on Electrode Processes, Boston, MA, May 1979, The Electrochemical Society, Princeton, NJ.

11. Many, A., Goldstein, Y., and Grover, NE, Semiconductor Surfaces, Wiley, New York, 1965.

12. Grove, AS, Physics and Technology of Semiconductor Devices, Wiley, New York, 1967.

13. Morrison, SR, The Chemical Physics of Surfaces, Plenum Press, New York, 1977.

14. Milnes, AG, Deep Impurities in Semiconductors, John Wiley and Sons, New York, 1973.

15. Hall, RN, Electron-hole Recombination in Germanium, Phys. Rev., 87, 387, 1952.

16. Skockley, W., and Read, WT, Jr., Statistics of the Recombinations of Holes and Electrons, Phys. Rev., 87, 835, 1952.

17. Goetzberger, A., Kalusmann, E., Schulz, MJ, Interface States on Semiconductor/Insulator Interfaces, CRC Critical Reviews in Solid State and Material Sciences, 6, 1, 1976.

18. Many, H., Relation Between Physical and Chemical Processes on Semiconductor Surfaces, CRC Critical Reviews in Solid State Sciences, 4, 515, 1974.

19. Balestra, CL, Lagowski, J., and Gatos, HC, Determination of Surface State Parameters from Surface Photovoltage Transients, Surface Science, 64, 457, 1977.

20. Deal, BE, Standardized Terminology for Oxide Charges Associated with Thermally Oxidized Silicon, J. Electrochem. Soc., 127, 979, 1980.

21. Sze, SM, Physics of Semiconductor Devices, John Wiley and Sons, New York, 1969.

22. Card, HC, Photovoltaic Properties of MIS-Schottky Barriers, Solid State Electronics, 20, 971, 1977.

23. Lax, M., Giant Traps, J. Phys. Chem. Solids, 8, 66, 1959.

24. Lax, M., Cascade Capture of Electrons in Solids, Phys. Rev., 119, 1502, 1960.

25.   Henry, CH and Lang, DV, Nonradiative Capture and Recombination by Multiphonon Emission in GaAs and GaP, Phys. Rev. B, 15, 989, 1977.

26.   Inoue, T., Watanabe, T., Fujishima, A., and Honda, K., Competitive Oxidation at Semiconductor Photoanodes, in Semiconductor Liquid--Junction Solar Cells, Heller, A., Ed., The Electrochemical Society, Princeton, NJ, 1977, 210.

27.   Wilson, RH, Analysis of Charge Transfer at Photoexcited CdS Electrodes with Na$_2$S in the Electrolyte, J. Electrochem. Soc., 126, 1187, 1979.

28.   Wilson, RH, Observation and Analysis of Surface States on TiO$_2$ Electrodes in Aqueous Electrolytes, J. Electrochem. Soc., 127, 228, 1980.

29.   Piacente, PA, Determination of Ground and Excited State pK$_a$ values of Several New Ru(II) Complexes, MS Thesis, Rensselaer Polytechnic Inst., Troy, NY, August 1979.

30.   Abruna, HD, Meyer, TJ, Murray, RW, Chemical and Electrochemical Properties of 2,2'-bipyridyl Complexes of Ruthenium Covalently Bound to Platinum Oxide Electrodes.

RECEIVED October 3, 1980.

Competing Photoelectrochemical Reactions on Semiconductor Electrodes

W. P. GOMES, F. VAN OVERMEIRE, D. VANMAEKELBERGH, and F. VANDEN KERCHOVE

Rijksuniversiteit Gent, Laboratorium voor Fysische Scheikunde, Krijgslaan 271, B-9000 Gent, Belgium

F. CARDON

Rijksuniversiteit Gent, Laboratorium voor Kristallograti en Studie van de Vaste Stof, Krijgslaan 271, B-9000 Gent, Belgium

1. Introduction

As part of the research on solar energy conversion by photo-electrochemical methods, a considerable amount of investigation has been carried out during the last half decade on the stabilization of illuminated n-type semiconductor electrodes by means of competing hole reactions with reducing agents added to the solution. However, very little attention has hitherto been paid to the influence of the light intensity upon the relative rates of the competing reactions. Recently, it has been reported that the ratio of the rate of anodic photodissolution of the ZnO electrode versus that of electrooxidation of dissolved halide ions is light-intensity dependent, and an interpretation has been given to this effect based upon the concept of the quasi-Fermi level ($\underline{1}$). Independently, we have also found that the stabilization of n-GaP by $Fe(CN)_6^{4-}$ ion in aqueous medium is light-intensity dependent, and here, the effect has been interpreted in kinetic terms ($\underline{2}$, $\underline{3}$). In the present work, we have extended the experimental data on light-intensity dependent stabilization of n-type III-V semiconductor electrodes, as well as the kinetic analysis of this phenomenon.

2. Experimental

The measurements were made by means of the rotating ring-disk technique. The ring consisted of Au (for the determination of $Fe(CN)_6^{3-}$) or of amalgamated Au (for the determination of $Fe(III)-EDTA$). The disk consisted of either n-InP ($N_D = 5 \times 10^{16}$ cm^{-3}) or n-GaP ($N_D = 1.6 \times 10^{17}$ cm^{-3}). The characteristic dimensions of the RRDE were $r_1 = 2.50$ mm, $r_2 = 2.65$ mm and $r_3 = 2.95$ mm (for InP) or $r_3 = 3.65$ mm (for GaP). For both semiconductors, the ($\bar{1}\bar{1}\bar{1}$) face was exposed to the electrolyte. Before each experiment, the electrode surface was etched, the etchant being 0.3 % Br_2 in CH_3OH for InP and 0.5 M KOH + 1 M $K_3Fe(CN)_6$

in H_2O for GaP, respectively. The indifferent electrolyte consisted of 0.25 M K_2SO_4 plus buffer in H_2O, the buffer being either phtalate (pH $\approx$ 3.8) or borax (pH $\approx$ 9.2). Reducing agents added to the solution were either Fe(II)-EDTA (at pH = 3.8) or $Fe(CN)_6^{4-}$ (at pH = 9.2). Three types of interfaces were studied, i.e. n-InP/Fe(II)-EDTA, n-GaP/Fe(II)-EDTA and n-GaP/$Fe(CN)_6^{4-}$. For the disk potential, a value within the saturation photocurrent region was chosen (0.0 V vs. the sulphate electrode for InP and 0.0 or + 1.0 V vs. the sulphate electrode for GaP). The electrode was illuminated with a Hg lamp, and the Fe(III) complexes formed by photoelectrochemical oxidation were reduced at the ring (at - 1.4 V and - 0.80 V vs. the sulphate electrode for Fe(III)-EDTA and $Fe(CN)_6^{3-}$, respectively). By changing the rotation velocity of the RRDE between 500 and 2000 rpm, it was certified that the effects further described were not due to rate-limiting transport of the reactants from the solution to the disk.

The results were expressed in terms of the stabilization ratio s, given by

$$s = I_Y/I_{ph} \tag{1}$$

where I_Y represents the current corresponding to the anodic oxidation of the reducing agent Y (the Fe(II) complex), and I_{ph} the total photocurrent measured at the disk. The value of I_Y was determined from the measured ring current increase ΔI_R by means of the relationship

$$I_Y = - \Delta I_R/N \tag{2}$$

N being the collection efficiency of the RRDE. For the latter, the theoretical value was used, as it follows from the characteristic dimensions of the RRDE. In one case (nGaP/$Fe(CN)_6^{4-}$) where, from independent measurements, the conditions are known under which s = 1, we were able to check that the experimental N value is in good agreement with the calculated one.

All results which will be reported pertain to steady-state values of both I_{ph} and s. At relatively high values of the ratio of photocurrent density vs. concentration at pH = 3.8, an initial decrease of the photocurrent was observed, indicating the formation of some surface layer. Such cases have not been taken into consideration here.

All results were obtained at room temperature and with the exclusion of oxygen.

3. Results

For the three systems considered, the stabilization ratio s was determined as a function of the total photocurrent density $j_{ph} = I_{ph}/A$ (A = surface area) and the concentration of reducing agent y. In all cases, s was found to increase with increasing y and to decrease with increasing j_{ph}. Typical examples are shown in Figures 1 and 2.

4. Discussion

4.1. Kinetic analysis. The observed light-intensity effect should permit a deeper insight into the mechanism of photoelectrochemical reactions. In this framework, a kinetic analysis of the stabilization process will now be made for an n-type compound semiconductor AB, which is being oxidized photoelectrochemically according to the over-all reaction equation

$$AB + n\ h^+ \longrightarrow \text{products} \tag{3}$$

with n = 6 for GaP (4, 5) and 6 < n < 8 for InP (6). The oxidation of the dissolved reducing agent Y will be taken to be one-equivalent, i.e.

$$Y + h^+ \longrightarrow Z \tag{4}$$

and, for simplicity, the reverse step of reaction (4) will be assumed to be negligible.

Reaction (3) must obviously be complex, and we will assume that in a first step, a hole arriving from the interior reacts at the semiconductor surface $(AB)_s$ to form some surface intermediate X_1, according to

$$(AB)_s + h^+ \underset{k_{-1}}{\overset{k_1}{\rightleftharpoons}} X_1 , \tag{5}$$

k_1 and k_{-1} being the forward and reverse rate constants respectively.

It will be further assumed that the competing reactions with the reducing agent Y either occurs with holes coming from the bulk directly, according to

$$Y + h^+ \xrightarrow{k_0} Z \qquad \text{case (H)} \tag{6}$$

or with the intermediate, according to

$$Y + X_1 \xrightarrow{k''_{-1}} Z + (AB)_s \qquad \text{case (X)} \tag{7}$$

Reactions (6) and (7) are denoted as the cases (H) and (X) respectively.

We will use the symbol j to indicate the number of holes consumed in electrochemical reactions per unit of surface area and per second. Hence, if e represents the elementary charge,

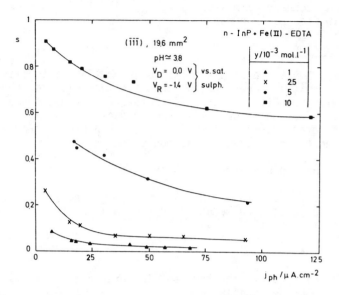

Figure 1.   Stabilization ratio s *as a function of the photocurrent density* j_{ph} *at different concentrations of solved reducing agent* y, n-InP/Fe(II)-EDTA *(rotation speed of the electrode 1000 rpm)*

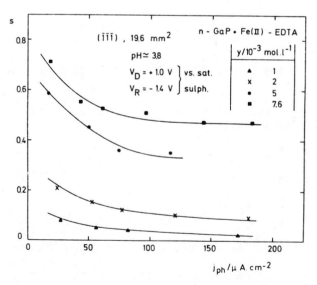

Faraday Discussions

Figure 2.   Plot of s *vs.* j_{ph} *for different* y, n-GaP/Fe(II)-EDTA *(7) (rotation speed 1000 rpm)*

$$j = j_{ph}/e = I_{ph}/Ae \tag{8}$$

Assuming that Y is not adsorbed, the following expressions can then be written for sj, the fraction of j associated with the anodic oxidation of Y:

- case (H) : $sj = k_o \; y \; p$ (9)

- case (X) : $sj = k''_{-1} \; y \; x_1$ (10)

with p the free hole concentration at the electrode surface and x_1 the surface concentration of X_1 intermediates.

We will further define a rate constant k'_{-1} which takes into account the consumption of X_1 species through the reverse reaction (5) as well as through reaction (7):

$$k'_{-1} = k_{-1} + k''_{-1} \; y \tag{11}$$

As we have assumed that the reactions (6) and (7) do not occur simultaneously, it follows that $k'_{-1} \equiv k_{-1}$ in the case (H). As for the case (X), we will assume that the consumption of X_1 predominantly occurs through reaction (7), so that here, $k''_{-1}y \gg k_{-1}$ and hence $k'_{-1} \simeq k''_{-1}y$.

We will now consider two possibilities for the nature of the intermediate X_1. The first possibility will be that X_1 is localized, i.e. represents an electron missing in a well-determined surface bond; this case will be denoted by (L). The alternative is that X_1 is mobile within the surface, in a manner analogous to that of a free hole in the bulk. Energetically speaking, X_1 might then belong to a surface energy band. This case will be denoted in what follows by the symbol (M). If one further takes into consideration the possibilities that reaction (5) may be either irreversible (I) or reversible (R), a series of different possible reaction mechanisms can now be discussed, constituting combinations of the alternatives described above. A survey of these mechanisms is given in Table I.

Table I. Expressions of the stabilization ratio for different reaction mechanisms.		$(AB)_s + h^+ \rightleftharpoons X_1$	
		(L) X_1 localized: $X_1 + h^+ \rightarrow X_2$	(M) X_1 mobile: $2X_1 \rightarrow X_2$
(H) $Y + h^+ \rightarrow Z$	(I) $(AB)_s + h^+ \rightarrow X_1$	(L)(H)(I) $s \neq f(j)$	(M)(H)(I) $s \neq f(j)$
	(R) $(AB)_s + h^+ \rightleftharpoons X_1$	(L)(H)(R) $\frac{s^2}{1-s} \sim \frac{y^2}{j}$	(M)(H)(R) $\frac{s^2}{1-s} \sim \frac{y^2}{j}$
(X) $Y + X_1 \rightarrow Z + (AB)_s$		(L)(X) $s = f(\frac{y}{j})$	(M)(X) $\frac{s^2}{1-s} \sim \frac{y^2}{j}$

Case (L). It is logical to assume here that the further oxidation of an AB unit occurs through consecutive hole-capture steps, so that reaction (5) is followed by the steps

$$X_1 + h^+ \xrightarrow{k_2} X_2$$

$$- - - -$$

$$X_{n-1} + h^+ \xrightarrow{k_n} products \tag{13}$$

(12) brace covers the two equations above.

which for simplicity will be taken to be irreversible. As a consequence of the steady-state which is established, the formation rates of all n oxidation products of AB will be equal and hence each given by $(1-s)j/n$, since $(1-s)j$ is the fraction of j corresponding to the photodecomposition of AB. For the formation rate of X_1, one has:

$$(1-s)j/n = k_1 p - k'_{-1} x_1 \; ; \tag{14}$$

for that of the higher intermediates:

$$(1-s)j/n = k_2 x_1 p = \ldots\ldots = k_n x_{n-1} p \tag{15}$$

Elimination of x_1 from eqns. (14) and (15) leads to:

$$k_1 p = \frac{(1-s)j}{n}\left(1 + \frac{k'_{-1}}{k_2 p}\right) \tag{16}$$

In the case (L), the two alternatives (H) or (X) for the oxidation of Y can be considered. For the alternative (H), distinction can further be made between the case where reaction (5) would be either irreversible (I) or reversible (R). This leads to the following possible mechanisms.

- Mechanism (L)(H)(I) : since here, $k'_1 \equiv k_{-1} = 0$, combination of eqns. (9) and (16) leads to an expression of the type

$$s/(1-s) \sim y \tag{17}$$

which is independent of j and hence of the light intensity.

- Mechanism (L)(H)(R) : assuming that the reverse reaction (5) is fast with respect to reaction (12), so that $k'_{-1} \equiv k_{-1} \gg k_2 p$, it follows from eqns. (9) and (16) that

$$\frac{s^2}{1-s} \sim \frac{y^2}{j} \tag{18}$$

- Mechanism (L)(X) : we recall that we have postulated in this case that the reverse reaction (5) can be neglected with respect to reaction (7), so that $k'_{-1} \approx k''_{-1}y$. By combination of eqns. (10), (15) and (16), one gets an expression of the type

$$\left(\frac{s}{n} + \frac{s^2}{1-s}\right) \sim \frac{y}{j} \tag{19}$$

Case (M). Let us now consider the alternative in which X_1 stands for a mobile surface intermediate. In this case, we will postulate that the two-equivalent oxidation product X_2 is formed by the encounter of two X_1 species, that the subsequent oxidation reactions occur with X_1 species and that all these reaction steps are irreversible.

$$2X_1 \xrightarrow{k_2} X_2 \tag{20}$$

$$\left.\begin{array}{l} X_2 + X_1 \xrightarrow{k_3} X_3 \\ \text{- - - -} \\ X_{n-1} + X_1 \xrightarrow{k_n} \text{products} \end{array}\right\} \begin{array}{c}(21)\\ \\ (22)\end{array}$$

(for simplicity, the same symbols for the rate constants have been used as for case (H)).

Under steady-state conditions, the following rate equations can be written here:

$$j = k_1 p - k_{-1} x_1 + k_o yp \qquad (23)$$

$$(1-s)j/n = k_2 x_1^2 = k_3 x_2 x_1 = \ldots = k_n x_{n-1} x_1 \qquad (24)$$

The last term in eqn. (23) differs from zero in the case (H) only.

Here again, the alternatives (H) and (X) will be discussed, and in the former case, distinction will be made between the possibilities of the first step in the decomposition reaction being either irreversible or reversible. Hence, the following cases can be considered.

- Mechanism (M)(H)(I) : putting $k_{-1}' \equiv k_{-1} = 0$ in eqn. (23) and combining this with eqn. (9), a relationship of the type (17) is obtained.

- Mechanism (M)(H)(R) : here we will assume for simplicity that a quasi-equilibrium establishes in reaction (5), so that

$$k_1 p = k_{-1} x_1 \qquad (25)$$

From eqns. (9), (24) and (25) then follows an expression for s of the type of relationship (18).

- Mechanism (M)(X) : a relationship of the type (18) is obtained from the appropriate expressions, i.e. eqns. (10) and (24).

The results of this kinetic analysis have been included in Table I. It can be seen that, if both the anodic decomposition of the semiconductor and the anodic oxidation of the competing reactant would occur by irreversible hole-capture steps ((L)(H)(I) or (M)(H)(I)), as was hitherto generally accepted, the stabilization should be independent of light intensity, in contradiction with the results described above. The mechanism in which the reducing agent reacts by donating an electron to a localized surface hole ((L)(X)) leads to an expression in which s is a function of the variable (y/j) only. The three other mechanisms considered lead to the relationship of the type (18), in which s is a function of (y^2/j).

4.2. Interpretation of experimental results. In order to compare the foregoing conclusions with experimental findings, we have replotted the data for InP/Fe(II)-EDTA of Figure 1 as a function of the variable (y/j_{ph}) in Figure 3 and of (y^2/j_{ph}) in Figure 4. It can be seen that, whereas s cannot be considered as a function of (y/j_{ph}) only, the data are reasonably well grouped on one single curve when plotted as s vs. (y^2/j_{ph}). The same conclusion is obtained for GaP/Fe(II)-EDTA (compare Figures 5 and 6) and for GaP/Fe(CN)$_6^{4-}$ (3). In Figure 7 finally, the results of Figure 2 have been replotted as log $[s^2/(1-s)]$ vs. log (y^2/j_{ph}), demonstrating that relationship (18) gives an acceptable des-

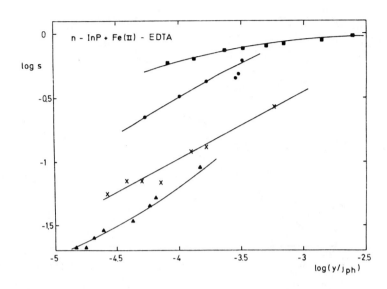

Figure 3. Plot of log s vs. log (y/j_{ph}); same data as in Figure 1; (y/j_{ph}) in arbitrary units

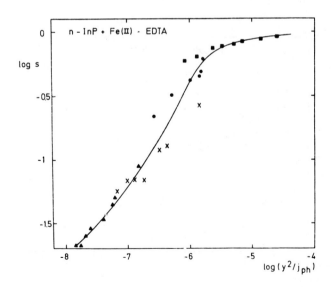

Figure 4. Plot of log s vs. log (y²/j_{ph}); same data as in Figure 1; (y²/j_{ph}) in arbitrary units

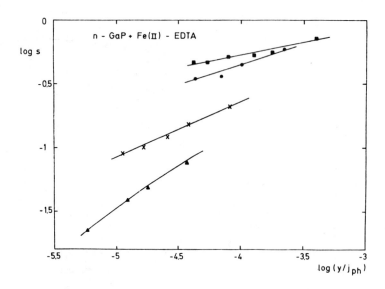

Figure 5.   Plot of log s *vs. log* (y/j_{ph}); *same data as in Figure 2;* (y/j_{ph}) *in arbitrary units (7)*

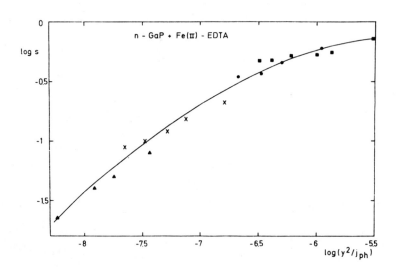

Figure 6.   Plot of log s *vs. log* (y²/j_{ph}); *same data as in Figure 2;* (y²/j_{ph}) *in arbitrary units (7)*

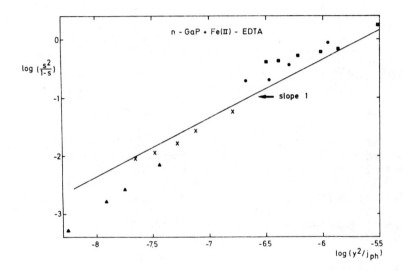

Figure 7.   Plot of log [s²/(1 − s)] vs. log (y²/jₚₕ); same data as in Figure 2; (y²/jₚₕ) in arbitrary units (7)

cription of the stabilization kinetics. (One has to consider that, with $s^2/(1-s)$ as the variable, experimental errors become relatively important for s either close to zero or to one).

According to the foregoing discussion, it must hence be concluded that either the first step in the decomposition process is reversible, or that the mechanism is that in which the decomposition involves a bimolecular step between two mobile surface holes X_1 and in which the reducing agent reacts by donating an electron to such a partially broken surface bond.

It must be finally noted that for all light-intensity dependent cases discussed here, the stabilization ratio was found to decrease with increasing light intensity. All reaction schemes proposed here to explain this effect are based upon the fact that the decomposition reaction is two- or multi-equivalent (see Table I). It might hence well be that this multi-equivalence is at the origin of the observed light-intensity effect on stabilization. This effect may constitute a problem for photoelectrochemical solar cells which for economic reasons may be used under concentrated sunlight, since this would lead to greater deterioration of the electrode under concentrated light.

Abstract.

The stabilization of illuminated n-type III-V semiconductor electrodes through competing hole capture by reducing agents added to the aqueous solution has been studied as a function of concentration and of the light intensity. The main result concerns the observed light-intensity dependence. From a kinetic analysis of the stabilization process, it follows that two types of reaction mechanisms can be held responsible for the observed kinetics.

Literature Cited.

1. Kobayashi, T.; Yoneyama, H.; Tamura, H.: Chem. Letters, 1979, 457.
2. Gomes, W.P.: Electrochem. Soc. Meetings Extended Abstracts, 1979, vol. 79-2, 1565.
3. Van Overmeire, F.; Vanden Kerchove, F.; Gomes, W.P.; Cardon, F.: Bull. Soc. Chim. Belg., 1980, 89, 181.
4. Madou, M.J.; Cardon, F.; Gomes, W.P.: Ber. Bunsenges. Phys. Chem., 1977, 81, 1186.
5. Kohayakawa, K.; Fujishima, A.; Honda, K.: Nippon Kagaku Kaishi, 1977, 6, 780.
6. Ellis, A.B.; Bolts, J.M.; Wrighton, M.S.: J. Electrochem. Soc., 1977, 124, 1603.
7. Cardon, F.; Gomes, W.P.; Vanden Kerchove, F.; Vanmaekelbergh, D.; Van Overmeire, F.: Faraday Discussions, 1980, nr. 70, in print.

RECEIVED October 3, 1980.

Factors Governing the Competition in Electrochemical Reactions at Illuminated Semiconductors

HIDEO TAMURA, HIROSHI YONEYAMA, and TETSUHIKO KOBAYASHI

Department of Applied Chemistry, Faculty of Engineering, Osaka University, Yamadakami, Suita, Osaka 565, Japan

Electrochemical reactions at metal electrodes can occur at their redox potential if the reaction system is reversible. In cases of semiconductor electrodes, however, different situations are often observed. For example, oxidation reactions at an illuminated n-type semiconductor electrode commence to occur at around the flat-band potential E_{fb} irrespective of the redox potential of the reaction E_{redox} if E_{fb} is negative of E_{redox} [1,2,3]. Therefore, it is difficult to control the selectivity of the electrochemical reaction by controlling the electrode potential, and more than one kind of electrochemical reactions often occur competitively. The present study was conducted to investigate factors which affect the competition of the anodic oxidation of halide ions X^- on illuminated ZnO electrodes and the anodic decomposition of the electrode itself. These reactions are given by Eqs 1 and 2, respectively:

$$ZnO + 2p^+ \rightarrow Zn^{2+} + 1/2O_2 \qquad (1)$$

$$2X^- + 2p^+ \rightarrow X_2 \qquad (2)$$

Until now, it has been pointed out that at least two factors play important roles in determining the degree of competition, neglecting kinetic factors. One is concerned with thermodynamics, and predicts that the most negative E_{redox} for a reaction will be the most reactive. The theoretical background for this concept was given by Gerischer [4,5] and by Bard and Wrighton [6], and experimental work to verify this concept has been done by Honda's group [7,8,9] which included ZnO electrodes. Another important factor, which was proposed by Gerischer [5,10], is illumination intensity, and we have already reported preliminary results to support this view [11] for the present electrode/electrolyte systems.

Experimental

The competitive reactions were studied by using the rotating

0097-6156/81/0146-0131$05.00/0

ring-disk electrode technique (7,12). A schematic illustration of the experimental setup is given in Fig. 1. A single crystal of 2 mm thickness was machined to 5.8 mm diameter and mounted in a Teflon electrode holder as the disk electrode. A platinum ring electrode having 7.0 mm inner diameter and 9.0 mm outer diameter was also mounted in the Teflon holder coaxial to the ZnO disk electrode.

Pretreatments of the electrode were carried out by soaking first in HCl for 10 s and then in H_3PO_4 for 5 min, followed by washing with de-ionized water. The RRDE was then set in a RRDE measuring system (Nikko Keisoku, model DPGS-1). The rotation rate of the electrode was usually 1000 rpm except where specially noted. Electrolytes were of reagent grade chemicals and twice distilled water. A 500 W super high pressure mercury arc lamp (Ushio Electric, model USH-500D) was used as a light source. A horizontal light path from the lamp was bent vertically by a mirror to illuminate the electrode from the bottom of the cell. All the potentials cited in this paper are referred to a SCE which was used as a reference electrode in this study.

According to Eqs 1 and 2, halogen molecules, zinc ions and oxygen molecules are produced by the competitive reactions. If the Pt ring electrode is set at a potential where only halogen molecules are selectively reduced, the disk current due to the oxidation of halide ions i_D can then be estimated by dividing the measured ring current i_R by the collection efficiency of the ring electrode N:

$$i_D(X^-) = i_R(X_2) \cdot \frac{100}{N} \qquad (3)$$

With estimated $i_D(X^-)$, the percentage of oxidation of halide ions $\Phi(X^-)$ can be determined as a ratio of i_D to the total disk photocurrent measured:

$$\Phi(X^-) = [i_D(X^-)/i_{total}] \cdot 100 \qquad (4)$$

Therefore, it is important to find out the potential of the Pt ring electrode where only halogen molecules are selectively reduced. The potential of the Pt ring electrode most suitable for detection of X_2 was determined as the potential where reduction of X_2 occurs under diffusion-limited conditions but oxygen and Zn^{2+} were not reduced at all. The potentials thus determined were 0 V for I_2, 0.05 V for Br_2, and 0.2 V for Cl_2 when 0.5 M K_2SO_4 was used as the base electrolyte. In cases where the base electrolyte was more acidic than 0.5 M K_2SO_4, these potentials were a little changed depending on the pH value of electrolytes.

Results

Figure 2 gives i_D and i_R as a function of the disk electrode potential in 0.5 M K_2SO_4 containing 10^{-2} M KI. It is seen that i_R

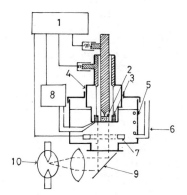

Figure 1. Schematic of experimental setup for measurements of the rotating ring-disk electrode: (1) dual potentiogalvanostat; (2) ZnO disk electrode; (3) Pt ring electrode; (4) Teflon electrode holder; (5) electrolytic cell; (6) N_2 gas inlet; (7) Pt counter electrode; (8) SCE; (9) mirror; (10) Hg arc lamp

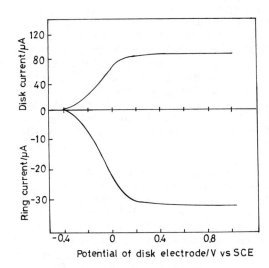

Figure 2.   Ring current and disk current as a function of potential of the ZnO disk electrode.   Potential of Pt ring: 0 V vs. SCE; rotation rate: 1000 rpm; solution: 10^{-2}M KI in 0.5M K_2SO_4.

was proportional to i_D, and both i_D and i_R became saturated at disk potentials positive of 0.5 V. The ratio of i_R to i_D, which was 0.35 in this case, did not vary with increasing concentration of KI and/or with decreasing pH values of electrolytes. Furthermore, under experimental conditions giving the results shown in this figure, no signal for reduction of O_2 could be seen in i_R-i_D curves at ring electrode potentials where oxygen can be reduced in addition to the reduction of I_2. Therefore, this ratio, 35%, was judged to show the collection efficiency of the ring electrode N. The value obtained was almost the same as that given theoretically (13).

Figure 3 shows $\Phi(X^-)$ (X = I, Br) as a function of the concentration of electrolytes for two different disk currents. According to this figure, a large $\Phi(X^-)$ was obtained for a high concentration of KI, as easily expected from general electrode kinetics. However, it is noticed that the concentration dependence of $\Phi(X^-)$ is different even for the same kind of halide ions if the i_D chosen is different. In order to investigate this point in detail, $\Phi(X^-)$ was obtained as a function of i_D for different concentration of halide ions.

Figure 4 shows results for iodide ions. It is seen in this figure that in a region of relatively small i_D, $\Phi(I^-)$ is relatively high and constant with concentration. However, it decreases with an increase in i_D if the magnitude of the disk photocurrent exceeds a certain critical value which depends on the concentration of iodide ions. The decrease in the percentage was judged to reflect a simple mass transport problem, because $\Phi(I^-)$ increased when the rotation rate of the electrode was increased. In the region where the constant $\Phi(I^-)$ was observed, it was not influenced by the rotation rate of the electrode but was varied by changing the pH values of electrolytes, as shown in Figure 5. According to this figure, $\Phi(I^-)$ increased with increasing acid concentration of the electrolytes.

Figure 6 shows $\Phi(Br^-)$ as a function of i_D for a variety of electrolyte concentrations. By comparing this figure with Figure 4, it is noticed that the dependency of $\Phi(Br^-)$ on i_D is quite different from the case of iodide ions. $\Phi(Br^-)$ increases with an increase in i_D in a region of relatively small disk currents, and complete suppression of the anodic dissolution could be achieved under certain critical i_D in solutions having a concentration about thousand times higher than that required for iodide solutions. In a region of photocurrents where an ascending trend of $\Phi(Br^-)$ was observed, the percentage was not changed by the rotation rate of the electrode, suggesting that $\Phi(Br^-)$ in this region is not controlled by any factor on the solution side. As in the case of iodide ions, $\Phi(Br^-)$ was distinctly affected by solution pH, and was high for solutions of low pH. The results to show this are given in Figure 7.

Figure 8 shows $\Phi(Cl^-)$ as a function of i_D. The ability of chloride ions to suppress the anodic decomposition of the electrode

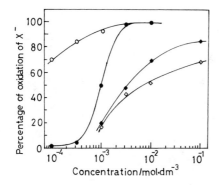

Figure 3.   Percentage of oxidation of I⁻ and Br⁻ as a function of the concentration of halide ions for two different disk currents: (−○−, −◇−) 10^{-5} A; (−●−, −◆−) 4.5×10^{-5} A; (−○−, −●−) KI in 0.5M K_2SO_4; (−◇−, −◆−) KBr in 0.5M K_2SO_4; rotation rate—1000 rpm

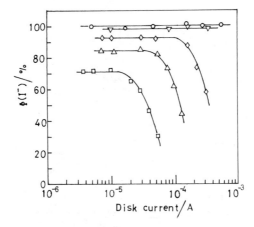

Chemistry Letters

Figure 4.   Percentage of oxidation of I⁻, $\Phi(I^-)$, as a function of disk current for a variety of concentrations of KI in 0.5M K_2SO_4. Concentration (M): (−○−) 1×10^{-2}; (−▽−) 3.16×10^{-3}; (−◇−) 1×10^{-3}; (−△−) 3.16×10^{-4}; (−□−) 1×10^{-4} (11).

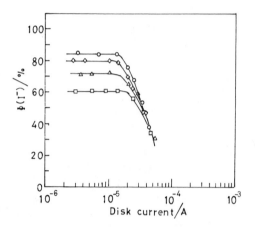

Figure 5.   As in Figure 4, but for a variety of solution pH (concentration of KI:
10^{-4}M; rotation rate: 1000 rpm; pH value: (–○–) 4.2; (–◇–) 5.5; (–△–) 6.2;
(–□–) 8.0)

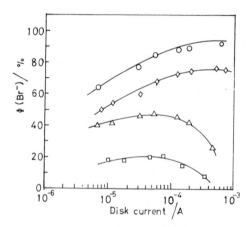

Chemistry Letters

Figure 6.   Plot of $\Phi(Br^-)$ as a function of disk photocurrent for a variety of concen-
trations of KBr in 0.5M K_2SO_4 (concentration (M): (–○–) 1 × 10^{-1}; (–◇–) 1 ×
10^{-2}; (–△–) 3.16 × 10^{-4}; (–□–) 1 × 10^{-4}; rotation rate: 1000 rpm (11)

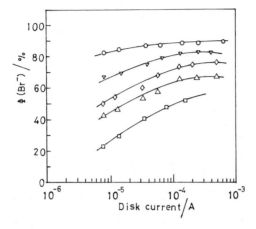

Figure 7. As in Figure 5, but for a variety of solution pH (concentration of KBr: 10^{-2}M; solution pH: ($-\bigcirc-$) 4.2; ($-\triangledown-$) 5.0; ($-\diamondsuit-$) 6.2; ($-\triangle-$) 7.0; ($-\square-$) 9.0)

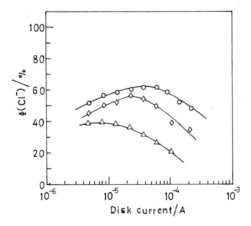

Figure 8. Plot of $\Phi(Cl^-)$ as a function of disk current for different concentrations of KCl in 0.5M K_2SO_4 (rotation rate: 1000 rpm; concentration (M): ($-\bigcirc-$) 1.0; ($-\diamondsuit-$) 0.562; ($-\triangle-$) 0.316)

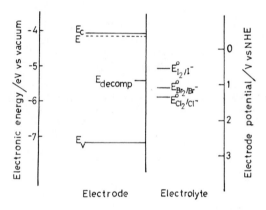

Figure 9. Energetic correlation between ZnO electrodes and halide solutions at pH = 6.0 E_c, E_v, and E_{decomp} denote the energy level of the conduction band edge, valence band edge, and the decomposition reaction of ZnO, respectively (11).

was much poorer than that of bromide ions, as judged from a comparison of Figures 6 with 8. Although ascending trends of $\Phi(Cl^-)$ with i_D were observed also in this case, the complete suppression of the electrode decomposition was not achieved even in solutions as high as 1.0 M.

Discussion

In Figures 4 to 8, $\Phi(X^-)$ is given as a function of i_D. Since i_D is the saturated photocurrent and is proportional to the illumination intensity, the results given in these figures can be read in such a way that $\Phi(X^-)$ is given as a function of the illumination intensity. From the results given above, therefore, the following are important factors affecting the degree of competition: the concentration of halide ions, the illumination intensity, the acidity of the electrolytes, the redox potential of the electrolytes, and kinetic factors.

The importance of the first factor, the concentration, is clear because the anodic photocurrent due to oxidation of halide ions should be proportional to the product of the concentration of positive holes at the electrode surface and that of halide ions in solution. When i_D becomes large, the supply of halide ions to the electrode surface by diffusion becomes unable to follow i_D, resulting in a decrease of $\Phi(X^-)$, which depends on the concentration of X^-.

Up to date, the importance of illumination intensity has been paid little attention. According to the results obtained, $\Phi(Br^-)$ and $\Phi(Cl^-)$ were noticeably affected by the illumination intensity, although this was not the case for iodide ions. The ascending trend of $\Phi(X^-)$ observed with increasing i_D, which was invariant with changing the rotation rate of the electrode, is believed to reflect an illumination intensity effect.

Figure 9 shows the energetic correlation of the ZnO electrode and the electrolyte at pH = 6. According to this figure, the energy level of the I^-/I_2 couple is higher than that of the anodic decomposition reaction of ZnO, while a reverse relation is seen for the cases of Br^-/Br_2 and Cl^-/Cl_2. By changing the pH value from 6 to 2, these energetic correlations are not upset, because the redox potentials of the competitive reactions are not changed by changing the pH values of electrolytes as long as the solution conditions are acidic. It is noticed from such energetic correlations that the illumination intensity effect is dependent on the relative positions of the energy levels of the competing reactions.

The energy level of the quasi-Fermi level of positive holes, $_pE_F^*$, which gives a measure of the average energy of positive holes, is given by Eq 5.

$$_pE_F^* = E_V - kT[ln(p_0 + p^*)/N_V]$$ (5)

In this equation, E_V is the energy level of the valence band edge, N_V the effective density of states of the valence band, p^* the concentration of photo-generated positive holes, and p_0 that of positive holes in thermal equilibrium in the dark which can be eventually ignored in a semiconductor of large band gap such as ZnO.

According to this equation, $_pE_F^*$ approaches the valence band with an increase in the illumination intensity. Under dynamic conditions in which electrochemical reactions are occring, photo-generated positive holes are consumed. In addition, some fraction of these holes recombine with electrons and are annihilated. Therefore, the determination of the concentration of positive holes at the illuminated surface of the electrode is quite difficult, making it also difficult to estimate the exact position of $_pE_F^*$. Even so, the increase of $\Phi(X^-)$ with an increase in i_0 can be qualitatively understood from the shift of pE_F^* with illumination intensity.

The energetic position of the anodic decomposition reaction of ZnO is higher than those of Br_2/Br^- and Cl_2/Cl^-. With an increase in the illumination intensity, $_pE_F^*$ at the surface shifts down first to reach the energy level of the anodic decomposition reaction of the electrode. The anodic dissolution of the electrode is therefore predominant under illumination of relatively low intensity. If the rate constant of the anodic decomposition reaction is so large that the reaction proceeds reversibly, the rate of positive hole consumption will be high enough not to bring about a further shift of $_pE_F^*$ towards the lower level with further increase in the illumination intensity. However, if this is not the case, the $_pE_F^*$ will then shift down to reach E_{Br_2/Br^-} and E_{Cl_2/Cl^-} depending on the nature of electrolytes. In this case, the anodic oxidation of X^- becomes feasible. If the rate constant of the anodic oxidation reaction of halide ions is larger than that of the anodic decomposition reaction of the electrode itself, the former reaction must occur preferentially. The increase in $\Phi(Br^-)$ and $\Phi(Cl^-)$ with increasing illumination intensity is believed to reflect such effects of the illumination intensity. If the rate constant of the anodic oxidation of X^- is smaller than that of the decomposition reaction, the percentage of the oxidation of halide ions will not be appreciably increased with increasing illumination intensity. In the case of iodide solutions, the energy level of the redox electrolyte is higher than that of the anodic decomposition reaction of ZnO, so that preferential oxidation of iodide ions occurs. The observed invariance of $\Phi(I^-)$ with illumination intensity seems to reflect that the rate constant of the anodic oxidation of iodide ions is large enough for $_pE_F^*$ to be pinned at a fixed energetic position of E_{I_2/I^-}. Under such circumstances, a relative position of $_pE_F^*$ to the energetic position of the decomposition reaction is also fixed, so that the ratio of the anodic oxidation of iodide ions to the decomposition reaction is invariable.

As Figures 5 and 7 show, $\Phi(X^-)$ was large in solutions of low pH. It is known from the energetic correlation between ZnO and electrolytes that the energy level of the redox electrolyte relative to that of the anodic decomposition reaction is invariant with changing pH values of the electrolytes. Therefore, the pH dependency observed cannot be explained from a point of thermodynamics. The energy separation of E_{X_2/X^-} from the valence band edge increases with an increase in acidity of electrolytes, making it difficult to explain the observed pH dependence from the point of view of overlaps of the density of states of electrolytes and the valence band.

It may be plausible to assume that the energy levels of surface states match well with those of halide ions with decreasing pH values of the electrolytes (8). However, similar pH dependencies of $\Phi(X^-)$ are also observed on other n-type semiconducting oxides such as TiO_2, WO_3, and $\alpha\text{-}Fe_2O_3$ (14). Furthermore, increasing trends of the activity for the anodic oxidation of iodide and bromide ions with an increase in acidity of electrolytes has been reported for SnO_2 (15). Similar pH effects are also observed for the anodic oxidation of chloride ions at metallic conductive RuO_2 electrodes (16). Considering such a broad pH dependence of the reactivity of halide ions, arguments based on surface states do not seem valid.

Based on surface chemistry arguments, the double layer structure of metal oxide surfaces is effected by the solution pH (17,18), and hydroxide groups of the surface becomes less abundant with a decrease in solution pH. The process may be represented by the following equation (18);

$$M^{m+}(OH)_m(H_2O)_n \underset{-\ H^+}{\overset{+\ H^+}{\rightleftharpoons}} [M^{m+}(OH)_{m-1}(H_2O)_{n+1}]^{+1} \qquad (6)$$

In this equation, M represents a surface metal cation and the sum of m and n fulfills the coordination of M. Such a change in surface conditions is reflected in the pH dependence of the flat-band potential of ZnO electrodes (19). Judging from the point of zero charge (pzc) of ZnO, which is at about pH=8.7 (20), the surface of the electrode must be charged positively in the soltuion chosen in the experiments. Then, there may arise specific adsorption of halide ions onto the cationic sites, and a mechanism is postulated that the observed pH effects of $\Phi(X^-)$ is due to contribution from the specifically adsorbed halide ions. Measurements of the flat-band potential of ZnO electrodes as a function of the concentration of iodide ions, however, gave no indication of the specific adsorption. Then, this model is ruled out.

Another model for giving an explanation of the pH dependence of the reactivity of halide ions may be that surface cations serve as effective sites for adsorption of reaction intermediates which are produced in the course of the anodic oxidation of halide ions. Usually, the anodic oxidation of halide ions is believed to proceed according to the following equation:

$$X^- + p^+ \rightarrow X_{ad} \tag{7}$$

followed by either

$$X_{ad} + X_{ad} \rightarrow X_2 \tag{8}$$

or

$$X_{ad} + X^- + p^+ \rightarrow X_2 \tag{9}$$

If the adsorption of neutral halogen atoms occurs on the surface cation sites with replacing surface-bound water, then the amount of the adsorbed intermediates will be high in acidic solutions, as implied by Eq 6, resulting in an increase in apparent reactivity of halide ions with decreasing pH values.

The importance of the thermodynamic correlation for competing reactions has already been discussed from theoretical and experimental points of view as described in the Introduction. The difference in reactivity created by difference in the redox potential of X_2/X^- systems is clearly reflected in the results obtained in the present study in that the concentration required to stabilize the electrode is different among the kind of halide ions (lowest for iodide but highest for chloride ions).

Conclusion

We have discussed the important factors which determine the degree of competition. These factors are usually intermixed and hence make the observed phenomena complicated. However, the present study has revealed that there are cases where the illumination intensity plays an important role in determining the degree of competition. Furthermore, it is suggested from the present study that stabilization of electrodes will be achieved by appropriate choices of electrolytes, solution pH, and illumination intensity even for electrode/electrolyte systems of thermodynamic instability.

Abstract

The anodic oxidation of iodide, bromide and chloride ions at illuminated ZnO electrodes, which occurs in competition with the anodic decomposition of the electrode itself, was studied as functions of halide ion concentration, illumination intensity and solution pH in order to investigate factors which affect the degree of competition. The reactivity of halide ions, obtained under fixed conditions, was in the order of $I^- > Br^- > Cl^-$, reflecting the importance of the redox potential in determining the reactivity. The illumination intensity was found to influence on the degree of competition, and the effects were different among the kind of halide ions. The observed phenomena are interpreted from the point of shift of the quasi-Fermi level of positive holes with illumination. The pH values of electrolytes was also found to have a marked effect on determining the degree of competition in such a way that the apparent reactivity of halide ions increased with decreasing pH values. The importance of the hydration structure of the electrode surface is suggested.

Literature Cited

1. Gerischer, H., in "Physical Chemistry," Vol. IX A, Eyring, H.; Henderson, D.; Jost, W., Ed. Academic Press, New York, N. Y., 1976; p. 463.
2. Memming, R., in "Electroanalytical Chemistry," Vol. 12., Bard, A. J., Ed., Dekker, New York, N. Y., 1979; p. 1.
3. Bard, A. J. J. Photochem.,1979, 10, 59.
4. Gerischer, H. J. Electroanal. Chem., 1977, 82, 133.
5. Gerischer, H. J. Vac. Sci. Technol., 1978, 15, 1422.
6. Bard, A. J.; Wrighton, M. S. J. Electrochem. Soc., 1977, 124, 1706.
7. Inoue, T.; Watanabe, T.; Fujishima, A.; Honda, K.; Kohayakawa, H. J. Electrochem. Soc., 1977, 124, 719.
8. Inoue,T.; Fujishima, A.; Honda, K. Bull. Chem. Soc. Jpn., 1979, 52, 3217.
9. Fujishima, A.; Inoue, T.; Honda, K. Chem. Lett., 1978, 377.
10. Gerischer, H., in "Solar Power and Fuels," Bolton, J. R., Ed. Academic Press, New York, N. Y., 1977; p. 77.
11. Kobayashi, T.; Yoneyama, H.; Tamura, H. Chem. Lett., 1979, 457.
12. Memming, R. Ber. Bunsenges. Phys. Chem., 1977, 81, 732.
13. Albery, W. J.; Bruckenstein, S. Trans. Farady Soc., 1966, 62, 1920.
14. Kobayashi, T.; Yoneyama, H.; Tamura, H. to be published.
15. Yoneyama, H.; Laitinen, H. A. J. Electroanal. Chem., 1977, 79, 129.
16. Arikado, T.; Iwakura, C.; Tamura, H. Electrochim. Acta, 1978, 23, 9.
17. Ahmed, S. M. J. Phys. Chem., 1969, 73, 3546.
18. Boehm, H. Discuss. Farady Soc., 1972, 52, 264.
19. Lohmann, F. Ber. Bunsenges. Phys. Chem., 1966, 70, 87, 428.
20. Block L.; DeBruyn, P. L. J. Colloid Interf. Sci., 1972, 32, 518.

RECEIVED October 3, 1980.

Electrochemical Behavior and Surface Structure of Gallium Phosphide Electrodes

Y. NAKATO, A. TSUMURA, and H. TSUBOMURA

Department of Chemistry, Faculty of Engineering Science, Osaka University, Toyonaka, Osaka, 560 Japan

The practical success of semiconductor electrochemical photo-cells depends on how to prevent the photo-corrosion of the electrode materials. The various electrochemical processes at the surface of semiconductor photo-electrodes — electron transfer as well as decomposition reactions — have been discussed so far mainly by taking account of the static electronic energy levels of the semiconductors and the solution; that is to say, the conduction band edge, E_c, the valence band edge, E_v, the redox potentials of the redox couple in the solution, $E°(Ox/R)$, the decomposition potential, E_d; etc. (1-6). However, the competition between the electron transfer process and the decomposition reaction paths should better be understood from a kinetic point of view (4). Namely, if the rate of the former is faster than the latter, the photoanode is maintained stable. Also, it seems vitally important to take into account the presence of the surface state whose energy and structure may be dynamically changed by the electrode reactions.

In this paper, we will report our experimental findings on the photo-anodic behavior of n-type gallium phosphide (GaP) in aqueous electrolyte and discuss them based on a picture of the reaction intermediates, which are to play an important role on the reaction pathway as well as on the creation of photo-voltages and photo-currents. The main point is that the surface band energies (depicted for an n-type semiconductor in Fig. 1), which play the most important role in the electrode processes, by no means remain constant, although this has been tacitly assumed to be the case in many previous papers, but change during the photoelectrode processes by the accumulation of surface intermediates and of surface charge (7,8).

For later discussions, we also define a potential U_s, which is a potential at which the inverse square of the differential capacitance $1/C^2$ tends to zero as determined from the $1/C^2$ vs potential plot (Mott-Schottky plot). It is related to E_c^s in the following way:

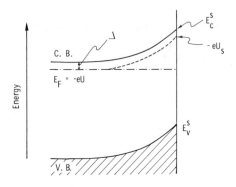

Journal of the Electrochemical Society

Figure 1.   Schematic of the band structure and energy terms of a semiconductor
(7)

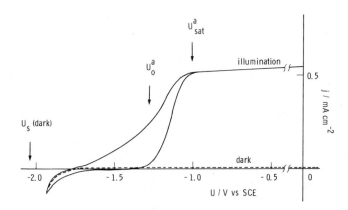

Journal of the Electrochemical Society

Figure 2.   Current–potential curve for the illuminated (111)-face of an n-GaP
electrode in a 0.1M NaOH solution (7)

$$E_c^S = -eU_s + \Delta$$

where e is the elementary charge and Δ a small energy difference
between the conduction band edge in the bulk and the Fermi level.

Experimental

The n-type GaP used was a single crystal in the form of wafers,
99.999 % pure and doped with sulfur to the concentration of 2 to 3
x 10^{17} cm^{-3} (Yamanaka Chemical Industries Ltd.). The p-type GaP
used was doped with zinc to 3.7 x 10^{17} cm^{-3} (Sanyo Electric Co.,
Ltd.). Both were cut perpendicular to the [111]-axis. The ohmic
contact was made by vacuum deposition of indium on one face of the
crystal, followed by heating at ca. 500 °C for 10 min. The side
connected with a wire was covered with epoxy resin. Before the
experiment, the crystals were polished and etched with warm aqua
regia. The (111)-face (Ga face) and the ($\bar{1}\bar{1}\bar{1}$)-face (P face) were
distinguishable by microscopic inspection of the etched surfaces,
the former very rough and the latter smooth.

The current-potential curve was obtained with a Hokutodenko
HA-101 potentiostat. The differential capacitance was measured
mostly with a Yokogawa-Hewlett-Packard Universal Bridge 4265B,
having a modulation frequency of 1 kHz and a modulation amplitude
of 20 mV (peak to peak). The frequency dependence of Mott–Schottky
plots were checked by connecting a function generator to the
bridge. The electrode was illuminated through a quartz window with
a 250 watt high pressure mercury lamp. The light intensity was
attenuated by use of neutral density filters made of metal nets.

Solutions were prepared from deionized water by using reagent
grade chemicals, in most cases, without further purification. They
were deaerated by bubbling nitrogen and stirred with a magnetic
stirrer. The pH of the solutions in the range of 3 to 11 was controlled by using buffer mixtures; acetate, phosphate, borate, or
carbonate, each at about 0.01 M, where M means mol/dm^3. This
abbreviation is used throughout the present paper.

Results and Discussions

1. The Effect of Illumination. In an alkaline solution, an
n-GaP electrode, (111) surface, under illumination shows an anodic
photocurrent, accompanied by quantitative dissolution of the electrode. The current-potential curve shows considerable hysterisis
as seen in Fig. 2; the anodic current, scanned backward, (toward
less positive potential) begins to decrease at a potential much
more positive than the onset potential of the anodic current for
the forward scanning, the latter being slightly more positive than
the U_s value in the dark, U_s(dark).

These results suggest that the GaP surface with backward
scanning develops an oxidized structure, which is acting as a pre-
cursor, or precursors, to the anodic dissolution reactions.

We have found that a good linear Mott-Schottky plot can be obtained for the electrode not only in the dark but also under weak illumination (Fig. 3). This means that, even under anodic polarization and illumination, the state of the surface is maintained at a constant condition. The U_s value determined from the intercept of the plot with the abscissa is somewhat less negative than the U_s value determined similarly in the dark. The U_s values in the dark and under illumination, respectively, at various pH are given in Fig. 4 for the (111) face of GaP, together with the U_o^a value, the latter defined as the potential at zero current for backward scanning (see Fig. 2). The linear dependence of U_s(dark) on pH is understood, as is generally the case for some semiconductors (9,10), to be the result of the acid-base equilibrium on the surface. The deviation of U_s(ill.) (the U_s value under illumination) from U_s(dark) increases slightly with the illumination intensity, and is explained by assuming the accumulation of surface intermediates. That is, when the electrode is under illumination, the holes generated by illumination are drawn to the surface by the built-in potential gradient, and cause anodic decomposition of the electrode. As Gerischer suggested for Ge or GaAs electrodes (5,9,11), intermediates are formed as the precursors in the decomposition or dissolution path:

$$
\begin{array}{cccc}
-P & P \cdot & P \diagdown_O & Ga-OH \\
-P^{\oplus}-Ga-OH & P-Ga\diagdown^{OH}_{OH} & P-Ga-OH & Ga-\overset{\cdot}{P}-OH \\
-P & P & P & Ga
\end{array}
$$

The extent of the accumulation of such intermediates depends on their rates of the formation and those of the ensuing decomposition (or dissolution) reactions. If the latter are not high, the total density of such surface intermediates becomes so high that an appreciable surface potential $\Delta\psi$ is created by the electric double layer formed by the charge unbalance of these intermediates, as well as by the approach of counter ions and the hydration around these intermediates. The experimentally obtained difference between U_s(dark) and U_s (ill.) can be attributed to this $\Delta\psi$.

The surface potential postulated above should affect the U_s and the U_o^a mentioned previously by an equal magnitude. However, the shift of U_o^a from U_s(dark) as seen in Fig. 4 is much larger than that of U_s(ill.). This large shift of the former may be explained by assuming that the surface intermediates capture electrons in the conduction band and act effectively as recombination centers.

2. The Surface Potential arising from the Interaction between the Surface "States" and the Redox Couples in the Solution. When the ferricyanide/ferrocyanide redox couple is present in a 0.1 N NaOH solution, the dark cathodic current of the n-GaP (111)-face sets out at -1.1 V (SCE), showing that an electron transfer occurs

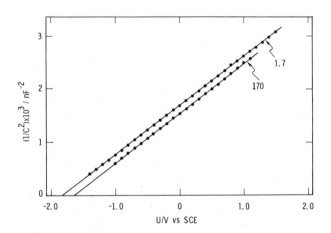

Figure 3. Mott–Schottky plots for the illuminated (111)-face of n-GaP (pH 12.9, modulation frequency 1 kHz). The anodic photocurrents flowing during measurements are shown in units of μAcm^{-2}.

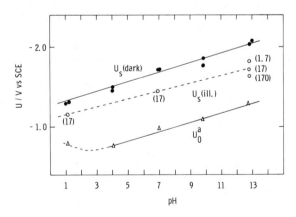

Figure 4. The U_s (dark), U_s (illuminated), and $U_o{}^a$ for the (111)-face of n-GaP in solutions of 0.05M Na_2SO_4 and buffers. The anodic photocurrents flowing during measurements are shown in units of μAcm^{-2}.

from the conduction band of n–GaP to the ferricyanide. This onset-
potential is ca. 0.6 V more anodic than the onset potential for the
cathodic current which corresponds to H_2 evolution observed in a
solution lacking the redox couple (Fig. 5), the latter being close
to U_s(dark). This quite large anodic deviation is analogous to
the anodic shift of U_o^a from U_s(dark) mentioned above, and indicates
the formation of surface intermediates which act as an effective
electron transfer mediator as well as being responsible for
shifting the surface band energy E_c^s. As Fig. 6 shows, the dark
cathodic current for a p–GaP electrode in the presence of the
ferricyanide/ferrocyanide couple sets out at ca. 0 V (SCE), indicating that hole injection to the valence band of GaP occurs from
the ferricyanide. This suggests that holes are injected by ferricyanide into GaP and supports the above idea that holes are chemically relaxed at the surface to form surface intermediates.
    A good linear Mott–Schottky plot was obtained for the (111)-
face of n–GaP in a solution of ferricyanide/ferrocyanide, as shown
in Fig. 7. The U_s values determined from such plots, in electrolyte solutions either with the absence or with the presence of the
redox couple (U_s and U_s(redox), respectively), are plotted at
various pH (Fig. 8). Although the U_s changes linearly with pH, the
U_s(redox) stays constant from pH 5 to pH 10, yielding a constant
difference of 1.7 V with the observed redox potential E(Ox/R) of
the ferricyanide/ferrocyanide couple. This result can be understood by again assuming that a surface potential, arising from the
accumulated surface intermediates, (including surface trapped
holes), brings down the electronic energy of the surface-trapped
hole (as well as the surface band energies) to the level of the
redox potential of the redox couple in the solution and achieves
electron exchange equilibrium. The surface trapped hole may be
visualized as the electron deficient Ga–P bond, stabilized by the
distortion of the surface Ga–P framework and the hydration or the
approach of OH^- ions in the solution. There also might possibly
be contributions from the intermediates formed by the chemical
relaxation processes thereof. However, in the present discussion,
we tentatively assume that only one species ⎯⎯ a surface trapped
hole ⎯⎯ is responsible for the electrochemical behavior. This
species, and the unreacted Ga–P bonds, would constitute a redox
system which is characterized by a redox potential E_h related to
the surface potential ψ:

$$E_h = E_h^\circ + \frac{kT}{e} \ln C_h + \psi + \text{const.}$$

where C_h is the density of the surface trapped hole which is under
reversible electronic equilibrium with the solute. In most cases,
the effect of ψ on E_h is much stronger than the second term of the
above equation.
    With this view, the energy intervals between the E_c^s, E_v^s and
$-eE_h$ becomes nearly constant, because these levels all change
equally with ψ. We can determine their relative positions by the

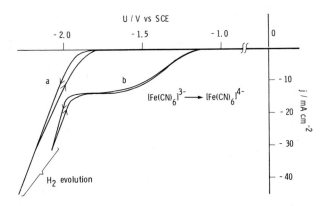

Journal of the Electrochemical Society

Figure 5. Current–potential curves for the (111)-face of n-GaP in a 0.1M NaOH solution, in the absence (a) and the presence (b) of 0.05M potassium ferrocyanide and 0.05M potassium ferricyanide (7).

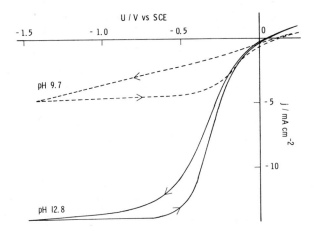

Figure 6. Dark cathodic currents at a p-GaP electrode in solutions of 0.05M potassium ferricyanide and 0.05M potassium ferrocyanide

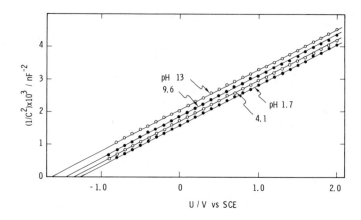

Figure 7. Mott–Schottky plots for the (111)-face of an n-GaP electrode in solutions containing 0.05M potassium ferricyanide and 0.05M potassium ferrocyanide

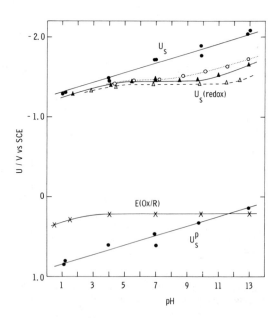

Figure 8. U_s values for the (111)-face of n-GaP (dark) at various concentrations of the ferricyanide/ferrocyanide couple (equal concentrations): (●) 0M; (○) 0.005M; (▲) 0.05M; (△) 0.4M; E(Ox/R) redox potential of the redox couple determined by the cyclic voltammetry; (U_s^p) the U_s for a p-GaP in the absence of the redox couple.

following argument. As mentioned before, at the equilibrium with the ferricyanide/ferrocyanide couple the difference between U_s and $E(Ox/R)$ is 1.7 V. Then, by taking Δ to be 0.1 V, the difference between E_c^s and $E(Ox/R)$ becomes 1.8 V. If we set E_h = E(ferricy-anide/ferrocyanide), which is observed to be 0.2 V (SCE), then from the band gap, E_g, of GaP (2.3 V), E_v^s is calculated to be 0.5 V below E_h. As an example, the U_s of n-GaP at pH 6 with no redox couple present in the solution is -1.6 V (SCE). This shows that E_h without a redox couple is ca. +0.1 V (SCE). It is then under-stood that the interaction of a ferricyanide/ferrocyanide couple shifts E_h, as well as U_s, by 0.1 V to the anodic direction at pH 6 (Fig. 11). At pH 10, the shift is 0.4 V.

The breakdown at a pH higher than 10 of the parallelism be-tween the U_s(redox) and the redox potential of the redox couple in solution can be explained by assuming that the rate of the dis-solution reaction, caused by the attack of H_2O or OH^- on the sur-face trapped hole, is so high in this pH range that the electron exchange equilibrium at the interface is no longer achieved.

For the ($\bar{1}\bar{1}\bar{1}$) face of n-GaP, the measured U_s(redox) changed almost linearly with the pH, showing that the surface-trapped hole is less stable than that for the case of the (111) face above men-tioned.

In the presence of a ferric/ferrous (Fe^{3+}/Fe^{2+}) couple, and that of tetraammine copper (II) $Cu(NH_3)_4^{2+}$ ion, the measured U_s values also shifted. The redox potential for the ferric/ferrous couple, both in equal concentrations, is +0.5 V (vs SCE) in a low pH region, and that for the tetraammine copper (II)/(Ⅲ) couple is 0 to -0.2 V, depending on the concentration of ammonia. In the presence of a vanadate/vanadite (V^{3+}/V^{2+}) couple at pH $\leq$ 3, whose formal redox potential is very highly negative (-0.5 V (SCE)), the U_s shifted very little and was the onset potential for the catho-dic current.

It was pointed out previously (12,13) that n-GaP emitted lumi-nescence under cathodic polarization in contact with a ferricyanide or other oxidant solution. We have also measured the electrochemi-luminescence spectrum in the presence of the ferricyanide/ferrocy-anide couple (Fig. 9). The spectrum is similar to that observed by Pettinger, Schöppel and Gerischer, while that measured by Beckman and Memming showed two peaks. The luminescence peak measured by us is at 1.6 eV, which is in rough agreement with the energy differ-ence between E_c^s and E_h (1.8 eV). The luminescence can be observed only when the cathodic current is fairly strong and only when the redox couple is present. Hence the luminescence can be probably assigned to an electronic transition from the conduction band to the surface trapped hole.

3. Stability of Illuminated n-GaP in Redox Solutions. Figure 10 shows the current-potential curves for the n-GaP electrode under illumination, in the presence of ferrous oxalate $Fe(C_2O_4)_2^{2-}$ and ferrocyanide $Fe(CN)_6^{4-}$, together with that for the solution without

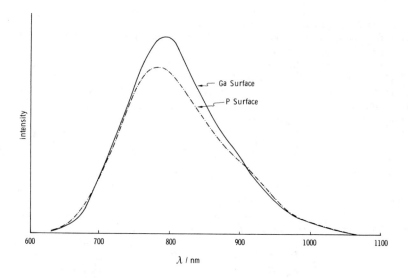

Figure 9. Electroluminescence spectra from an n-GaP electrode under about −2.5 V (SCE) in a 0.3M ferro- and ferricyanide solution at pH 13

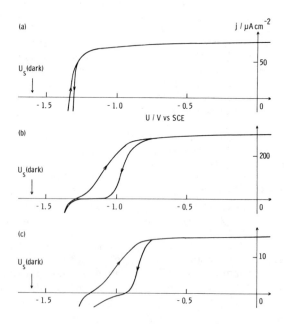

Figure 10. Current–potential curves for the (111)-face of n-GaP under illumination at pH 6.0 in the presence of 0.05M ferrous oxalate (a) and 0.05M ferrocyanide (b), together with that of a solution containing only 0.05M Na_2SO_4 as a supporting electrolyte

a reducing agent. For the case of ferrous oxalate, (pH 6.0), the onset potential for the photoanodic current is largely moved to the negative, and the i-U curve has lost its strong hysteretic behavior, indicating that the surface trapped holes are effectively quenched by ferrous oxalate and that the electrode is kept from corrosion. The sharp increase of the cathodic current at -1.3 V is undoubtedly due to hydrogen evolution. For the case of ferrocyanide, the onset potential of photoanodic current is somewhat more cathodic but the hysteresis still remains, indicating that there are still some surface trapped holes. The electrode is, however, kept intact from corrosion, the photocurrent showing no decay even at the magnitude of 100 $\mu A/cm^2$, in contrast to the case of the solution containing no reducing agent, where the photocurrent decays quickly by the oxide film formation if the initial value of the photocurrent is higher than 30 $\mu A/cm^2$.

All these results can be explained in terms of the model proposed above (cf. Fig. 11). Namely, with ferrous oxalate having a standard redox potential $E°(Ox/R)$ of -0.2 V (SCE), which is a little more negative than the E_h of the surface trapped hole located ca. 0.5 V above E_v^S, the surface trapped hole is effectively quenched by the rapid reduction, and the photoanodic current flows without decomposition. With ferrocyanide, having an $E(Ox/R)$ of 0.2 V (SCE), which is more positive than the E_h of the surface trapped hole, the surface trapped holes are accumulated to the extent that the surface potential created will level it down to the $E(Ox/R)$ of the redox couple. At this point, the rates of nucleophillic attack of H_2O and OH^- to the surface trapped holes are still low and the electrode decomposition is prevented.

Concluding Remarks

In the discussions by many authors of the energy conversion efficiency of semiconductor photoelectrochemical systems, it has been tacitly assumed that the maximum theoretical photovoltages produced is the difference between E_c^S (in units of eV) and $E(Ox/R)$. The best conversion efficiency should then be obtained with a redox couple whose standard redox potential is as low as possible, with a reasonable margin x, say 0.3 V, above E_v^S (Fig. 11). From this it follows that the maximum photovoltage obtainable is equal to the band gap, E_g, in an eV unit, minus a small margin x plus Δ.

It has been pointed out by some authors (1,2) that for a semiconductor having a thermodynamic decomposition potential, E_d, in between E_c^S and E_v^S, a redox couple with a standard redox potential, $E°$, more negative than E_d is needed in order to operate the photoanode without decomposition. Then, the maximum photovoltage attainable is $U_s - E_d$, which is often much lower than $E_g - \Delta - x$. For GaP, this is only 0.8 V (4) (Fig. 11).

In this regard, the main conclusion of the present paper is as follows:
1) The n-GaP photoanode can be operated in a stable condition with

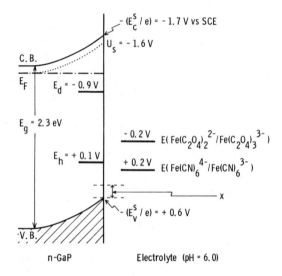

Figure 11.   Energy diagram of the interface between n-GaP *and an electrolyte solution of pH 6.0*

a reducing agent (e.g., ferrocyanide) having E° more positive than the decomposition potential E_d. This means that the theoretical limit of the output photovoltage can be higher than $U_s - E_d$.
2) The best photovoltage can be obtained with a redox couple having E° slightly higher (say — 0.2 V) than the E_h of surface trapped hole (0.1 V), e.g., ferrous oxalate. The use of a redox couple having E° much more positive than that has no merit because in that case the surface trapped holes are accumulated and the surface potential moves up so that not only E_h but also U_s are shifted more positive, thus reducing the theoretical limit of the photovoltage.

Finally it is pointed out that these conclusions for n–GaP can be extended to other various n-type semiconductors for general criteria of the performance of photoelectrochemical systems.

Abstract

The current-potential curves of the n-GaP electrode were studied in aqueous solutions in the dark and under illumination, in connection with the surface conduction band energy E_c^S of the GaP, which is equivalent to the electrode potential U_s determined from the Mott-Schottky plots. From the hysteresis in the current-potential curves, and the change of U_s under anodic polarization caused by the action of light or by an oxidizing agent in the solution, it has been concluded that the surface trapped holes or surface intermediates of the anodic decomposition reactions are rather stable in the region of pH between 5 and 10, and, by the accumulation of these species, a surface potential is built up, causing the shift of U_s to the positive. In such a pH region, the U_s values are fixed by the presence of a redox couple, e.g., ferricyanide/ferrocyanide. It is deduced that the redox couple is in an electron transfer equilibrium with the surface trapped holes having a 'standard' redox potential E°(trapped hole) of 0.5 eV above the valence band edge. Electrochemiluminescence spectrum was observed and attributed to the electron transition from the conduction band to the surface trapped hole. Based on these results, it has been theoretically concluded that a photoelectrochemical cell can be operated as a stable system when a redox couple is present in the aqueous phase which has E°(redox) slightly more negative than E° (trapped hole). This conclusion has been experimentally varified. The photovoltage of such a photocell can have the theoretical limit defined by the difference between U_s and E°(trapped hole), which is much higher than those previously quoted rather pessimistically, on the basis of the thermodynamic decomposition potentials.

Literature Cited

1. Gerischer, H. *J. Electroanal. Chem.*, 1977, *82*, 133.
2. Bard, A. J.; Wrighton, M. S. *J. Electrochem. Soc.*, 1977, *124*, 1706.

3. Fujishima, A.; Inoue, T.; Watanabe, T.; Honda, K. Chem. Lett. 1978, 375; Inoue, T.; Watanabe, K.; Fujishima, A.; Honda, K. Bull. Chem. Soc. Jpn., 1979, 52, 1243.
4. Memming, R. J. Electrochem. Soc., 1978, 125, 117.
5. Gerischer, H.; Ed. "Topics in Applied Physics, Solar Energy Conversion"; vol. 31, Springer: Berlin, Heidelberg, New York, 1979.
6. Yoneyama, H.; Tamura, H. Chem. Lett., 1979, 457.
7. Nakato, Y.; Tsumura, A.; Tsubomura, H. J. Electrochem. Soc., 1980, 127, 1502.
8. Nakato, T.; Tsumura, A.; Tsubomura, H. ibid., submitted for publication.
9. Gerischer, H.; Hoffmann-Perez, M.; Mindt, W. Ber. Bunsenges. phys. Chem., 1965, 69, 130.
10. Lohmann, F., ibid., 1966, 70, 428.
11. Gerischer, H.; Mindt, W. Electrochim. Acta, 1968, 13, 1329; Gerischer, H. Surf. Sci., 1969, 13, 265.
12. Beckmann, K. H.; Memming, R. J. Electrochem. Soc., 1969, 116, 368.
13. Pettinger, B.; Schöppel, H. -R.; Gerischer, H. Ber. Bunsenges. phys. Chem., 1976, 80, 849.

RECEIVED October 3, 1980.

Surface Aspects of Hydrogen Photogeneration on Titanium Oxides

F. T. WAGNER, S. FERRER, and G. A. SOMORJAI

Materials and Molecular Research Division, Lawrence Berkeley Laboratory, and
Department of Chemistry, University of California, Berkeley, CA 94720

Strontium titanate and titanium dioxide have received considerable attention as materials for photoanodes and photocatal-
ysts in the dissociation of water (1,2,3), and in other photoassisted reactions. Knowledge of how the surface composition and
electronic structure of these materials change under illumination
when in contact with gases or liquid electrolytes is essential if
detailed understanding of the mechanisms of semiconductor photochemistry is to be achieved. Although these wide bandgap oxides
do not exhibit gross photocorrosion under most reaction conditions
(2) and would appear less susceptable to possible Fermi-level
pinning than many semiconductors with smaller bandgaps (4), more
subtle surface chemical effects have been documented. Evidence
for photocorrosion (5,6,7), surface-state mediation of electron
and hole transfer to electrolyte species (8,9), and the dependence
of quantum efficiencies on surface preparation techniques (10),
indicate important roles for surface species on wide bandgap
materials.

Most detailed studies of water photodissociation on $SrTiO_3$
and TiO_2 have concentrated on photoelectrochemical cells (PEC
cells) operating under conditions of optimum efficiency, that is
with an external potential applied between the photoanode and
counterelectrode. We have become interested in understanding and
improving reaction kinetics under conditions of zero applied
potential. Operation at zero applied potential permits simpler
electrode configurations (11) and is essential to the development
of photochemistry at the gas-semiconductor interface. Reactions
at the gas-sold, rather than liquid-solid, interface might permit
the use of materials which photocorrode in aqueous electrolyte.
The gas-solid interface is also more amenable to the application
of ultrahigh vacuum surface analytical techniques.

In this paper the hydroxide concentration dependence of the
rate of hydrogen production in $SrTiO_3$ systems (12) is discussed
in light of surface analytical results. The surface elemental
composition before and after illumination in various aqueous
electrolytes has been monitored with Auger electron spectroscopy

0097-6156/81/0146-0159$05.00/0

and is compared with the composition obtained by ultrahigh vacuum surface preparation techniques. Auger spectroscopy, while less sensitive than photoelectron spectroscopies to subtle changes in the oxidation states of surface species, is more easily applied to the imperfectly clean surfaces obtained in basic aqueous electrolytes using present technology. Carbon and silicon impurities are found on surfaces exposed to electrolytes; the carbonaceous species have some filled states which may make them effective for the mediation of charge transfer across the interface. The effects of surface platinization on photoactivity are discussed and evidence for a thermal reaction between Pd and TiO_2 surfaces is given. ([13])

A hydrogen–producing stoichiometric photoreaction occurs between pre–reduced $SrTiO_3$ and 10^{-7} Torr water vapor.([14]) At these low pressures surface TI^{3+} and hydroxyl species can be observed by photoelectron spectroscopies. Comparison of the reaction conditions required for hydrogen photogeneration from low pressure water vapor and from aqueous electrolyte allows some speculation as to the roles of hydroxyl species.

II. Experimental

II.1. Liquid–solid Interface Experiments. Single crystal wafers for experiments in liquid electrolyte were cut to within 2% of the (111) face and etched 3–5 minutes in molten NaOH held in a gold lined crucible. Wafers were then rinsed in water, soaked 5 minutes in aqua regia, rinsed, soaked 5 minutes in high purity 35% aqueous NaOH (Apache SP 7329), rinsed in 7M-Ω triply distilled water, and air dried.

Liquid phase hydrogen photogeneration experiments were carried out with a gas chromatographic detection system described in more detail elsewhere.([12]) Crystals rested in a 2–10 ml pool of electrolyte within a borosilicate glass vacuum flask. The detection system was sensitive to rates of at least 5×10^{15} molecules H_2/hr-cm^2 $SrTiO_3$ ($\equiv 5$ monolayers/hr), but slow rates of oxygen production could not be followed. A 500W high pressure mercury lamp provided a flux of bandgap photons of $10^{16} cm^{-2} s^{-1}$.

The electrolyte for most experiments was compounded from reagent–grade materials and low conductivity water. However, in some experiments high purity (Apache SP7329 35% NaOH) or ultrapurity (Alfa 87864 30% NaOH) solutions were employed. Glassware for these experiments was prepared by soaking in 1:1 H_2SO_4:HNO_3, thoroughly rinsing in 7M-Ω water, and rinsing in the electrolyte to be used.

After being illuminated, the crystals were rinsed in low conductivity water, allowed to dry in air, and transferred into either a Physical Electronics 590 scanning Auger microprobe or into a UHV chamber equipped with a sample enetry lock, a Varian cylindrical mirror Auger analyzer, and a glancing incidence electron gun. Unless otherwise noted, crystals received no argon sputtering, heating, or other cleaning treatments *in vacuo*.

II-2 Gas-solid Interface Experiments. Low pressure photoreactivity experiments were carried out in a UHV chamber previously described (15) equipped with an electron analyzer for Auger,
photoelectron, and low resolution energy loss spectroscopies.
D_2O vapor was admitted into the chamber through a variable leak
valve. Photogenerated projects were detected with a UTI quadrupole mass spectrometer. The system was sensitive to H_2 generation
rates as low a 0.3 monolayers/hour and had still higher sensitivity to oxygen.

A Physical Electronics single-pass CMA and a Phi 4-grid LEED
Optics unit were used in the studies of metal films on TiO_2. Pd
and Au films were evaporated from high-density alumina effusion
sources.

$SrTiO_3$ and TiO_2 crystals for gas-phase studies were polished
with 1μ diamond paste. "Pre-reduced" $SrTiO_3$ crystals were baked
four hours in flowing hydrogen at 1270 K. TiO_2 crystals were reduced *in vacuo*. Clean surfaces were produced by Ar^+ bombardment
and thermal annealing.

III.   Results

III-1. Hydrogen production and surface stoichiometry.
Hydrogen production was observed upon illumination of both stoichiometric and pre-reduced $SrTiO_3$ crystals in concentrated NaOH
solutions. On metal-free crystals, rates of hydrogen production
of 20 and up to 100 monolayers per hour (1 monolayer $\equiv 10^{15}$
molecules/cm^2 $SrTiO_3$) were commonly observed, corresponding to a
quantum efficiency for photons with $h\nu > 3.2$ eV of 0.03-0.15%.
Hydrogen production required bandgap radiation and could be maintained for hours (Figure 1). Rates of hydrogen evolution from
platinum-free crystals in various electrolytes are listed in
Table I. The rate of hydrogen production increases at high OH⁻
concentrations.(12) Similar rates were observed in 10M NaOH
prepared from reagent-grade NaOH (heavy metal impurities 1 ppm
sensitivity, Pt not reported). No hydrogen production was observed in 10M $NaClO_4$ compounded from a reagent containing 5 ppm
heavy metal impurities.

Figure 2 shows Auger spectra of water-rinsed $SrTiO_3$ crystals
(A) before illumination, (B) after illumination in 30% NaOH, and
(C) after illumination in $NaClO_4$. (Spectrum D will be discussed
later.) These spectra were taken with high beam current densities to allow more ready detection of impurities. Figure 3 shows
Ti-O Auger spectra previously taken at low beam currents (<0.1µA)
to minimize beam damage to the surface. The O(507 eV/Ti(380 eV)
peak ratio of 2.6 is the same before and after hydrogen-producing
illumination in 30% NaOH, but is higher (3.8) after illumination
in 10M $NaClO_4$. Crystals illuminated in other electrolytes (pure
water, 1 M H_2SO_4) gave spectra similar to those for illumination

Table I. Hydrogen Photogeneration from Stoichiometric, Initially Metal-Free, SrTiO$_3$ in Various Electrolytes (prereduced crystal; 30% NaOH = 10M)

Electrolyte	Run duration/ Hr	Normalized H$_2$ yield (monolayers)	Average yield (monolayers/hr)
30% reagent NaOH	12	1300	110
30% ultra-pure NaOH	17	940	55
30% high purity NaOH, acetate added to 8x10^{-2}M	18	1250	70
30% reagent NaOH Pt (IV) added to 4x10^{-4}M	12	19,200	1600
10M reagent NaClO$_4$	24	0	0
7M-Ω water	13	36	3
1M H$_2$SO$_4$	15	0	0
40% reagent NaOH	10	600	60

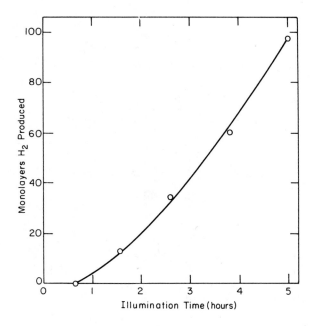

Figure 1.   Hydrogen accumulation as a function of time on a stoichiometric, metal-free SrTiO₃ crystal in 20M NaOH (41)

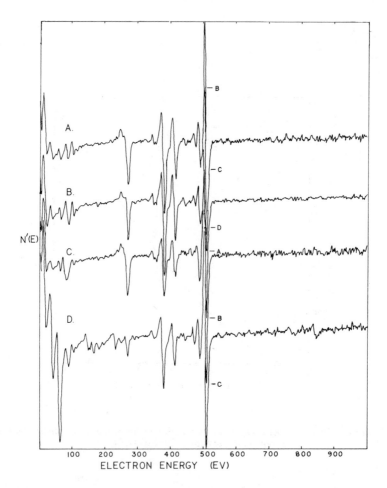

Figure 2. Auger spectra of water-rinsed stoichiometric crystals (A) after pretreatment described in Experimental section; (B) after illumination in 30% (10M) NaOH (H$_2$ produced); (C) after illumination in 10M NaClO$_4$ (no H$_2$ produced); (D) after illumination in 30% (10M) NaOH with Pt(IV) added to 4 × 10^{-4}M

in NaOH, although no hydrogen was produced. The Sr (66 eV)/Ti (380 eV) peak ratios before and after illumination in NaOH were also identical at 0.11. Oxygen mapping with the scanning Auger microprobe showed no regions of crystal illuminated in NaOH to be significantly depleted in oxygen, with a spatial resolution of $\sim 3\mu$. Thus no obvious development of macroscopic anodic and cathodic regions of the crystal occurs.

After crystal illumination, spectrophotometric examination of the electrolyte by the pertitanate method (16) showed no dissolved titanium with a sensitivity on the order of 10 monolayers. It appears that no irreversible change in the surface stoichiometry of constituent elements accompanies the slow photogeneration of hydrogen on metal-free crystals in aqueous NaOH. Although a change in stoichiometry occurred in $NaClO_4$, no such change occurs in other electrolytes which are also ineffective for hydrogen production.

III-2. Surface Contaminants. Most discussions of the role of surface states in the photochemistry of metal oxides have considered primarily states in the metal-oxygen system (8), induced by adsorption of major species from the electrolyte (17), or generated by intentional derivitization of the surface.(18) Surface impurities may also play a role in mediating charge transfer across the interface. A major carbon impurity (272 eV) was observed on all crystals contacted with liquid water. Typical carbon coverage was on the order of one monolayer, as estimated from Auger sensitivity factors based on TiO_2 and graphite. The carbon peak shape was more characteristic of graphite or a partially hydrogenated carbon layer than of a carbide (19) and the signal was too intense to be due to a carbonate.(20,21) Figure 9 shows Auger spectra of $SrTiO_3$ after Ar^+ sputtering (A), exposure to room air (B), and exposure to triply distilled water. Even on this highly reactive (22) sputtered surface the majority of carbon contamination arises from exposure to liquid water rather than to air.

UPS spectra of clean Ar^+ sputtered and *in vacuo* carbon-contaminated surfaces are shown in Figure 4. On the clean, sputtered surface a filled state due to Ti^{3+} lies 0.6 eV below the conduction band.(22) Carbon-induced filled states lie in a broad peak with considerable intensity between the valence band edge and the Ti^{3+} peak. Frank *et al.*(8) reported evidence that a state lying about 1.2 eV above the valence band mediates electron transfer from TiO_2 electrodes. Although these carbon states are as of now poorly defined and have not been directly implicated in any aqueous photochemistry, their nearly ubiquitous presence should be considered in discussions of charge transfer at real oxide surfaces.

The presence of carbon contamination on $SrTiO_3$ surfaces raises the question of whether a carbonaceous species, rather than water, is oxidized during hydrogen photogeneration on Pt-free

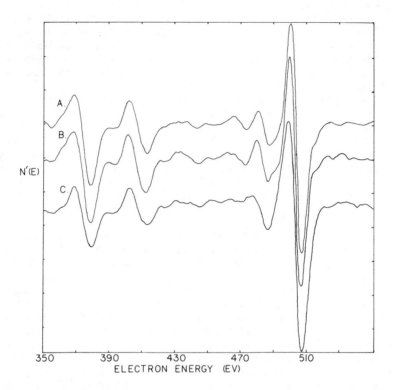

*Figure 3.   Low-beam current ($< 0.1\mu A$) Auger spectra: (A) before illumination;
(B) after illumination in 30% (10M) NaOH; (C) after illumination in 10M CaClO$_4$*

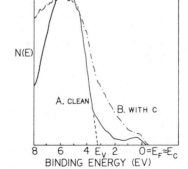

*Figure 4.   UPS spectra of clean, Ar$^+$-sput-
tered SrTiO$_3$ before (A) and after (B)
contamination with about 1 monolayer C
in vacuo*

crystals, where difficulties with detection of slow oxygen production and the low photoactivity leave the issue unresolved. Several authors (23,24) have reported photoassisted reactions between carbon and water yielding some hydrogen and CO_2. Photo-Kolbe reactions of carboxylates have also been demonstrated (25). However, neither addition of acetate to 0.08M nor an unintentional gross contamination of the 10M NaOH electrolyte with charred epoxy residue caused significant acceleration of hydrogen production in our experiments. The presence of carbon monolayers on $SrTiO_3$ shows the need for caution in evaluating photoreactions where the total product yield is on the order of one monolayer or less.

Silicon (Auger peaks at 92 and 1613 eV) appeared on $SrTiO_3$ surfaces after illumination in both NaOH (where hydrogen was produced) and in $NaClO_4$ (where no hydrogen evolution occurred). Less Si deposited from NaOH than from $NaClO_4$, but still less Si appeared after illumination in other electrolytes from which no hydrogen was evolved. It appears that high alkali cation concentrations accelerate the etching of the borosilicate reaction vessel. Surfaces illuminated in $NaClO_4$ exhibited an unusual oxygen peak shape, most clearly visible in the higher-resolution spectra of Figure 5 as the feature with an inflection point at 484.5 eV. Knotek (26) has reported a similar peak shape on a TiO_2 surface exposed to water vapor and aged *in vacuo*. The feature may be due to a peculiar form of hydroxylation or peroxidation of the surface. However, Legaré et al.(27) reported a strikingly similar spectrum for air oxidized silicon and ascribed this peak to a bulk plasmon in SiO_2. The ambiguity in this data produced by silicon contamination is indicative of the problems encountered in the use of glassware with highly concentrated electrolytes. Some advantages may be found in the use of adherent thin films of electrolytes.(12)

III-3. Metallic Impurities and Surface Metallization.
Kraeutler and Bard (28) have demonstrated that metal ion impurities in aqueous electrolytes readily plate out on illuminated TiO_2. No heavy metal contaminants were observed on $SrTiO_3$ surfaces illuminated in the electrolytes used here, though surface coverages less than 1% of a monolayer would have remained undetected. When H_2PtCl_6 was added to a concentration of $4x10^{-4}M$ (2000 monolayers Pt in solution) in 30% NaOH and a stoichiometric crystal was illuminated for 12 hours therein, platinum deposited unevenly onto the illuminated surface. As the Pt plated out, the rate of hydrogen evolution increased from an initial 50 monolayers /hr. to 2000 monolayers/hr. Figure 1D shows the spectrum of the thickest part of the Pt deposit, which formed around a hydrogen-containing bubble trapped under the crystal. The Auger equivalent of 3 or 4 monolayers of platinum is present. The carbon/titanium peak ratio appears significantly smaller than observed on metal-free surfaces, possibly due to carbon-consuming photoreactions (23, 24) or selective deposition of Pt on carbon.

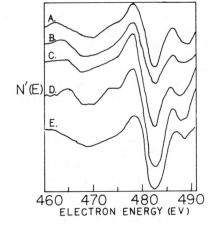

Figure 5. Details of oxygen Auger spectra of SrTiO₃ crystals: (A) Ar⁺-sputtered; (B) illuminated in 7M Ω H₂O; (C) illuminated in 30% NaOH; (D) illuminated in 10M NaClO₄; (E) illuminated in 1M H₂SO₄. Inflection point of main oxygen peak at 501 eV.

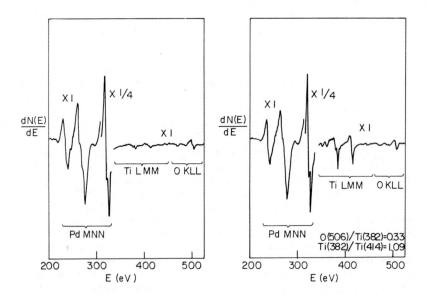

Figure 6. Auger spectra of Pd film evaporated onto TiO₂ before and after annealing ((left) ~ 12 monolayers Pd deposited on TiO₂ (110) at room temperature, $E_i = 3.1$ KeV; (right) after 5 min anneal at 700°C)

Reducible oxides such as TiO_2 and $SrTiO_3$ have been shown to exhibit strong metal-support interaction effects in a number of catalytic reactions(29).   It is possible that direct metalliza-tion of oxide semiconductors may lead to somewhat different chem-istry from that obtainable with discrete oxide and metal elec-trodes.   Bahl *et al.*(30) have undertaken photoemission studies of platinized $SrTiO_3$ surfaces and have found *in vacuo* evidence of partial negative charge transfer from $SrTiO_3$ to each surface Pt atom.   Under more severe conditions intermetallic compounds may form.(13)   Figure 6 shows the Auger spectrum of about 12 monolayers of Pd evaporated onto vacuum-reduced TiO_2.   Both the titanium and oxygen Auger signals are almost completely masked by the overlying Pd.   Upon annealing above 700°C the Ti, but not the O, signal grows indicating a diffusion of Ti to the surface.   The Ti peak shape is moremetallic in character than that seen on clean TiO_2.   The Ti diffusion is more pronounced on Ar-sputtered surfaces with a high Ti^{3+} concentration and is not observed for Au on TiO_2 or Pd on $\alpha-Al_2O_3$.

Figure 7 shows a schematic representation of low energy electron diffraction patterns on the TiO_2(110) and the vicinal (320) "stepped" surfaces.   The pattern in Figure 7A was obtained after annealing a Pd film on either surface at ∿500°C; spots due to the substrate and to (111)-faced Pd crystallites were observed.The pattern in 7B developed on the stepped surface after annealing at higher temperature;   superimposed on the patterns due to the sub-strate and Pd(111) is another set of spots whose lattice parameter is consistent with the hexagonal basal plane of a known inter-metallic compound, Pd_3Ti.(31)   It is not yet clear whether such metal-metal oxide reactions can be stimulated photochemically. Such reactions could modify the photoelectrochemical properties of the system, as platinum-niobium intermetallics have proven superior to pure platinum for the electrocatalysis of oxygen re-duction at elevated temperatures.(32)

III-5. Comparison of Hydrogen Production with Current Mea-surements.

To allow direct comparison of hydrogen evolution re-sults from stoichiometric metal-free crystals in aqueous electro-lyte with rates obtained from a photoelectrochemical cell, dis-crete $SrTiO_3$ and platinized platinum electrodes were mounted in the vacuum reaction flask.   The $SrTiO_3$ electrode was prepared from a crystal polished with 1μ diamond paste, etched in molten NaOH, and reduced 4 hours in hydrogen at 1270 K.   Contact was made through Ga-In eutectic and silver epoxy, and the contact was in-sulated with UHV-grade epoxy.   Current measurements were made via the voltage drop across an 11Ω resistor placed across the elec-trical vacuum feedthroughs to which the electrodes were attached. Figure 8 shows the results of simultaneous measurements of inte-grated photocurrent and hydrogen accumulation, as measured with the gas chromatograph, in NaOH electrolytes of varying concentration.

Structure A (110 and stepped)

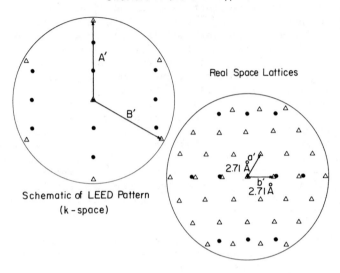

Structure B (stepped surface)

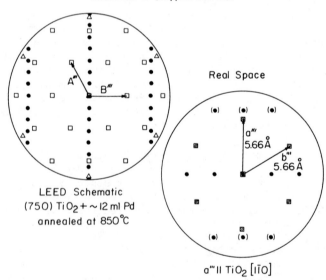

Figure 7. Schematics of LEED patterns and their real-space interpretations for Pd on TiO₂ (110) and (320). Key to spots: (●) substrate; (△) Pd(111)-faced crystallites; (□) intermetallic crystallites.

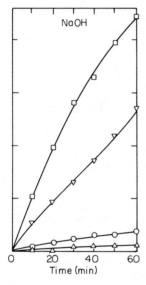
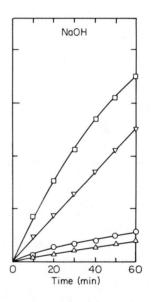

Figure 8. Simultaneous measurements of H_2 yield and integrated current passed as a function of NaOH electrolyte concentration for a n-$SrTiO_3$/Pt PEC cell ((left) gas chromatography; (right) coulometry). Each vertical division corresponds to the equivalent of 5000 monolayers H_2 produced. (($\triangle$) 0.01 F; ($\bigcirc$) 1 F; ($\bigtriangledown$) 10 F; ($\square$) 20 F)

The same trend of increasing photoresponse with increasing hydroxide concentration is seen in both measurements, showing that the hydroxide dependence of hydrogen production is not simply an artifact of the greater solubility of hydrogen in less concentrated solutions preventing escape of photogenerated hydrogen to the gas phase. However, some of the discrepancies between current and hydrogen measurements in dilute solutions may be due to this solubility factor. The role of dissolved oxygen and hydrogen in altering the effective "Fermi level" of the solution (33) has not yet been thoroughly investigated. Although oxygen is much less soluble in concentrated than in dilute NaOH solutions, it also diffuses much less rapidly at higher OH^- concentrations.(34) The relative oxygen concentrations in solutions of variable OH^- concentration deaerated by vacuum pumping (as used here) or by argon or nitrogen purging for finite times remain undetermined.

III-6. Stoichiometry of Vacuum Prepared vs. Water–dipped $SrTiO_3$. All of the Auger spectra in Figures 2, 3, and 5 were taken after rinsing the crystal in triple distilled water to remove electrolyte residue which would interfere with Auger analysis. Figure 9 shows the changes wrought upon an Ar^+ sputtered $SrTiO_3$ surface prepared in vacuo (A), upon exposure to room air for two minutes (B), or after a one minute rinse in triple distilled water (C). The vacuum-prepared surface, with or without air exposure, shows a higher Sr (68 eV/Ti 383 eV) ratio (0.5) than is seen in the surface exposed in liquid water (0.2). Tench and Raleigh (35) showed that immersion in water also removed Sr from $SrTiO_3$ cleaved in air. The effect is too large to be accounted for as differential attenuation by the carbon monolayer. The higher Sr concentration of the vacuum-prepared surface should lead to a lower surface electron affinity, giving electrons in the conduction band under flat-band conditions (which are likely on strongly illuminated, Pt-free crystals) greater reductive power than would be attainable at the water-dipped surface. Since rinsing in water can change the stoichiometry of vacuum-prepared crystals, it is possible that this rinsing step may obscure reversible changes in surface stoichiometry which occur during immersion and/or illumination in diverse electrolytes. The use of volatile electrolytes or gas-phase reactants is thus desirable.

III-7. Hydrogen Photogeneration from Low-Pressure Water Vapor. Water vapor can react with oxygen vacancies of illuminated pre-reduced $SrTiO_3$ surfaces to yield hydrogen and lattice oxide.(14) Vacuum-prepared (111) surfaces of pre-reduced and stoichiometric $SrTiO_3$ were heated to 400°C in 10^{-7} Torr D_2O in a UHV system equipped with a quadrupole mass spectrometer. Illumination of the pre-reduced crystal caused an increase in the D_2 pressure of the system equivalent to D_2 production of 3 monolayers/hr. No such effect was seen on the stoichiometric crystal.

Electron energy loss spectra showed that the surface Ti^{3+} concentration was smaller during illumination in water vapor than during dark exposure to water vapor. In 10^{-8} Torr D_2O photoproduction of HD, but not D_2 was observed. At crystal temperatures below 200°C photoproduction of neither D_2 or HD could be measured. No molecular or atomic oxygen photogeneration was observed under any conditions. In the absence of D_2O, heating the pre-reduced crystal to 400°C caused the slow evolution of hydrogen from the pre-reduced bulk, but no photoeffects were seen in the absence of water vapor. Both oxygen vacancies and hydrogen appear to diffuse towards the surface at 400°C. An oxygen diffusion coefficient of $7x10^{-12}cm^2S^{-1}$ (obtained by extrapolating the data of Paladino (36) to 400°C) would allow $3x10^{13}$ oxygen vacancies-cm^2S^{-1} to reach the surface initially from a crystal with a bulk vacancy concentration of $10^{19}cm^{-3}$. This rate seems adequate to account for the lack of gas-phase oxygen production in this experiment. No hydrogen photoproduction was observed on pre-reduced crystals near room temperature in water vapor at up to 20 Torr pressure.(12)

IV. Discussion

IV.1 Hydrogen Production on Metal-free Crystals. Hydrogen is photogenerated on metal-free, as well as 'platinized' $SrTiO_3$ crystals in aqueous alkali electrolytes. The use of electrolytes of higher purity did not significantly decrease the rate of hydrogen production. Similar results were obtained in the adherent electrolyte films in which the total amount of metallic impurities was several orders of magnitude less than in the bulk liquid electrolyte.(12) Hydrogen photoproduction was observed from the more rigorously Pt-free crystals exposed to water vapor. While a catalytic role for very low levels of metallic impurities can not be ruled out, it appears that the clean oxide surface does have some residual activity. Weber (37) has reported Tafel plots for hydrogen evolution on anodized sodium tungsten bronze, another perovskite semiconductor, under conditions carefully designed to prevent platinum contamination. On this material currents equivalent to hydrogen photoproduction on metal-free $SrTiO_3$ ($\sim2\mu A$) required an overpotential of 150 mV. As this is less than the difference between the flatband potential of $SrTiO_3$ and the hydrogen redox level (2) it is quite possible that the SrTiO3 itself has sufficient catalytic activity to account for the observed production.

During hydrogen evolution on metal-free $SrTiO_3$ photogenerated electrons must diffuse to the surface against an electric field within the depletion layer which tends to drive electrons into the crystal bulk. Krauetler and Bard (25) have proposed the existence of shallow surface electron traps to account for reductive chemistry on n-type oxides. The only electron-trapping surface species as yet identified by UPS on $SrTiO_3$ or TiO_2 is the

Ti^{3+} state. Though it has been impossible to monitor this state
at the liquid-solid interface, the Ti^{3+} concentration <u>decreases</u>
during hydrogen producing illumination in water vapor. One
would expect an increase in Ti^{3+} upon illumination if photopopu-
lation of this electron trap state controlled the reaction rate.
It thus appears that even on metal-free $SrTiO_3$ conduction-band
electrons are the primary reductants. Since similar reaction
rates occur on pre-reduced and stoichiometric crystals with disparate depletion layer widths, the electrons do not tunnel through
the depletion layer. With no Pt to provide an outlet for electrons at potentials far positive of the flatband potential,
strong illumination would flatten the bands almost completely and
allow electrons to reach the semiconductor surface. The presence
of both electrons and holes at the surface could lead to unique
chemistry as well as high surface recombination rates.

On stoichiometric crystals diffuse platinization of the
illuminated surface accelerated hydrogen production, while platinization of dark areas had little effect. Short diffusion distances for electrons (and/or possibly hydrogen atoms) between
$SrTiO_3$ and Pt centers are beneficial to hydrogen production.
Since addition of platinum, a good hydrogen evolution catalyst,
does not alter the hydroxide concentration effect, the effect
bears on oxygen, rather than hydrogen production.

V-2. The Hydroxide Concentration Effect. The higher
rates of hydrogen evolution obtained in highly concentrated alkaline electrolytes are not predicted by simple photoelectrochemical theory since the pH dependence of the flatband potential
is expected to be the same as that of the hydrogen and oxygen redox potentials.(2) Although the photoresponse of $SrTiO_3$ PEC cells
have been shown to be independent of electrolyte pH (2), these
measurements were taken at a high applied potential. Kawai and
Sakata (38) have found that pH-related differences in the dynamic
response to pulsed light of TiO_2 electrodes disappear upon application of an external anodic potential. At zero applied potential surface recombination is more rapid, and the speed of transfer of charge carriers to the solution may become critical to the
conversion efficiency of the system. Small changes in band bending due to specific chemical effects would also be more important
at zero applied potential.

pH-Dependent changes in surface stoichiometry could increase
band bending by decreasing the electron affinity of the surface,
as probably occurs during Ar^+ sputtering. However, no such
stoichiometry changes have been observed by Auger, and more purely
kinetic explanations of the hydroxide effect should be considered.

That rate-limiting step in oxygen production may be (1) absorption of a facile hole-acceptor species, (2) hole transfer to
an absorbed or electrolyte species, or (3) desorption of oxidized
products. These steps must compete with bulk and surface recombination processes. Williams and Nozik (39) have shown that

step (2) is likely to be highly irreversible and the solid-liquid interface, casting doubt onto whether step (3) could be rate controlling. Comparison of the conditions required for hydrogen photogeneration from aqueous electrolytes and from low pressure water vapor may shed some light on steps (1) and (2).

Hydrogen photogeneration from 10^{-7} Torr water vapor occurs only on pre-reduced crystals at elevated temperatures where a stoichiometric reaction with reduced centers (Ti^{3+}-Vo^-) from the crystal bulk is possible. These oxygen vacancies could react with zero valent oxygen produced at the surface in a reaction analogous to that yielding $O_{2(g)}$ in aqueous solution. Alternately, the oxygen from water may never be oxidized above the -II or -I state, and holes may be directly accepted by the reduced center. Figure 10 shows the energies of known surface species (from UPS spectra (22)) and estimates of the filled state distribution of several aqueous species (17) relative to the $SrTiO_3$ band edges. All of the UPS-detectable hydroxide states formed upon adsorption of low pressure water vapor lie well below the band edge, where neither thermalized nor somewhat hot holes could be accepted. On the clean, water vapor exposed surface only the Ti^{3+} species could accept holes, though considerable relaxation of the hole must occur. Oxidation of Ti^{3+} is, in fact, observed upon illumination. The filled state distribution of aqueous OH^- is believed to overlap the $SrTiO_3$ valence band edge, and rapid isoenergetic hold transfer is possible, leading to the catalytic oxidation of water. At high electrolyte concentrations hydroxide-hydroxide reactions may increase the overlap, thereby facilitating hole transfer. If the aqueous energy levels are as shown, no adsorbed surface species would be needed to mediate charge transfer, though carbon states are available at the proper energy.

It should be noted that while UPS could not detect an adsorbed hydroxide species with surface coverage less than 10% of a monolayer, such a species could be chemically active. The same hole-acceptor could be present at both the gas-solid and liquid-solid interface, but the rate of its formation may be inadequate to compete with oxygen diffusion into the bulk from the gas-solid interface. Munuera (40) has found that measurable rates of restoration of a hydroxyl species linked to the photoactivity of TiO_2 powders for oxygen desorption require treatments harsher than immersion in liquid water.

IV-3. Summary. No Auger detectable changes in surfaces composition correlate with the higher rates of hydrogen photogeneration on metal-free $SrTiO_3$ observed in highly concentrated aqueous alkaline electrolyte. Platinized crystals and PEC cells show similar enhancement of photoactivity at high hydroxide concentrations. Although no hydrogen photogeneration is seen from water vapor at pressures up to 20 Torr on crystals near room temperatures, hydrogen photogeneration does occur on pre-reduced crystals at temperatures where oxygen diffusion into the bulk is

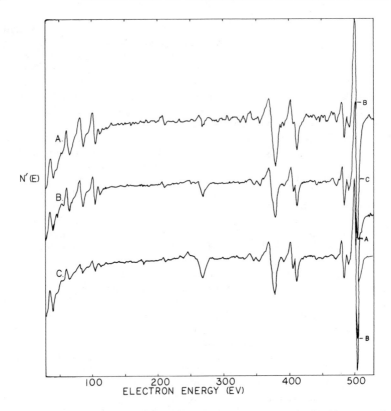

Figure 9.   Auger spectra of (A) freshly Ar⁺-sputtered surface; (B) (A) exposed to room air 2 min; (C) (A) exposed to 7M-Ω water for 1 min

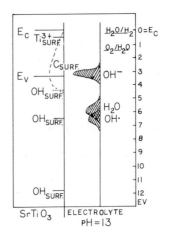

Figure 10.   Filled levels of surface and aqueous species referenced to SrTiO₃ band edges and H₂, O₂ redox couples

rapid. Kinetics of hole-transfer or hydroxylation may determine the overall reaction rate at zero applied potential. Carbon states lying within the $SrTiO_3$ forbidden gap may mediate charge transfer in some photochemical reaction systems, and intermetallic compound formation may alter the catalytic properties of directly metallized titaniferous semiconductors.

Acknowledgement

This work was supported by the Division of Chemical Sciences, Office of Basic Energy Sciences, United States Department of Energy under Contract No. W-7405-ENG-48. The authors would like to thank Dr. John Wang for assistance with the scanning Auger microprobe and Dr. Phillip Ross for the use of his Auger apparatus.

Literature Cited

1. Fujishima, A., Honda, K., Bull. Chem. Soc. Jpn., 1971, 44, 1148
2. Wrighton, M.S., Ellis, A.B., Wolczanski, P.T., Morse, D.L., Abrahamson, H.B., Ginley, D.S., J. Am. Chem. Soc. 1976, 98, 2774.
3. Mavroides, J.G., Kafalas, J.A., Kolesar, D.F., Appl. Phys. Lett. 1976, 28, 241.
4. Bard, A.J., Bocarsly, A.B., Fan, F.F., Walton, E.G., Wrighton, M.S., J. Am. Chem. Soc., 1980, 102, 3671.
5. Schrager, M., and Collins, F.C., J. Appl. Phys., 1975, 46, 1934.
6. Hurlen, T., Acta Chem. Scan., 1959, 13, 365.
7. Harris, L.A., Wilson, R.H. J. Electrochem. Soc.,1976,123,1010.
8. Frank, S.M., Bard, A.J. J. Am. Chem. Soc.1975, 97, 7427.
9. Frank, S.M., Bard, A.J. J. Am. Chem. Soc. 1977, 99, 303.
10. Wilson, R.H., Harris, L.A., Gerstner, M.E., J. Electrochem. Soc. 1979, 126, 844.
11. Nozik, J.A., Appl. Phys. Lett. 1977, 30, 567.
12. Wagner, F.T., Somorjai, G.A. J.Am. Chem. Soc. 1980, 102, 5494
13. Wanger, F.T., Somorjai, G.A. to be published.
14. Ferrer, S., Somorjai, G. A. to be published.
15. Lo, W.J., and Somorjai, G.A., Phys. Rev. B, 1978, 17, 4842.
16. Mühlebach, J., Müller, K., Schwarzenbach, G., Inorg. Chem. 1970, 9, 2381.
17. Sakata, T., Kawai, T., Ber. Bunsenges. Phys. Chem., 1979, 83 486.

18. Tomkiewicz, M., J. Electrochem. Soc., 1980, 127, 1518.
19. Haas, T.W., Grant, J.T., Dooley, G.J.III, J. Appl. Phys. 1972, 43, 1853.
20. Weber, R.E., R-D Magazine, Oct. 1972.
21. Ross, P.N., unpublished data.
22. Ferrer, S., and Somorjai, G.A., Surface Sci. 1980, 94, 41.
23. Sato, S., White, J.M., Chem. Phys. Lett., 1980, 70, 131.
24. Kawai, T., Sakata, T., Nature, 1979, 282, 283.
25. Kraeutler, B., Bard., A.J., J. Am. Chem. Soc., 1978, 100, 5985.
26. Knotek, M.L., in "Proc. Symp. on Electrode Materials and Processes for Energy Conversion and Storage," eds. McIntyre, J.D.E., Srinivasan, S., Will, G. (Proceedings, Vol. 77-6, The Electrochem. Soc., Princeton, NJ, 1977).
27. Legaré, P., Maire, G., Carriére, B., DeVille, J.P., Surface Science, 1977, 68, 348.
28. Kraeutler, B., Bard, A.J., J. Am. Chem. Soc., 1978, 100, 431F.
29. Tauster, S.J., Fung, S.C., Garten, R.L., J. Am. Chem. Soc. 1978, 100, 170.
30. Bahl, M.K., Tsai, S.C., Chung, Y.W., Phys. Rev. B., 1980, 21, 1344.
31. Nishimura, J., Hiramatsu, T., J. Japan Inst. Metal, 1958, 22, 38.
32. Ross, P.N., National Fuel Cell Seminar Abstracts, 1980, 42.
33. Memming, R., Electrochimica Acta, 1980, 25, 77
34. Gubbins, K.E., Walker, R.D., J. Electrochem, Soc. 1965, 112, 469.
35. Tench, D.M., Raleigh , D.O., in "Electrocatalysis on Non-Metallic Surfaces," NBS Spec. Pub. 455, 1976, p.229.
36. Paladino, A.E., Rubin, L.G., Waugh, J.S., J. Phys. Chem. Solids, 1965, 26, 391.
37. Weber, M.F., "Electrocatalytic Activity and Surface Properties of Tungsten Bronzes," Ph.D. Thesis, Iowa State University, 1977, p.76.
38. Kawai, T., Sakata, T., Chem. Phys. Lett. in press.
39. Williams, F., Nozik, A.M., Nature, 1978, 271, 137.
40. Mumuera, G., Rives-Arnau, V., Saucedo, A., J. Chem. Soc. Faraday Trans. I, 1979, 75, 736.
41. Wagner, F.T., Somorjai, G.A., Nature, 1980, 285, 559.

RECEIVED October 9, 1980.

Photocorrosion in Solar Cells

The Enhanced Effectiveness of Stabilization Agents Due to Oxide Films[1]

S. ROY MORRISON, MARC J. MADOU, and KARL W. FRESE, JR.

SRI International, Menlo Park, CA 94025

Studies of the corrosion of the n-type silicon electrode show that Fe(II) EDTA, Fe(II) cyano and ferrocene (the last in ethanol) stabilize silicon temporarily against photocorrosion, but only do so after a thin oxide layer is formed. Analysis of voltammetric C/V and I/V curves suggests the reason for this behavior: a voltage develops across the oxide that raises the energy levels of the reducing agent in solution, relative to the energy of the valence band edge, and increases their hole capture efficiency. This effect may also be present with other systems, e.g. GaAs and GaP, where for some stabilizing agents the greatest stability is found at intermediate pH where the Ga_2O_3 is relatively insoluble. An important side benefit of the observation is the possibility of analyzing the voltammetry data for the reorganization energy, λ. Preliminary values of λ are given. Finally the disadvantages of requiring a thin oxide on a photoelectrochemical solar cell are briefly discussed.

In the development of photoelectrochemical (PEC) solar cells, one of the most difficult problems is the corrosion problem. In any solvent, but particularly in solvents with water present, anodic currents flowing from the solid to the solution will usually lead to corrosion. Specifically the corrosion will take the form of anodic oxidation of the semiconductor, with the products remaining as a film, dissolving into the solution, or evolving as a gas. Any such action will degrade the solar cell.

We have been studying silicon as a PEC cell, partly because silicon would be an exceptionally valuable material for solar cells if the corrosion could be controlled, and partly because the corrosion product, SiO_2, has particularly simple characteristics: it is generally insoluble in aqueous solutions, and it

---

[1] Supported by the Solar Energy Research Institute

is a good insulator. Thus silicon not only has potential for a
solar cell material, but it is of great interest for studies to
generate new basic information regarding photocorrosion.

   In this report we will try to summarize our observation (1)
that the presence of a thin oxide layer on the surface of silicon
has a dominating effect on the effectiveness of stabilizing agents
in the solution. The mechanism is simple. As photoproduced holes
are captured by interface states, a voltage develops across the
oxide layer and the energy levels of the ions in solution become
higher relative to the edge of the valence band. Then the photo-
produced holes are more easily captured by the energy levels of
the stabilizing agent. Now the increased ability of the reducing
agent to capture holes arises because the valence band edge becomes isoenergetic with the higher density part of the Gaussian
distribution of energy levels (2,3). Thus if we plot the current
to the stabilizing agent as a function of the voltage across the
oxide layer, the most important parameter in the relationship is
the reorganization energy, λ. Then another feature of interest
in this oxide layer model is that from such measurements we can
evaluate the parameter λ. In the discussion to follow, first we
will consider the experimental observations, secondly, we will
go into the theory of the measurement of λ and of how the presence
of the oxide improves the effectiveness of the stabilizing agent.

   As mentioned above, silicon has been found to be a very
valuable material with which to study the influence of the oxide
layer. This is because the oxide layer on silicon is an excellent
insulator, i.e. there is no problem with carrier transport through
impurity bands. This makes the interpretation much simpler with
silicon than it would be, for example, with the oxide on gallium
arsenide (4) where the oxide seems to have sufficient conductance
to complicate the interpretation.

Results and Discussion

   A lack of stabilization of the silicon means that when holes
reach the surface the oxide continually grows. In our measurements on silicon, the loss of stability is thus observed by the
decrease in photocurrent at a given voltage as the oxide grows.
In a typical run we cycle the electrode potential of the silicon
from strongly cathodic to strongly anodic and back. If the holes
oxidize the silicon during the anodic part in the cycle, then the
current at a given voltage becomes lower during the next cycle
because of the growth of the insulating silicon oxide. The loss
of current is assumed due to lower tunneling probability. If on
the other hand, the stabilizing agent is effective, the current
at a given voltage is maintained at a constant level through
several cycles.

   In the bottom of Figure 1 we show a series of sweeps from
cathodic to anodic, with a saturated solution of ferrocene in
acetonitrile present in the solution. The effect of the growing

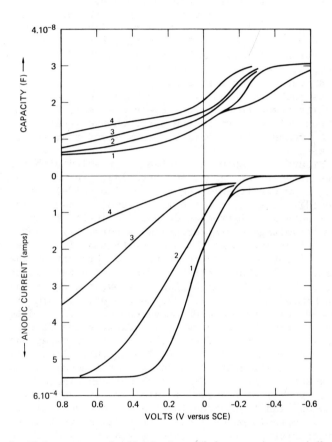

Figure 1. Voltammetry under illumination with ferrocene in acetonitrile for stability. The system does not stabilize the silicon, as shown by the decrease in current in consecutive Curves 1–4.

oxide on the current voltage characteristics is clear. The photocorrosion at a given voltage is lower in each successive cycle.
We show results where the stabilizing agent is ineffective rather
than results with no stabilizing agent present because the intent
is to show the slow change in successive cycles due to a slowly
increasing oxide thickness. With no stabilizing agent present
the current decreases after one anodic sweep.

In the top half of Figure 1 the behavior of the capacity is
shown, as we sweep from the accumulation layer with its very high
capacity toward a depletion layer with its rather low capacity.
On the first sweep (#1) a substantial band bending can be induced
by the positive electrode potential as indicated by the low capacity attained. In successive sweeps however we are unable to
develop as much band bending at the same electrode potential.
The capacity found to be frequency-independent, is assumed to
reflect the voltage drop in the space charge region, and the in-
ability to reach a low capacity simply means that the voltage
drop appears across the oxide.

In Figure 2 we show the equivalent results with a reasonably
effective stabilizing agent, ferrocyanide, present. In this
figure we show results with HF present (no oxide) for comparison
as curve (d). The first cycle of a freshly etched electrode in
the 0.1 M potassium ferrocyanide/water solution is indicated in
the current/voltage characteristics as curve (a). Now in curve
(a) the current begins at the same voltage as it begins with HF
present. In both cases there is no oxide present on the surface
so the current is characteristic of an oxide-free silicon sample.
However, as anodic current begins to flow, the curve (a) deviates
from the HF curve. Curve (a) resembles curve (1) in Figure 1,
and the deviation from the HF curve can be attributed to corrosion (oxide growth). So curve (a) indicates the development of
an oxide at the surface of the silicon. As we sweep back cathodic, we sweep along curve (b). Following this first cycle, the
onset of current is shifted as shown in the anodic sweep to curve
(b). Now the significant difference between Figure 1 and Figure 2
is that curve (b) of Figure 2 is similar for the anodic and cathodic sweep, and is reproducible in successive cycles. There is
apparently no rapid oxide growth; there is no further change in
the characteristics of the sample as was observed in Figure 1.

After the oxide is grown in the first cycle, the results of
Figure 2 thus indicate stability against photocorrosion. Although not shown in the figure, it is found that the photocurrent
rises to a saturation value close to the saturation current observed with the same light intensity with HF present in the solution. As further indication that the changes from curve (d) to
curve (b) is due to an oxide layer, we have intentionally formed
an oxide. After a thin oxide is grown on silicon (in a 0.1 M KCl
electrolyte), and then the sample is transferred to a solution
containing a stabilizing agent, curve (d) is observed directly,
bypassing the "oxide growing" step of curve (a).

The mechanism by which the oxide promotes stability can be determined by observing the capacity/voltage relationship in the top half of Figure 2. The curve again shows the case with HF present in the solution. Curve (b) in the C/V plot corresponds to curve (b) in the current/voltage plot. The initial part of the C/V characteristics under illumination, more cathodic than about -0.2 volts (SCE), looks very much the same as the characteristic with the HF present. However, as the electrode potential is made more positive, a plateau in the C/V curve (b) appears. The plateau in the capacity extends from an applied voltage of about -0.3 to +0.1 volts (SCE). Because the capacity reflects the voltage in the space charge region, it is clear that a constant capacity means that the voltage in the space charge region is constant. Thus the change in applied voltage must appear elsewhere. Specifically, the changes in applied voltage must appear across the oxide layer.

Thus as we sweep the applied voltage from -0.3 to +0.1 volts, the voltage appears across the oxide layer, raising the energy level of ions in solution relative to the energy level of the valence band in the silicon. Eventually the energy levels of the ferrocyanide come close enough to the energy level of the valence band so that holes can tunnel through the oxide, reaching the reducing agent energy levels in solution.

In Figure 3 we show a case to indicate that the sample is indeed changed in a way consistent with such a development of an oxide layer. Curve (i) indicates that the capacity/voltage relationship before any oxide is grown. Curve (ii) is representative of a capacity/voltage curve with the light on, which can be repeated several times. Curve (iii) is again in the dark with no holes present, after these repeated cycles under illumination. It is observed that the cycles under illumination caused a shift in the flat band potential of the sample that is consistent with the presence of another phase or an adsorbed species at the surface.

It is noted that in Figure 3 stabilization is found with ferrous (EDTA) ions in the solution. Ferrocene in ethanol is a third effective stabilizing agent (5). The effectiveness of various stabilizing agents can be approximately compared by comparing the number of cycles that can be applied to the sample before degradation due to excessive oxide growth is observed. With equivalent voltage and photocurrent, such as the values shown in Figure 2, and with a sweep speed of 0.01 V/s, ferrocyanide ions stabilize the silicon for about half a dozen cycles. Typically, with ferrous (EDTA) present, the curves are unchanged for about 10 to 20 cycles and with ferrocene in ethanol present the curves are unchanged for significantly more. For any of these systems, however, exceeding some photocurrent limit (the limit depending on the stabilizing agent) or some voltage limit (close to + 0.16 SCE) will result in rapid degradation of the characteristics as typified by Figure 1.

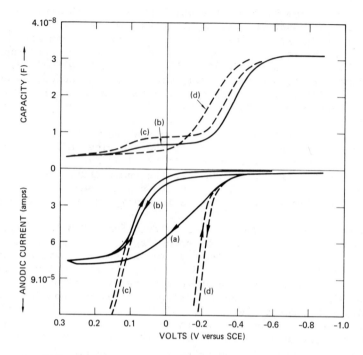

Figure 2. *Voltammograms of capacity and current for* n-*silicon under illumination under various conditions: (a)* $Fe(CN)_6^{-4}$, *no HF, first cycle; (b) same solution, subsequent cycles; (c) higher illumination intensity; (d) HF added*

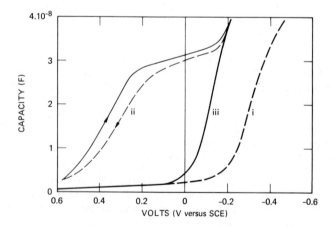

Figure 3. *Shift in dark capacity due to oxide growth in ferrous-EDTA: (i) first curve in dark; (ii) one of several curves in light; (iii) dark again*

   In Figure 4 we suggest a band model for the silicon under
three bias conditions: (a) strongly cathodic, (b) flat band, and
(c) strongly anodic.   The energy levels in solution are considered
a constant reference system and are shown together with the silicon band diagram in Figure 4(b).   Now with a strongly cathodic
bias, as in Figure 4(a), a significant negative charge may accumulate in the interface states between the silicon and the silicon
oxide.   Such charging is suggested by the flat band shift in
Figure 2 between curve (d) and curve (b) or (c) as is observed in
the cathodic region ($<$ -0.2 V SCE) of the C/V curves.   With the
negative charge in interface states a voltage appears across the
oxide as is indicated in Figure 4(a) and the valence band of the
semiconductor will be substantially raised from the energy levels
of the reducing agent in solution.   We have indicated in curve
(b) (the flat band case), that here also the energy level of the
valence band is too high for effective hole transfer to the reducing agent in solution.   Now in the strongly anodic case, under
illumination, Figure 4(c), hole trapping will occur at the interface states.   The switch from negative to positive charge on the
interface states will be observed as a plateau on the C/V characteristics.   The positive charge will lead to a substantial lowering of the valence band relative to the ions in solution, as
indicated in Figure 4(c), and now holes not only can come to the
surface, but when they reach the surface they are at an energy
highly favorable for tunneling through to the energy levels of
the reducing agent.
   We can easily develop the theory in more mathematical detail,
especially the equations for low surface barrier V_s, and use the
results to estimate λ, the reorganization energy of the reducing
agent.   The reorganization energy is a parameter in the expression
describing the probability that the energy level of the ion in
solution has the energy E.   Because of fluctuations in the polarization of the medium and in the bond lengths between the ion and
its ligands, the energy of the ion fluctuates widely in energy
with a normal distribution function, as indicated in Figure 4,
dominated by λ.   Figure 5 indicates the symbolism used.   We assume
that the current to the surface, J_a, is proportional both to the
density of holes at the surface and to the density of ions in
solution with a donor level isoenergetic with the valence band
edge.   The proportionality constant includes
                              a tunneling probability exp $(-\gamma x_o)$, with γ
constant.

$$J_a = Bp_s \exp\{-(E_{vs}-E_{red})^2/4\lambda kT\} \exp(-\gamma x_o) \tag{1}$$

From Figure 5, with E_F (SCE) defined as zero, and (as common with
the band model) the potentials more positive toward the bottom of
the figure, we find:

$$- (E_{vs} - E_{red}) = qV_m - qV_s - (Eg-\mu) + qE^o + \lambda \tag{2}$$

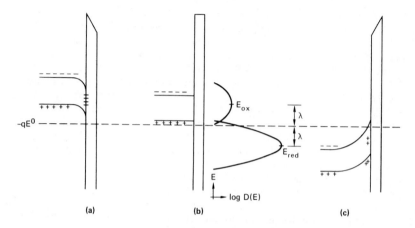

(a)   (b)   (c)

Figure 4.   Band model of oxidized silicon.

The energy levels in the solution are kept constant, and the applied voltage shifts the bands in the oxide and the silicon. The Gaussian curves in Figure 4b represent the ferrocyanide/ferricyanide redox couple with an excess of ferrocyanide. E° is the standard redox potential of iron cyanide. With this, one can construct (a) to represent conditions with an accumulation layers, (b) with flatbands, where for illustration, we assume no charge in interface states, and (c) with an inversion or deep depletion layer (high anodic potential).

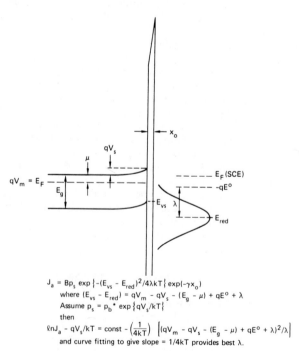

$$J_a = Bp_s \exp\left\{-(E_{vs} - E_{red})^2/4\lambda kT\right\} \exp(-\gamma x_0)$$

where $(E_{vs} - E_{red}) = qV_m - qV_s - (E_g - \mu) + qE^0 + \lambda$

Assume $p_s = p_b{}^* \exp\left\{qV_s/kT\right\}$

then

$$\ln J_a - qV_s/kT = \text{const} - \left(\frac{1}{4kT}\right)\left[(qV_m - qV_s - (E_g - \mu) + qE^0 + \lambda)^2/\lambda\right]$$

and curve fitting to give slope = $1/4kT$ provides best λ.

Figure 5.   Determination of λ

Now at low V_s we can assume:

$$P_2 = p_b^* \exp\{qV_s/kT\} \qquad (3)$$

where p_s is the surface, p_b^* the bulk hole density. Then

$$\ell nJ_a - qV_s/kT = const - (4kT)^{-1}\{qV_m - qV_s - (Eg-\mu) + qE^o + \lambda\}^2/\lambda$$
$$(4)$$

The physical picture is as follows. As the valence band edge is lowered relative to the energy levels of the reducing agent, that is, $E_{vs} - E_{red}$ becomes smaller, the anodic current will change. The parameter that dominates this change is the parameter λ. All the slope-determining parameters in Equation 4 except λ are measurable. The anodic current is measured directly, of course; the surface barrier V_s is measured by measuring the capacity of the silicon surface; the energy gap of the silicon and its Fermi energy μ are known from resistivity measurements, and E^o is the standard electrode potential of the couple. From a plot according to Equation (4) we find the value of λ that gives a slope of $(4 kT)^{-1}$. By curves such as shown in Figure 2 we have been able to analyze λ in preliminary measurements, finding the order of 0.45 eV for ferrocene and the order of 0.9 eV for ferrocyanide. Careful experiments intended to determine λ more accurately using this method are in progress.

Concluding Remarks

The ability to measure λ for various reducing agents in solution is of course critical in photocorrosion. The understanding of the role of thin passivating (possibly oxide) layers is also important. It is observed in the present study that the presence of the oxide may be desirable and in some cases is probably necessary to make the stabilizing agents effective in preventing photocorrosion. This observation is not limited to silicon. It is well known that gallium arsenide (5) and gallium phosphide (6) show most stability under the conditions of pH where the gallium oxide is insoluble in an aqueous solution. By analogy to the present work we suggest that in these cases one also apparently needs a thin oxide to promote the greatest stability against photocorrosion. To illustrate in slightly more detail how the oxide can be effective in photocorrosion, Equation 1, giving the anodic current to the stabilizing agent as a function of the thickness of the oxide, is plotted in Figure 6. Actually we plot the maximum current that the stabilizing agent can accept. The value of $\Delta E = E_{vs} - E_{red}$, with zero voltage across the oxide, and λ are indicated as the parameters for the curves of Figure 6. We have used 0.5 $\mathring{A}^{-1}$ for the tunneling coefficient γ and $10^{13} cm^{-2}$ for the interface state density (which determines the maximum voltage obtainable across the oxide) and 3 for the dielectric constant of the oxide.

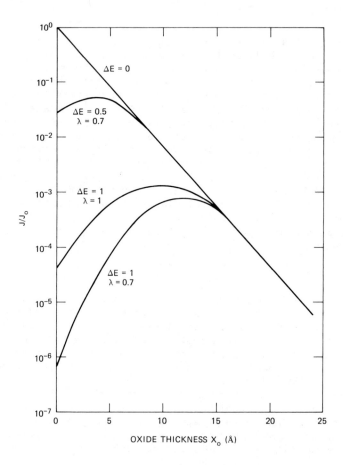

Figure 6. Maximum hole current to a reducing agent as a function of oxide thickness. Equation 1 is plotted, with p_s assumed constant and incorporated into J_o. $\Delta E = E_{vs} - E_{red}$ with no oxide, λ is the reorganization energy, and other constants have been chosen as described in the text.

In Figure 6, where we plot the maximum current that the stabilizing agent can accept, if the current rises above the value indicated, the surface is by definition unstable. Consider for example the curve where $\Delta E = 1$ and $\lambda = 0.7$. If we consider a current
of, say 10^{-4} J_0 A/cm^2, it is observed that with no oxide there is
no stability, so the oxide will grow. The oxide will grow, according to this curve, until it is 5 Å in thickness. Then it will
provide stability. If, however, the oxide grows beyond 18 Å in
thickness then the stabilizing agent can no longer capture the
holes, primarily because with the value of γ chosen, the tunneling current is too low. Then again the silicon will no longer
be stable. Of course in practice 100% stability cannot be expected for any value of oxide thickness--even if 99.99% stability
is achieved, the 0.01% of the holes causes irreversible oxide
growth in this system where there is no mechanism for oxide
removal. Thus in the experiments reported above, a lasting stability was never reached.

In conclusion it should be pointed out that the stability of
the silicon or other semiconductor against photocorrosion is not
the only important parameter in PEC solar cells. Free current
flow from the silicon valence band to the reducing agent in solution must be possible in order to have an efficient solar cell.
Thus with stabilizing agents such as we have been discussing in
the present experiments, where an oxide is needed, a compromise
must be reached between the improved corrosion resistance of the
materials and the poor current flow characteristics. An oxide
may or may not affect the open circuit voltage or the short circuit current adversely, but the fill factor of the solar cell will
suffer most due to the increased voltage across the oxide as the
photocurrent increases.

Literature Cited

1. Madou, M. J., Frese, Jr., K. W., and Morrison, S. R.,
    J. Phys. Chem. (submitted for publication).
2. Gerischer, H., Z. Phys. Chem., 1961, NF 27, p. 48
3. Morrison, S. R., "Electrochemistry at Semiconductor and
    Oxidized Metal Electrodes," Plenum: New York, 1980 (in
    press).
4. Frese, Jr., K. W. and Morrison, S. R., J. Vac. Sci. Tech.,
    1980, 17, 609.
5. Wrighton, M. S., Bolts, J. M., Bocarsley, A. B., Palazzotto,
    M. C., and Walton, E. G., J. Vac. Sci. Tech., 1978, 15, 1429.
6. Madou, M. J., Frese, Jr., K. W., and Morrison, S. R.,
    J. Electrochem. Soc., 1980, 127, 987.
7. Madou, M. J., Cardon, F., and Gomes, W. P., Ber. Bunsenges,
    Phys. Chem., 1977, 81, 1186.

RECEIVED October 3, 1980.

Conditions for Rapid Photocorrosion at Strontium Titanate Photoanodes

R. E. SCHWERZEL, E. W. BROOMAN, H. J. BYKER, E. J. DRAUGLIS,
D. D. LEVY, L. E. VAALER, and V. E. WOOD

Battelle Columbus Laboratories, 505 King Avenue, Columbus, OH 43201

In 1912, the great Italian photochemist, Giacomo Ciamician, published a remarkable paper entitled "The Photochemistry of the Future" (1) in which he considered the wealth of benefits which might be gained by the photochemical utilization of solar energy for the production of useful chemical materials. In discussing the role of plant crops (or biomass, as we would say now) as solar energy transducers, he suggested that: "The harvest, dried by the sun ought to be converted, in the most economical way, entirely into gaseous fuel, taking care during this operation to fix the ammonia (by the Mond process, for instance) which should be returned to the soil as nitrogen fertilizer together with all the mineral substances contained in the ashes". This elusive goal of efficient, economical fuel production from renewable biomass resources has stimulated research efforts around the world since Ciamician's time. While much progress has been made, the problems involved are far from solved, and the production of gaseous fuels from biomass is still too expensive to be economically feasible on a large scale. Nonetheless, recent developments in several aspects of photoelectrochemistry have offered renewed promise that the production of useful fuels with solar energy might indeed become a viable process.

Of particular interest in this context has been the finding that the Kolbe reaction, the anodic oxidation of carboxylic acids (Equation 1) (2), can be made to occur at n-type oxide semiconductor photoanodes to the virtual exclusion of oxygen formation (3,4,5).

$$2CH_3CO_2H \longrightarrow C_2H_6 + 2CO_2 + H_2 \qquad (1)$$

$$\Delta G^O = -22.3 \text{ kJ/mole } (-5.3 \text{ kcal/mole})$$

0097-6156/81/0146-0191$05.00/0

While this reaction is substantially exothermic (6), it provides an intriguing approach to the production of fuels from
renewable resources, as the required acids (including acetic
acid, butyric acid, and a variety of other simple aliphatic
carboxylic acids) can be produced in abundant yields by the
enzymatic fermentation of simple sugars which are, in turn,
available from the microbiological hydrolysis of cellulosic
biomass materials (7). These considerations have led us to
suggest the concept of a "tandem" photoelectrolysis system,
in which a solar photoelectrolysis device for the production
of fuels via the photo—Kolbe reaction might derive its
acid—rich aqueous feedstock from a biomass conversion plant
for the hydrolysis and fermentation of crop wastes or other
cellulosic materials (4).

     As one aspect of our recent studies in this field, we
have sought to extend the range of conditions under which the
photo—Kolbe reaction could be conducted, so as to explore the
sensitivity of the process to such variables as light intensity, pH, concentration of carboxylic acid, and so on. It
has gradually become apparent that under at least some of our
experimental conditions, the strontium titanate photoanodes
can undergo severe photodegradation. When this occurs, visible pits or craters are formed in the illuminated portion of
the electrode after several hours or days of exposure to
focused light from an Eimac 150W xenon arc lamp. This observation is totally unprecedented; after all, strontium
titanate is one of the few materials that has been unanim-
ously reported in the literature to be a robust, stable
photoanode (8-16).

     We have conducted numerous replicate control experiments
to determine whether a procedural error or an equipment mal-
function (such as, for example, the leakage of alternating
current from the line circuits into the dc circuits of the
photoelectrolysis apparatus) could have been responsible for
the observed effects. Ultimately, these control experiments
led us to add an oscilloscope to the diagnostic equipment (to
check for ac leakage), to replace each of the major electronic components (including the potentiostat, voltage scan
unit, and electrometer) with alternative components of comparable quality, and to replace the simple, one—compartment
cell we had been using with a newly designed two—compartment
cell, so as to minimize the possibility that cathodic products could somehow affect the stability of the strontium
titanate photoanodes. In addition, the electrodes used in
the new cell were mounted to their Pyrex support tubes using
heat—shrinkable Teflon tubing rather than epoxy cement. As
before, the new cell had a Pyrex window through which the
semiconductor electrode could be irradiated, a Luggins
capillary positioned close to the surface of the illuminated
electrode, and integral gas burets above each electrode for
the collection of gas samples.

Despite these precautions, marked corrosion was still observed on some, but not all, of the n-SrTiO₃ photoanodes obtained from four different sources. The corrosion appeared to be most severe after several experiments (totalling typically 20 hours or more of use as an electrode) had been conducted under photo-Kolbe reaction conditions. A fine white film was also observed to form gradually on the irradiated areas of n-SrTiO₃ when the acid electrolyte (typically 2N H_2SO_4) was used in the absence of added acetic acid.

Characteristics of the Photocorrosion Process

The magnitude of the problem can be appreciated by comparing the crystals marked A and B in Figure 1. Crystal A is typical of the condition of our n-SrTiO₃ electrodes just prior to etching and mounting; it is smooth, shiny (after polishing with a 1.5μ diamond paste cloth), and somewhat transparent. Crystal B illustrates the degree of photocorrosion which occurred in a similar n-SrTiO₃ electrode after approximately 50 hours of irradiation (during several experiments) under typical photo-Kolbe conditions, in this case aqueous sulfuric acid containing acetic acid. The severely eroded area is located where the light beam was focused on the electrode; the residual pitting on the surface is probably due to scattered light striking the electrode. Virtually no photo-Kolbe products have been observed by mass spectrometry in the gas evolved from the decomposing electrodes; the primary gaseous product is oxygen, although variable amounts of CO_2 have been observed at times. Thus, there appears to be a competition between electrode decomposition and the photo-Kolbe reaction under these conditions.

While we have not yet carried out detailed kinetic measurements on the rate of photocorrosion, our impression is that the process is relatively insensitive to the specific composition of the strontium titanate. Trace element compositions, obtained by spark-source mass spectrometry, are presented in Table I for the four boules of n-SrTiO₃ from which electrodes have been cut. Photocorrosion has been observed in samples from all four boules. In all cases, the electrodes were cut to a thickness of 1-2 mm using a diamond saw, reduced under H_2 at 800-1000 C for up to 16 hours, polished with a diamond paste cloth, and etched with either hot concentrated nitric acid or hot aqua regia. Ohmic contacts were then made with gallium-indium eutectic alloy, and a wire was attached using electrically conductive silver epoxy prior to mounting the electrode on a Pyrex support tube with either epoxy cement or heat-shrinkable Teflon tubing.

Some information about the nature of the photocorrosion process is provided by a comparison of the UV/visible ab-

TABLE I.   TRACE ELEMENT COMPOSITION OF n–SrTiO$_3$ BOULES[a]

Element	Sample			
	1[b]	2[c]	3[d]	4[e]
W	2.0	0.2	0.3	1.0
Ta	<0.03	30.0	0.3	<1.0
Nb	<0.05	<0.05	<0.05	<0.3
Ba	20.0	10.0	50.0	Not reported
Fe	∿3	∿3	∿3	∿5

(a)   Analyses are reported as parts per million (by weight), and were obtained by spark–source mass spectrometry. The identity of the bulk material as SrTiO$_3$ was confirmed by x-ray powder diffraction measurements.

(b)   Obtained from National Lead Company some 5 or 6 years ago for another project at Battelle–Columbus, and made available to us in 1977 when our research in photoelectrolysis began. Our initial photo-Kolbe experiments were carried out using electrodes cut from this boule.

(c)   This sample was kindly provided by Mr. Fred Wagner, of the Materials Science Laboratory, The University of California at Berkeley (1978).

(d)   Purchased from Commercial Crystal Laboratories (1978).

(e)   Purchased from Atomergic Laboratories (1979).

sorption spectra of the electrodes before and after extended
irradition. As Figure 2 illustrates, the photocorrosion
process is accompanied by a dimunition of the band edge below
400 nm and a pronounced broadening of the visible absorbance
around the Ti^{3+} band at 505 nm. The magnitude of the visible absorbance does not change appreciably, remaining about
0.7 to 0.8 at 505 nm before and after corrosion for these
electrodes. To obtain these spectra, a thin (<1 mm) single-
crystal wafer was used as an electrode after it had been
reduced to a bright blue color (as viewed by transmitted
light), polished, and etched as described above. After the
electrode had been corroded by irradiation under photo-Kolbe
conditions for some 20 hours, it was demounted from its
holder and polished again. These spectra therefore reflect
changes in the bulk semiconductor, which are manifested by a
change in color from the original blue to greyish-tan.

Further insight is provided by the Fourier-transform
infrared absorption spectrum of the corrosion residue which
could be scraped from the surface of a n-SrTiO₃ electrode
after partial photocorrosion in the mixed H_2SO_4:HOAc
electrolyte. As Figure 3 shows, the spectrum contains strong
absorption bands at 460 cm^{-1} and 620 cm^{-1}, which are
characteristic of SrTiO₃. However, it also contains strong
bands at 620 cm^{-1}, 650 cm^{-1}, 1000 cm^{-1}, 1100 cm^{-1}, 1140
cm^{-1}, and 1205 cm^{-1}, which are characteristic of $SrSO_4$,
and two weaker bands at 410 cm^{-1} and 425 cm^{-1}, which appear to be diagnostic for $Sr(OAc)_2$. There is no striking
evidence for the presence of TiO_2, although its infrared
spectrum is relatively featureless. The assignments reported
here have been confirmed by recording the infrared spectra of
authentic samples of $SrSO_4$ and $Sr(OAc)_2$. Since $Sr(OAc)_2$
is significantly more soluble in water than is $SrSO_4$ (which
precipitates immediately when a dilute solution of $Sr(OAc)_2$
is added to dilute sulfuric acid) it is plausible that the
photocorrosion process may be accelerated under these experimental conditions by the formation of $Sr(OAc)_2$ and the
subsequent precipitation of $SrSO_4$ at the electrode surface.
In sulfuric acid solutions alone, the formation of an insoluble layer of the sulfate on the electrode surface may serve
to inhibit the rapid bulk corrosion which has been observed
in the presence of acetic acid.

Finally, we note that the photocorrosion process is
strongly pH-dependent, occurring most readily in strongly
acid solutions, and that the presence of a carboxylic acid is
required for the occurrence of severe photocorrosion. In
Table II we present analytical results, based on inductively
coupled argon plasma (ICP) emission spectroscopy, for representative electrolyte solutions after 6-8 hr. of photo-Kolbe
electrolysis with n-SrTiO₃ anodes. It can be seen that the
formation of soluble strontium and titanium species is

Figure 1. Comparison of n-*SrTiO₃ photoelectrodes (A) before and (B) after corrosion under photo-Kolbe reaction conditions*

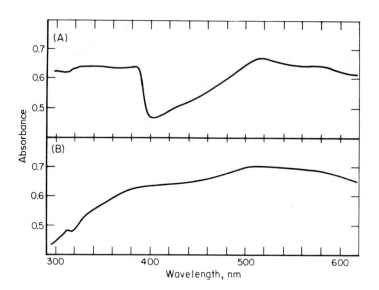

Figure 2. UV/visible absorption spectra of n-*SrTiO₃ (A) before and (B) after partial photocorrosion in* 2N H₂SO₄ *containing 0.5N HOAc (about 20 h irradiation)*

TABLE II. ANALYSIS OF ELECTROLYTES FOR PHOTOCORROSION
PRODUCTS FROM n-SrTiO$_3$

Sample	Electrolyte	pH	[Sr][a]	[Ti][a]
1	0.2$\underline{M}$ NaOH + 0.5$\underline{M}$ NaOAc[b]	13	0.10	<0.05
2	0.5$\underline{M}$ HOAc + 0.5$\underline{M}$ NaOAc[c]	5	0.25	<0.05
3	1.0$\underline{M}$ H$_2$SO$_4$ + 0.5$\underline{M}$ HOAc[c]	0	4.10	2.20

(a) Concentration of strontium and titanium, respectively, in μg/ml, as determined by inductively coupled argon plasma (ICP) emission spectroscopy (Jarrell-Ash ICP spectrometer).

(b) Bias potential, 0.0 V vs. SCE.

(c) Bias potential, +1.0 V vs. SCE.

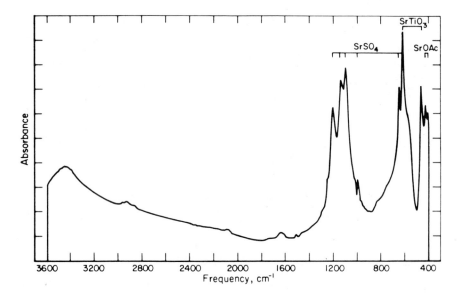

Figure 3. Fourier transform IR absorption spectra of corrosion residue from n-SrTiO₃ electrode formed after about 20 h irradiation in 2N H₂SO₄ containing 0.5N HOAc

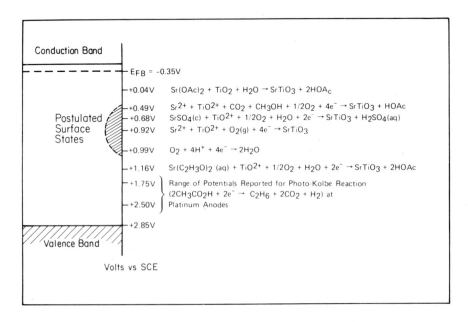

Figure 4. Estimated redox potentials of possible photocorrosion processes in strontium titanate at pH = 0 (V vs. SCE)

greatly accelerated in the strongly acid (pH=0) solution, and
that only trace amounts of Sr^{2+} can be detected in the other
solutions. It is not yet clear whether the presence of sulfuric acid is required for rapid corrosion, or whether the
rapid corrosion observed in the solutions containing sulfuric
acid simply reflects the effect of low pH. The same electrode was used all for these experiments, and all irradi-
ations were carried out with the sample apparatus for approximately equal periods of time; thus, the relative amounts
of strontium found in the solutions provide an indication of
the relative rates of photocorrosion in the different electrolytes. As noted before, little or no photocorrosion could
be detected visually in either acidic or basic electrolytes
under comparable conditions in the absence of added acetic
acid (or acetate).

Proposed Mechanism of Photocorrosion

The electrochemical redox potential of several possible
decomposition reactions at pH = 0 (relative to the potential
of the saturated calomel electrode), which have been estimated from thermodynamic parameters (6,17-21), are shown
schematically in Figure 4. The band levels are shown for
open-circuit conditions. The standard electrode potentials
were calculated from the free energies of formation, which
are summarized below in Table III.

All but one of the reactions in Figure 4 lead to the
formation of the soluble TiO^{2+} ion; this seems consistent
with the observed changes in the visible absorption spectrum
of the solid electrode. It may also be that other titanium
species are formed in solution, such as peroxytitanium complexes like H_4TiO_5. We have no direct evidence as to
the identity of the solution species at this time, and have
limited the candidate corrosion reactions shown in Figure 4
to those for which thermodynamic data are readily available.
Nonetheless, the fact that titanium is observed in the electrolyte only upon extensive photocorrosion (and then in
smaller amounts than strontium) suggests that the initial
photocorrosion process involves the loss of strontium from
the $SrTiO_3$, with the formation of $Sr(OAc)_2$ or $SrSO_4$.
However, we cannot really choose among the possible photocorrosion reactions on the basis of the information presently
available.

Some additional insight as to possible mechanisms of the
photocorrosion process can be gained from a more detailed
consideration of the effects of pH on the band levels in
$SrTiO_3$ and on the redox potentials of oxygen formation and
the photo-Kolbe reaction. These data, along with the band
levels for TiO_2, are shown in Figure 5. It is important
to remember that the photocorrosion process occurs in com-

TABLE III. STANDARD FREE ENERGIES OF FORMATION

Compound	ΔG_f (kcal/mole)	Reference
$SrTiO_3$ (c)	−378.8	17
$SrSO_4$ (c)	−318.9	6
TiO_2 (c)	−203.8	6
$Sr(OAc)_2$ (aq)	−311.8	6
Sr^{2+} (aq)	−133.2	6
TiO^{2+} (aq)	−138.2	18, 19
H_2SO_4 (aq)	−177.34	6
CO_2 (g)	−94.26	6
H_2O (l)	−56.69	6
CH_3OH (aq)	−42.12	20
HOAc (aq)	−96.19	21

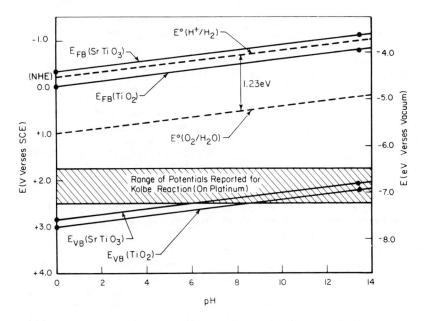

Figure 5.   Effects of pH on the energy levels of n-SrTiO₃ and n-TiO₂

petition with both the photo-Kolbe reaction and normal photoelectrolysis (although oxygen is generally evolved in parallel with photocorrosion), and that this competition is dominated not by thermodynamics (as all three processes are energetically allowed) but by surface kinetic effects which can only be speculated on at present. The crucial observation, however, is that the photo-Kolbe reaction occurs cleanly and infallibly on n-TiO$_2$, particularly in acid solution (5), with little if any degradation of the titanium dioxide, while on n-SrTiO$_3$ the photo-Kolbe reaction sometimes occurs (3,4) and sometimes is overwhelmed by photocorrosion and/or oxygen evolution.

As was indicated by the results in Table I, the success or failure of the photo-Kolbe reaction to occur on SrTiO$_3$ appears to be unrelated to the concentration of doping impurities present in the SrTiO$_3$. Similarly, there seems to be no obvious correlation with the extent of reduction of the SrTiO$_3$, as similar behavior has been observed on crystals which ranged in appearance from bright blue to nearly black. We speculate, then, that the behavior we observe may rely on the presence of defect surface states which can provide optimal overlap with the SrTiO$_3$ corrosion potentials, and most likely with the oxygen evolution potential as well, for those electrodes which undergo photocorrosion.

If this is true, then the oxidation of carboxylic acids should be the preferred process on SrTiO$_3$ photoanodes, in the absence of such defect surface states. We see from Figure 5 that the range of potentials reported for the normal Kolbe reaction (at platinum) actually crosses the valence band levels of both SrTiO$_3$ and TiO$_2$ in the neutral pH region. It may well be that at high pH, the photo-Kolbe potential lies at or below the valence band edge for these semiconductors, consistent with the observation that photo-Kolbe products are not observed under these conditions. Where there is direct overlap with the valence band edge, the electron transfer process may be so facile as to give rise to the Hofer—Moest reaction (2), in which the intermediate alkyl radical is itself oxidized (while it is still adsorbed to the electrode surface) to give a carbonium ion. The reaction of this carbonium ion with the aqueous electrolyte would then yield water-soluble products such as methanol, in keeping with our observation that anodic gas evolution is suppressed under these conditions. In acidic solutions, where the Kolbe reaction is energetically allowed, its kinetic competition with the other reactions on SrTiO$_3$ thus depends on the absence of defect surface states which are present in some electrode crystals and not in others.

Summary and Conclusions

According to this picture, then, the photo-Kolbe reaction should be the preferred reaction for defect-free SrTiO$_3$ photoanodes in acidic solutions, while the Hofer-Moest reaction should be preferred in alkaline solutions. The presence of defect surface states on some electrodes would thus deflect the anodic chemistry of these electrodes toward photocorrosion and oxygen evolution, to the detriment of the desired fuel-forming reactions. We do not yet know what the origin or nature of such surface states might be, nor indeed can we offer direct evidence that they exist. However, they provide a plausible rationale for the behavior we observe, as well as a logical target for future exploration. They also call attention to a crucial problem which is common to all of photoelectrochemistry at present: the understanding and kinetic control of energetically allowed reactions at the semiconductor-electrolyte interface. It is toward this end that future experiments must be directed if photoelectrochemistry is to become a viable means of solar energy utilization.

Acknowledgement

This research was supported by the Solar Energy Research Institute, U.S. Department of Energy. We thank Dr. Fred Wagner (University of California, Berkeley) for assistance with the preparation of several samples of strontium titanate.

Abstract

We report here that strontium titanate photoanodes, which previously have been thought to be impervious to electrolytic attack, can undergo rapid photocorrosion under photo-Kolbe reaction conditions in acidic electrolyte containing both sulfuric acid and acetic acid. Little, if any, corrosion is observed in aqueous sulfuric acid alone, and virtually no corrosion occurs in alkaline solutions, with or without added acetate. Furthermore, the corrosion process is independent of the trace element composition of the SrTiO$_3$, and occurs only on some, but not all, photoelectrodes cut from any given boule of SrTiO$_3$ under the same experimental conditions. We suggest that the observed behavior is consistent with the presence of defect surface states which provide good overlap with the potentials of several plausible SrTiO$_3$ corrosion reactions. The photo-Kolbe reaction is thus considered to be the preferred reaction on defect-free SrTiO$_3$ electrodes in acid solution, with photocorrosion dominating for kinetic reasons on those electrodes which have been etched insufficiently or otherwise treated so as to provide appropriately located surface states.

Literature Cited

1. G. Ciamician, Science, 36, 385 (1912).

2. A. K. Vijh and B. E. Conway, Chemical Reviews, 67, 623 (1967).

3. R. E. Schwerzel, "Methods for the Photochemical Utilization of Solar Energy", paper presented at the 29th Southeast Regional Meeting of the American Chemical Society, Tampa, Florida (November 9-11, 1977).

4. R. E. Schwerzel, E. W. Brooman, R. A. Craig, D. D. Levy, F. R. Moore, L. E. Vaaler, and V. E. Wood, in "Solar Energy: Chemical Conversion and Storage", R. R. Hautala, R. B. King, and C. Kutal, Eds., The Humana Press, Inc., Clifton, N.J. (1979), pp. 83-115.

5. B. Kraeutler and A. J. Bard, J. Amer. Chem. Soc., 100, 5985 (1978), and references cited therein.

6. "CRC Handbook of Chemistry and Physics", 53rd Ed., Chemical Rubber Publishing Company, Cleveland, Ohio (1972).

7. See, for instance J. E. Sanderson, D. L. Wise and D. G. Augenstein, "Liquid Hydrocarbon Fuels from Aquatic Biomass", Paper No. 27 presented at the Second Annual Fuels from Biomass Symposium, Rensselaer Polytechnic Institute, Troy, New York (June 20-22, 1978).

8. M. S. Wrighton, A. G. Ellis, P. T. Wolczanski, D. Morse, H. H. Abrahamson, and D. Ginley, J. Amer. Chem. Soc., 98, 2774 (1976).

9. J. Mavroides, J. Kafalas, and D. Kolesar, Appl. Phys. Lett., 28, 241 (1976).

10. T. Watanabe, A. Fujishima, K. Honda, Bull. Chem. Soc. Japan, 49, 355 (1976).

11. H. H. Kung, H. S. Jarret, A. W. Sleight, and A. G. Ferretti, J. Appl. Phys., 48, 2463 (1977).

12. A. B. Bocarsly, J. M. Bolts, P. C. Cummins, and M. S. Wrighton, Appl. Phys. Lett., 31, 568 (1977).

13. M. S. Wrighton, P. J. Wolczanski, and A. B. Ellis, J. Solid State Chem., 22, 17 (1977).

14. K. Honda, and T. Watanabe, Elektrokhimiua, 13, 924 (1977).

15. H. P. Maruska and A. K. Ghosh, Solar Energy, 20, 443 (1978).

16. F. Vanden Kerchove, J. Vandermolen, and W. P. Gomes, Ber. Bunsenges, Phys. Chem., 83, 230 (1979).

17. L. A. Zharkova, Zh. Fiz. Khim., 36, 985 (1962).

18. A. J. Bard and M. S. Wrighton, J. Electrochem. Soc., 124, 1706 (1977).

19. Estimated from the cell potential for:
$$TiO^{2+} + 1/2\ O_2 + 2e^- \rightarrow TiO_2,$$
where E^O = +1.422V vs H^+/H_2 as given by A. J. Bard and M. S. Wrighton, J. Electrochem. Soc., 124, 1706 (1977), and ΔG_f^O for TiO_2 = -203.8 kcal/mole.

20. From ΔG_f^O for $CH_3OH(1)$ and ideal solution correction to 1 molal aqueous solution.

21. Same as Reference 19. At pH = 0 essentially all of the acetic acid should be present as HOAc(aq).

RECEIVED October 21, 1980.

Photoelectronic Properties of Ternary Niobium Oxides

K. DWIGHT and A. WOLD

Department of Chemistry, Brown University, Providence, RI 02912

Ferric oxide is known to have an optical band gap of about 2 eV, and the literature contains many reports of photoelectrochemical measurements on "conducting", n-type Fe_2O_3 ([1], [2], [3], [4], [5]). Unfortunately, any attempt to reduce ferric oxide results in the formation of magnetite as a distinct separate phase, and there is no solubility of this spinel in the corundum structure ([5], [6]). Thus, all the properties reported above for Fe_2O_3 were measured either on multiphase samples or on samples which contained impurities.

The ternary iron oxides, as exemplified by the iron-niobium system, offer an opportunity to obtain single-phase, conducting n-type iron oxides; in which the conductivity can be controlled by means of chemical substitution. At first glance, $FeNbO_4$ and $FeNb_2O_6$ might appear to be very different materials. Yet as $MM'O_4$ and MM'_2O_6 they merely represent superstructures of the basic $\alpha-PbO_2$ structure obtained under the conditions of preparation ([7]). Consequently, they form a solid solution in which the two valence states of iron are uniformly distributed throughout a single homogeneous phase ([8]).

However, many ternary systems incorporate a second photoactive center in addition to the $[FeO_6]$ octahedra: in the present case, $[NbO_6]$ octahedra. The interaction between such multiple centers has not previously been investigated. In the present work, interband transitions are observed which appear characteristic of niobium centers, together with other transitions characteristic of the iron centers. Since these are homogeneous, single-phase materials, this result suggests that caution should be exercised when applying the conventional band model to such oxide semiconductors.

For materials with a single photoactive center, it is generally observed that the optical band gap and flat-band potential are interrelated, so that lower band gaps appear to be accompanied by more positive flat-band potentials ([4],[9]). Nevertheless, the non-active A-site ions in such ternary compounds as $BaTiO_3$, $SrTiO_3$, $Ba_{0.5}Sr_{0.5}Nb_2O_6$ and $Sr_2Nb_2O_7$ do have a perturbing effect

0097-6156/81/0146-0207$05.00/0

(4, 7, 9, 10). Consequently, if multiple photoactive centers can
maintain sufficiently independent existence in a single compound,
it would be conceivable that significant deviations from the usual
correlation of high flat-band potentials with low band-gap energies might occur.

Effects of Composition and Structure

Before proceeding to ternary oxides with multiple photoactive
centers, the effects of composition and structure upon such photoelectronic properties as optical band gap and flat-band potential
for a given active center should be considered. It will be seen
that composition appears to primarily affect the flat-band potential, whereas the band gap is more sensitive to structure.

$Sr_2Nb_2O_7$ is a pyrochlore; $Ba_{0.5}Sr_{0.5}Nb_2O_6$ is a defect perovskite. In both materials, the $[NbO_6]$ octahedra are the only photoactive centers. As shown in Figure 1, the flat-band potential
of the pyrochlore is more negative by 0.4 volts, and its band gap
is correspondingly larger, as would be expected. But the respective roles of structure and composition cannot be deduced from
this comparison alone.

$BaTiO_3$ and $SrTiO_3$ are both perovskites and have nearly the
same optical band gaps. Yet the flat-band potential of $SrTiO_3$ is
0.6 volts more negative than for the barium analog, a difference
comparable in magnitude to that noted above for the niobates.
Furthermore, it can be seen from Figure 1 that the band gap in the
rutile TiO_2 is significantly lower than in these perovskite ti-
tanates.

Thus, the behavior in both the titanium and niobium systems
is consistent with the hypothesis that the A-site cation is primarily responsible for variation in flat-band potential while the
structure is primarily responsible for variation in optical band
gap. Of course, it has been noted elsewhere that other properties
such as the magnitude of the quantum efficiency also depend upon
structure (10).

From Figure 1 it is evident that Fe_2O_3, $FeNbO_4$, and $FeTiO_3$
all have relatively positive flat-band potentials, which is presumably a characteristic of the iron. The band gap in the titanate appears to be associated with the $[TiO_6]$ octahedra; that in
the niobate appears to match ferric oxide within structural vari-
ability. From such a cursory analysis, there would appear to be
no effect from the presence of a second photoactive center in
these two materials.

However, the existence of such an effect can be demonstrated
by the application of a recently proposed technique for the study
of interband transitions having energies greater than the "optical" band gap (11). Standard procedures exist for the extraction
of band-gap information from measurement of the optical absorption coefficient, which has been shown to be proportional to the
quantum efficiency (photocurrent density divided by the incident

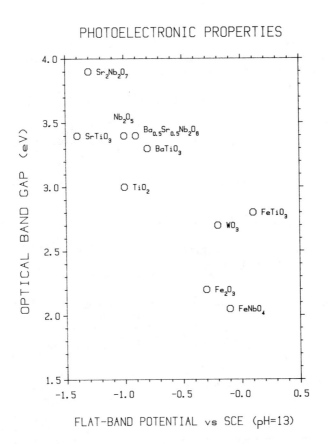

Figure 1. *Optical band gaps and flat-band potentials (adjusted to pH = 13) for some photoanode materials (4, 7, 9, 10)*

light flux) under conditions applicable to the materials considered here (11, 12). The photoelectrolysis experiment provides an
effective sampling region much thinner than can be obtained by
polishing crystals, thereby extending the range of measurement to
much higher energies. Since this technique is not yet widely
known, an outline of its principal features is presented in the
following section.

Band–Gap Analysis

Under moderate irradiation, the reaction rate in a photoelectrolysis cell is limited by the arrival rate of holes at the anode
surface (12), in which case the quantum efficiency η is given by:

$$\eta = 1 - [\exp(-\alpha W)]/(1 + \alpha L_p)$$

where α is the optical absorption coefficient, L_p is the hole diffusion length, and W is the width of the depletion layer (12).
Also,

$$W = [2\varepsilon\varepsilon_o (V - V_{fb})/eN_o]^{1/2}$$

and

$$L_p \lesssim [\varepsilon\varepsilon_o (kT/e)/eN_o]^{1/2}$$

since the hole diffusion length is determined by bulk recombination in highly defective oxides (11, 12).
    The dielectric constant ε can be estimated to be of the order
of 100, and the donor concentration N_o can be estimated from the
measured conductivity, activation energy and Hall mobility to be
of the order of 10^{20} cm^{-3}. Then $W \approx 10^{-6}$ cm and L_p is even
smaller, so that expansion of the exponential yields a quantum
efficiency porportional to the optical absorption coefficient even
for large values of α.
    The optical absorption coefficient for a single interband
transition is related to the photon energy by $\alpha \sim (h\nu)^{-1}(h\nu - Eg)^n$
where Eg is the band gap and n depends upon the character of the
transition (n = 0.5 for allowed direct transitions; n = 2 for
allowed indirect ones). Thus, if experimental values for α are
multiplied by $h\nu$ and are plotted as $(\alpha h\nu)^{(1/n)}$ against $h\nu$, then a
straight line intersecting the energy axis at Eg will be obtained
when n correctly characterizes the transition. Since the total
optical absorption coefficient α for a compound comprises the sum
of such contributions from successive interband transitions, its
complete analysis must proceed in stages. Each transition is
characterized in turn, starting from the lowest energy, whereupon
its contribution to the absorption is extrapolated to higher energies and subtracted from the total α. However, the simple determination of the interband transition energies does not require
this elaborate process, the onset of each additional contribution

to the total α being clearly visible as an abrupt increase in the
slope of the graph of $(\alpha h\nu)^{(1/n)}$ vs hν. Furthermore, higher-energy direct transitions can often be identified unambiguously without subtracting the contributions from lower-energy indirect ones.

The absorption coefficient increases with increasing photon
energy, and each successive transition adds to the rate of increase. Consequently, the analysis of higher-energy transitions
is limited by the maximum value of α which can be measured, which
is inversely proportional to the thickness of the sample. In the
photoelectrolysis experiment, the depletion layer forms a very
narrow sampling region, so that the maximum value of α measurable
by η is large. This permits the determination of interband transitions well above the energy of the lowest band gap (11).

In order to illustrate the power and reliability of this analytical procedure, the quantum efficiency η measured for $SrTiO_3$,
being proportional to α, has been multiplied by the photon energy
hν and is plotted in Figure 2 as $(\eta h\nu)^{0.5}$ vs hν. The linearity of
the lowest-energy section of this graph (with n = 2) characterizes
the transition as indirect. The energy intercept yields the value
of 3.2 eV for the lowest band gap, which is in good agreement with
previous absorption measurements (13) and with the calculation of
Kahn and Leyendecker (14).

The abrupt increase in slope at 3.37 eV signals the presence
of a higher-energy transition, in accord with the increased absorption found by electromodulation measurement (15). The decrease in slope at 3.5 eV corresponds to a saturation of this contribution to the total absorption and is not understood. Nevertheless, several other materials give evidence of similar
behavior.

Finally, the transition at 3.74 eV agrees both with the electromodulation spectra (15) and with the band structure calculation
(14). This higher-energy section does not appear greatly different from the rest of Figure 2, although there is some curvature of
the data. However, this region becomes truly linear when replotted as $(\eta h\nu)^2$ versus hν, which establishes the direct character of
the high-energy transition.

Results and Discussion

The quantum efficiency data for the defect pyrochlore
$Ba_{0.5}Sr_{0.5}Nb_2O_6$ is presented in Figure 3 (10). It shows an indirect band gap at 3.4 eV with a "tail" extending to nearly 2.6 eV.
The higher-energy transition at 4.4 eV shows some curvature of the
data, and indeed, corresponds to a direct transition when replotted as $(\eta h\nu)^2$ versus energy (10).

Similar data for the pyrochlore $Sr_2Nb_2O_7$ is plotted in Figure
4. Here the principle indirect band gap occurs at 3.9 eV with a
"tail" to nearly 3.4 eV. The data beyond 4.3 eV cannot be interpreted quantitatively because of a breakdown in the conditions
required for the band-gap analysis, but there is an indication of

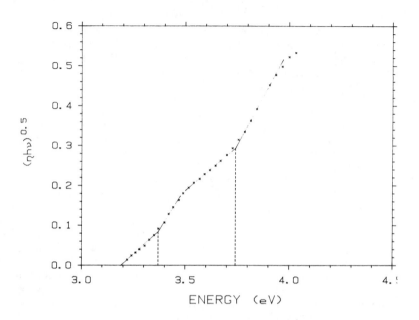

Figure 2.   Band-gap analysis for SrTiO$_3$ (11) showing transitions at 3.2, 3.4, and 3.75 eV

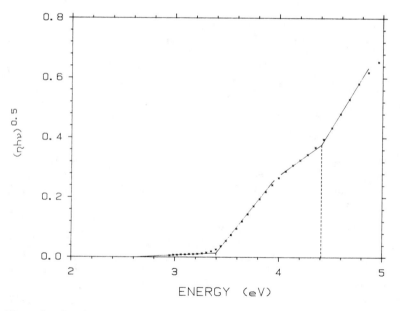

Figure 3.   Band-gap analysis for Ba$_{0.5}$Sr$_{0.5}$Nb$_2$O$_6$ (10) showing transitions at 2.6, 3.4, and 4.4 eV

a transition in the vicinity of 4.7 eV. Thus the behavior is qualitatively similar to that observed in the perovskite niobate, only shifted to higher energies.

The analagous results for the corundum Fe_2O_3 are given in Figure 5. This shows an indirect band gap at 1.85 eV together with a direct band gap at 2.5 eV (11). Such simple behavior is in sharp contrast with the complex succession of transitions shown in Figure 6 for $FeNbO_4$ (7). Here the lowest-energy transition at 2.05 eV is clearly indirect. It is followed by several higher-energy transitions at 2.68, 2.93, 3.24, and 4.38 eV, each giving rise to a sudden increase in the slope of the curve, but so close together as to preclude reliable determination of direct or indirect character.

The locations of these additional interband transitions are highly suggestive. That at 2.68 eV appears to correlate with the 2.58 eV transition for Fe_2O_3 shown in Figure 5; those at 3.24, 2.9, and 4.38 eV are reminiscent of the indirect transition at 3.4 eV, its "tail," and the direct transition at 4.4 eV shown in Figure 3 for $Ba_{0.5}Sr_{0.5}Nb_2O_6$. Thus the data for $FeNbO_4$ show all the characteristics of the $[NbO_6]$ octahedra in addition to all the characteristics of the $[FeO_6]$ centers. The greater similarity to the perovskite niobate can be attributed to closer agreement between their Nb-O bond strengths as compared with those in the pyrochlore structure.

Summary and Conclusions

When only a single species of photoactive center is present in a compound, the presence of a non-active, A-site cation produces a characteristic shift in the flat-band potential. A change in structure, however, will in general produce a shift in the optical band-gap energy. This is accompanied by corresponding shifts in any other, higher-energy interband transition, but the qualitative features remain the same, and hence appear to be characteristic of the particular photoactive center.

When two species of photoactive centers are simultaneously present, the higher flat-band potential appears to dominate. But it is evident that both species contribute their characteristic sets of interband transitions to the ensemble. In this respect, these oxide semiconductors behave differently than the conventional, broad-band semiconductors. It would appear that different photoactive centers remain at least partially independent.

However, further experimentation embracing a variety of ternary systems will be required to determine the degree of interaction between such multiple centers. Preliminary results for Fe_2WO_6 confirm the superposition of two characteristic sets of interband transitions. The optical band gap and flat-band potential are essentially the same as in $FeNbO_4$, but the quantum efficiency is considerably greater. This suggests that there may be some enhancement of the photoresponse due to interaction between the iron and tungsten centers.

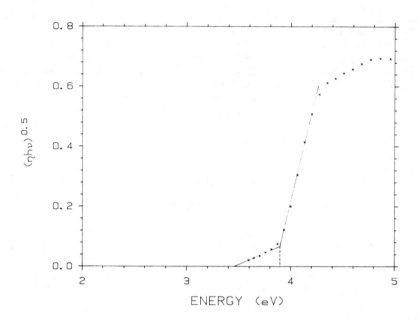

Figure 4. Band-gap analysis for $Sr_2Nb_2O_7$ (10) showing transitions at 3.4 and 3.9 eV

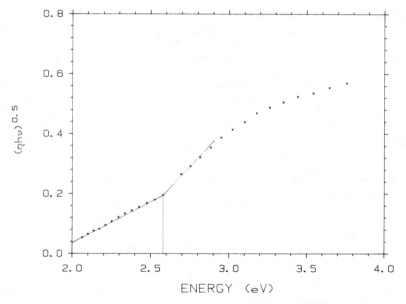

Figure 5. Band-gap analysis for Fe_2O_3 (11) showing transitions at 1.85 and 2.58 eV

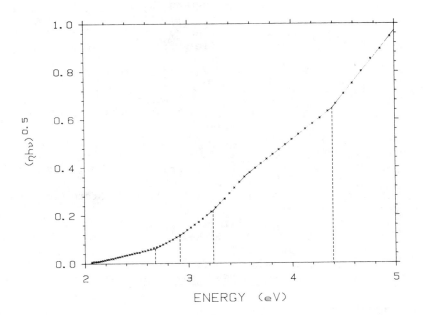

Figure 6.   Band-gap analysis for FeNbO$_4$ (7) showing transitions at 2.05, 2.68, 2.9, 3.24, and 4.38 eV

Acknowledgements

The authors would like to acknowledge the support of the Office of Naval Research, Arlington, Virginia for the support of Kirby Dwight.
The authors would also like to acknowledge the Solar Energy Research Institute, Golden, Colorado as well as the Materials Research Laboratory Program at Brown University for their support.

Literature Cited

1. Hardee, K. L.; Bard, A. J. J. Electrochem. Soc., 1976, 123, 1024 and J. Electrochem. Soc., 1977, 124, 215.
2. Quinn, R. K.; Nasby, R. D.; Baughman, R. J. Mat. Res. Bull., 1976, 11, 1011.
3. Yeh, L. R.; Hackerman, N. J. Electrochem. Soc., 1977, 124, 833.
4. Kung, H. H.; Jarrett, H. S.; Sleight, A. W.; Ferretti, A. J. Appl. Phys., 1977, 48, 2463.
5. Merchant, P.; Collins, R.; Kershaw, R.; Dwight, K.; Wold, A. J. Solid State Chem., 1979, 27, 307.
6. Salmon, O. N. J. Phys. Chem., 1961, 65, 550.
7. Koenitzer, J.; Khazai, B.; Hormadaly, J.; Dwight, K.; Wold, A. J. Solid State Chem., to be published.
8. Turnock, A. C. J. of the Am. Ceramic Soc., 1966, 49, 177.
9. Nozik, A. J. Ann. Rev. Phys. Chem., 1978, 29, 189.
10. Hormadaly, J.; Subbarao, S. N.; Kershaw, R.; Dwight, K.; Wold, A. J. Solid State Chem., to be published.
11. Koffyberg, F. P.; Dwight, K.; Wold, A. Solid State Communications, 1979, 30, 433.
12. Butler, M. A. J. Appl. Phys., 1977, 48, 1914.
13. Weakliem, H. A.; Burk, W. J.; Redfield, D.; Korsum, V. R.C.A. Review, 1975, 36, 149.
14. Kahn, A. H.; Leyendecker, A. J. Phys. Rev., 1964, 135, A1321.
15. Mack, S. A.; Handler, P. Phys. Rev., 1974, B9, 3415.

RECEIVED October 3, 1980.

14

Electrode Band Structure and Interface States in Photoelectrochemical Cells

JOHN G. MAVROIDES, JOHN C. FAN, and HERBERT J. ZEIGER

Lincoln Laboratory, Massachusetts Institute of Technology, Lexington, MA 02173

Efficient utilization of solar energy by means of photo-electrolysis requires an electrode material that is not only stable but also has an electron affinity that is small enough to give sufficient band bending. In addition, satisfactory material must have an energy gap that is well matched to the solar spectrum. The energy gaps of both TiO_2 and $SrTiO_3$, which have values of 3.0 and 3.2 eV, respectively, are considerably too large to satisfy this latter requirement.

Furthermore no better single chemical compound has been found. This suggests trying a combination of compounds for electrodes. We will describe our program to develop electrodes, with the desired electrochemical properties, by such an approach. In particular, we will discuss composite structures and solid solutions. This is an ongoing program and what will be presented are mainly the concepts with only a few preliminary results.

Composite Electrodes

Beginning with composite electrodes the simplest scheme here is to coat the surface of a small gap semiconductor that is well matched to the solar spectrum but which is electro-chemically unstable with a thin film of a wide gap, electrochemically stable semiconductor. To demonstrate the feasibility of using such a composite electrode, the film must be thin enough - of the order of 50-100 Å or less - so that at least some of the photogenerated carriers in the small bandgap material can tunnel through to the electrolyte. Furthermore the film must not have any cracks or pinholes since these would

allow the electrolyte to penetrate through to the small bandgap
material and cause dissolution. The formation of such a thin
defect-free film by conventional techniques is very difficult
and although many attempts have been reported, success has not
been achieved (1,2).
       We are investigating a slightly more complex type of
composite electrode, in which the small bandgap semiconductor is
protected by a much thicker, two-phase film such as shown in
Figure 1. Here the low bandgap substrate material is CdSe,
which has an ε_g = 1.7 eV, and the large bandgap material is
SrTiO$_3$. In this arrangement the protective coating, which is of
the order of 500Å thick, consists of small grains of CdSe (~ 50Å
in size) in a matrix of SrTiO$_3$. During photoelectrolysis, the
hole-electron pairs are generated in the CdSe and separated by
the depletion field. The holes move to the substrate-film
interface and then tunnel through SrTiO$_3$ to the nearest grain of
CdSe. After a few jumps between grains, the holes reach the
electrolyte. This film should be effective in protecting this
substrate provided the grains are isolated from each other. In
this case the grains that are initially in contact with the
electrolyte will be dissolved away until a continuous protective
surface of SrTiO$_3$ is left in contact with the electrolyte.
       Figure 2 is a schematic energy band diagram of this
composite electrode indicating how the minority carriers
from the bulk CdSe valence band tunnel through the SrTiO$_3$ to
CdSe grains via surface states. The easy passage of minority
carriers through the interface to the electrolyte requires that
the valence band edge of CdSe overlap the surface states in the
bandgap of the SrTiO$_3$. For free passage of the majority
carriers between the two seminconductors, their electron
affinities should be equal so that the potential barrier at the
interfaces between them is negligible. These conditions are
fairly well met in the CdSe-SrTiO$_3$ system.
       In connection with this system a number of questions arise.
First of all, because of interaction and scattering effects, the
optical absorption of the two-phase film could be significantly
greater than that of the individual components. In this case it
would be necessary to shine the light onto the back surface of
the substrate and the substrate would have to be very thin, of
the order of the diffusion length; secondly, even if the
carriers are excited through the protective layer and within the
depletion region of the substrate, how efficiently would they
transfer to the electrolyte? What would the nature of the
states at the interfaces be and how efficiently would
charge transfer between the grains to the electrolyte. What
would the tunneling efficiency be and would carrier scattering
be a problem? Thirdly, would the film fabrication process cause
work damage to the substrate?

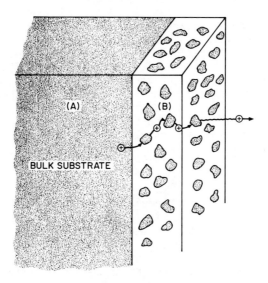

Figure 1.   Schematic of a CdSe–SrTiO₃ composite electrode (A refers to CdSe and B to SrTiO₃)

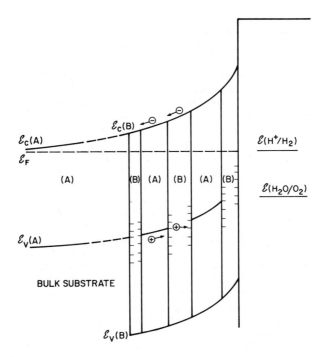

Figure 2.   Energy band diagram of the CdSe–SrTiO₃ composite electrodes (A refers to CdSe and B to SrTiO₃)

Finally how would one make such a film? One way is shown in Figure 3. What is shown here schematically is an rf sputtering system. The substrate is mounted on a rotating disc and a microprocessor-controlled unit is used to produce sequential sputtering of two different targets, one for each component, with provisions for adjusting the proportions of the two components in the film to any desired value. Annealing or some other treatment after the sputtering may be necessary to insure that the $SrTiO_3$ forms a continuous matrix with the CdSe embedded in it.

Alloys

Another approach to the electrode problem is that of alloying either TiO_2 or $SrTiO_3$ with analogous compounds to form solid solutions which could have lower bandgaps and still retain the other desirable properties.

Alloying is widely used for adjusting the bandgaps in solid solutions of such semiconductors as Ge and Si, and GaAs and GaP, where the outer electrons occupy s and p orbitals. In these broadband materials an adequate description of the electronic states is provided by the band theory approximation; as a result alloying shifts the energy gap approximately linearly with composition although there are some cases in which an intermediate composition can have either a maximum or a minimum bandgap.

Transition metals such as Ti, however, have outer d electrons which exhibit a strong tendency to be localized and the simple one-electron band model is frequently not applicable to transition-metal compounds. Consequently these compounds and their solid solutions are generally not so well understood, although some experimental data and theoretical work on them exists. For these alloys not only can you get bandgap changes, but you can also introduce separate bands of levels due to the individual constituents of the alloy, and these can be partially responsible for lower energy transitions. For example, in the case of metallic alloys of Ni with Cu, one finds not just a single set of d band states, but two sets of states, one corresponding to Ni d states and the other to Cu states (3).

It is well known that the valence and conduction bands of TiO_2 are derived mainly from the oxygen 2 p states and titanium 3 d states, respectively, (4) as shown in Figure 4(a). Less is known of the compounds which are completely miscible with TiO_2, namely TaO_2, WO_2, NbO_2, VO_2 and MoO_2. The first three of these compounds are semiconductors with a band structure similar to TiO_2 and energy gaps close to 3 eV. VO_2 on the other hand, is a semiconductor below 65°C where it undergoes a semiconductor-to-metal transition. Its bandgap of 0.7 eV separates two-subbands both derived from vanadium d-states as shown in Figure 4(b). X-ray photoemission data indicate that

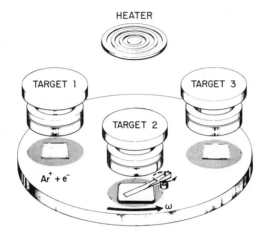

Figure 3.   Simultaneous RF sputtering module

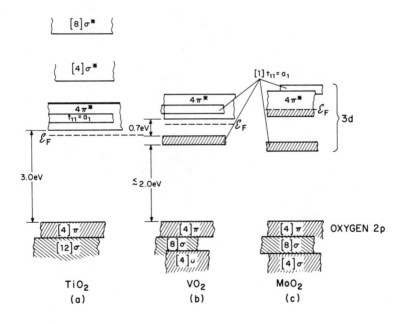

Figure 4.   Energy band diagrams (after Ref. 4) for TiO₂, VO₂, and MoO₂

the gap in VO_2 analogous to the p-to-d gap in TiO_2 is less than 2 eV. Therefore one might expect that solid solutions containing a limited concentration of VO_2 will have the same basic band structure as TiO_2 but with a smaller bandgap. Even less is known about MoO_2, except that it is a semimetal; its theoretical energy band diagram, as proposed by Goodenough, is shown in Fig. 4(c).

Another interesting system is based on alloys of $SrTiO_3$ with other ABO_3 compounds of the perovskite structure and is suggested by the work of Tributsch and co-workers, who have found stable photogalvanic action by using optical excitation between nonbonding d orbitals of transition metal dichalcogenides such as $MoSe_2$ (5,6,7,8). The same mechanism explains the photoelectrochemical stability of Fe_2O_3 (9,10,11,12) and $YFeO_3$ (13). However, a bias voltage must be applied to obtain photoelectrolysis with electrodes of the latter compounds, since their electron affinities are too high. $SrTiO_3$-$LaFeO_3$ solid solutions appear interesting because we feel that such alloys could exhibit substantial solar absorption due to transitions between the Fe^{3+} d-levels of $LaFeO_3$ while retaining a large proportion of the favorable band bending of $SrTiO_3$. In terms of the energy diagrams of Figure 4 $SrTiO_3$ would be represented by (a)and $LaFeO_3$ by (b) where the d-to-d gap is now ~ 2 eV. The preparation and characterization of perovskite based solid solutions, including a 50/50 $LaFeO_3$-$SrTiO_3$ alloy, has been reported very recently by Rauh et al (14). For the latter compound, these researchers observed photoresponse at photon energies in the 2 eV range, but with very low photocurrents.

Experiments

Preliminary experiments with the composite electrodes indicate that optical scattering does not appear to be a problem; the action spectra of the composite films are similar to those of the bare substrate electrodes. The current-voltage curves, such as shown at the bottom of Figure 5 for a 50/50 composition of CdSe and $SrTiO_3$, indicate that carrier scattering is a problem. Under illumination, the initial composite electrodes display very low photocurrents (~ 10^{-3} of the substrate material) and a linear dependence of current on electrode potential under both cathodic and anodic conditions, with no saturation effect in the anodic direction. Such behavior is suggestive of a barrier in which the carrier flow is diffusion limited, which would be the case if collisons at the CdSe-$SrTiO_3$ grain boundaries were significant (15). Annealing the electrode in an Ar atmosphere to ~ 800°C and then slowly cooling increases the photoresponse by one order of magnitude but how much improvement will ultimately be obtained is not yet known.

Most of the TiO_2 and $SrTiO_3$ alloys were made by firing mixtures of powders of the appropriate oxides. Stoichiometric mixtures of the starting materials, which had a purity of at least 99.995% as obtained from Johnson Matthey Chemicals Limited, are finely ground and sintered in air at ~ 1100°C for about 12 hours. The materials are then pressed into discs, about 1 cm in diameter and 1 mm thick, and then reduced, depending on the alloy and composition, at temperatures between 1000°C and 1400°C and times ranging from 3 to 12 hours in an atmosphere of either forming gas or Ar in order to obtain conductivities which range between 10^{-1} and 10^{-3} mho/cm. Although we have fabricated solid solutions of TiO_2 with TaO_2, NbO_2, VO_2 and MoO_2, the discussion here will be limited to the Nb alloys. In the case of $SrTiO_3$, only results of alloys between this semiconductor and $LaFeO_3$ will be given.

Figure 6 shows the variation of integrated photocurrent versus photon energy for several electrodes of Nb composition, namely for 5, 10 and 20% NbO_2 in TiO_2. Also shown for comparison is the result for a TiO_2 electrode which was reduced so as to be conducting. In these measurements the electrodes, which are all n-type, were held at a potential of + 1.0 V rel to SCE. A 150 W Xe lamp in combination with low-energy-pass color filters was used to obtain these action spectra. Such a technique, while only yielding approximate action spectra, is useful in cases where the photoresponse is low. It will be noted that all the electrodes have maximum response in the UV but that the Ti-Nb electrodes have a response which extends below 3 eV into the visible region. Unalloyed TiO_2 also showed some response into the visible, but it was two orders of magnitude lower than for the alloys. Starting with the dashed curves, an increase in the low energy photoresponse is obtained as the composition is increased from 5 to 10% Nb and a small decrease as the composition is further increased to 20% The decrease in the latter alloy is no doubt due to the fact that it is a two-phase material as identified by x-ray diffraction patterns. At the high energy end of the spectrum, on the other hand, a decrease in photoresponse is found in going from TiO_2 to increasing amounts of Nb. This tends to suggest that the same mechanism which produces the low energy sensitization also causes the lowered high energy response in agreement with the observations of Maruska and Ghosh for doped TiO_2 electrodes (16). This is not completely true, however, as the solid curve labelled (B) on this plot indicates. This curve is the action spectrum of the original 5% Nb sample labelled (A) taken after it was reoxidized in air at 1000°C for 10 minutes and then, reduced again. This heat treatment has improved not only the low energy end of the spectrum but also the high energy

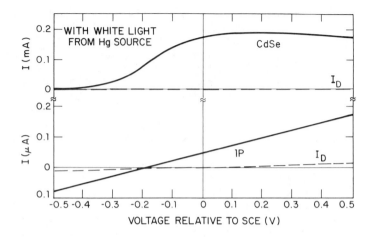

Figure 5.   *Comparison between I–V curves of a single-crystal, bare CdSe elctrode (top) and a composite electrode (bottom) with light (———) and in the dark (– – –)*
(pH = 4.7)

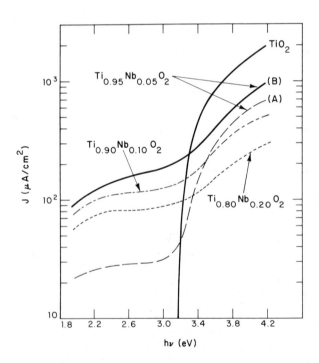

Figure 6.   *Integrated photocurrent vs. photon energy obtained with a 150-W Xe lamp in combination with low-energy-pass color filters. Electrodes are biased at +1.0 V relative to SCE (pH = 10). See text for difference between (A) and (B).*

end. The result suggests the need for more detailed and careful characterizations (both physical and chemical) of the basic material properties in order to optimize the photoresponse.

Figure 7 is a comparison of the responsivity (that is, the variation of the normalized photocurrent with photon energy), of the 5% Nb alloyed sample (labelled (B) in Figure 6), of TiO_2 single crystal and of a NbO_2 powder disc electrode. The peak response of the alloy in the ultraviolet is one order of magnitude down from that of TiO_2; the alloy also has a response in the visible which, however, is several orders down from that in the UV.

The sintered NbO_2 electrode response peaks in the UV but has a tail in response which extends into the visible. The photoresponse in the UV is fast which suggests that this response arises from electronic transitions between the valence and conduction bands, and gives a value for the energy gap of NbO_2 of about 3.3 eV. The visible photoresponse on the other hand is slower and probably arises from states within the bandgap. The response of the $Ti_{0.95}Nb_{0.05}O_2$ alloy is almost a replica of the NbO_2 response but with an apparent shift of the energy gap to lower energies (~ 2.7 eV) and also a higher responsivity. This represents a shift of the energy bands rather than a tailing of the TiO_2 response which takes place with doping (16). Interestingly, this extrapolation suggests that the alloy might have a bandgap which is less than that of the two starting materials. Such effects have been observed in wide band semiconductors, e.g., the ZnSe-ZnTe system (17). The fact that the structure in the alloy action spectrum is similar to that of pure reduced NbO_2 implies that similar states exist in the alloy as in the reduced NbO_2 and that the low energy response in the alloy is not due mainly to strains arising from inhomogeniety or alloying. Conductivity versus temperature plots indicate that relatively shallow energy levels below the conduction band are introduced by the reduction process. Thus these states are bulk and not surface states.

Figure 8 compares the responsivities of a pure $LaFeO_3$ powder disc, single-crystal $SrTiO_3$ and 20% $LaFeO_3$ in $SrTiO_3$. Curve A, the responsivity of the alloy, indicates response both in the visible as well as in the UV. Recently other methods of electrode fabrication have been reported with relatively high efficiencies; for example, Augustynski et al (18) have prepared thin films of TiO_2 doped with metals such as Al and Be on Ti substrates by a spraying technique. Another fabrication technique which is quite simple in principle is to fire oxide mixtures on Ti metal in the method recently described by Guruswamy and Bockris (19) who studied $LaCrO_3$-TiO_2 electrodes. Here again a very thin film of material is formed and a significant photoresponse is obtained. Fabrication technique is

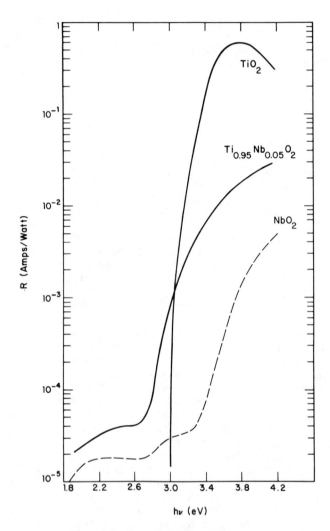

Figure 7.   Comparison between responsivities of single-crystal n-TiO₂, *pressed-disc* n-NbO₂, *and pressed-disc* n-Ti₀.₉₅Nb₀.₀₅O₂. *Electrodes are biased at* +1.0 V *relative to SCE (pH = 10).*

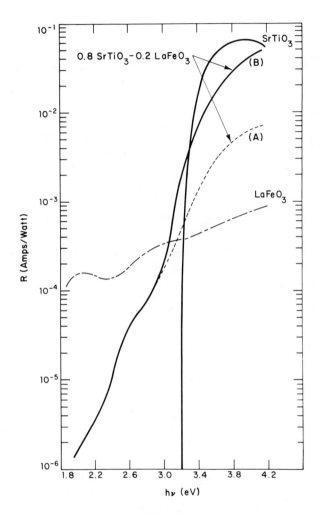

Figure 8.   Comparison between responsivities of single-crystal n-SrTiO₃, pressed-disc p-LaFeO₃, and n-0.8 SrTiO₃–0.2 LaFeO₃. SrTiO₃ and the alloys are biased at +1.0 V relative to SCE while p-LaFeO₃ is biased at−0.7 V relative to SCE (pH = 10). See text for difference between (A) and (B).

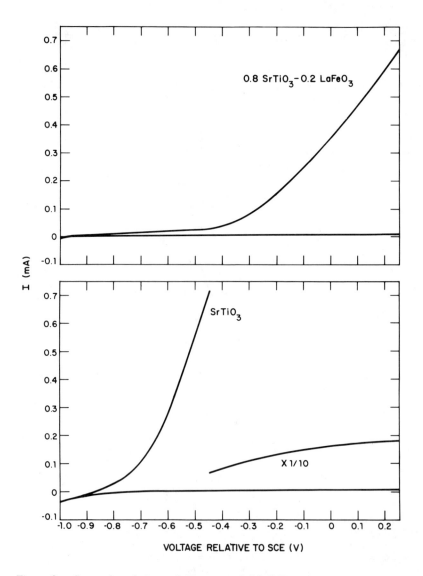

Figure 9.   Comparison between I–V curves of 0.8 SrTiO₃–0.2 LaFeO₃ film and single-crystal SrTiO₃ (pH = 10)

thus very important in determining photoresponse, especially for alloys.

Curve (B) of Figure 8 gives the photoresponse for a thin electrode prepared essentially by the method of Guruswamy and Bockris. It will be noted that although both electrodes (A) and (B) have the same visible photoresponse, which is intermediate between that of the two end materials but very low, the film, electrode (B), has a much higher response from the UV into the visible. If these alloys are to prove useful, their low energy response needs to be raised, perhaps by a combination of improved fabrication techniques and identification of the optimum alloy concentration.

Figure 9 compares the I-V curves of electrodes formed from the $SrTiO_3$-$LaFeO_3$ alloy film and the single-crystal $SrTiO_3$ at pH = 10. It will be observed that (with a 150 W Xe lamp) the onset of photocurrent for the alloy is at ~ -0.95 V relative to the SCE just as negative as for $SrTiO_3$. However, with a little bias the $SrTiO_3$ electrode gives a much larger photocurrent, indicating much less hole-electron recombination.

In summary, we have described the main concepts of two components of our program, one on composite materials and the other on alloys. The experiments on composite materials are still not advanced enough to tell how effective such materials will be as electrodes. The results on the alloys show some promise, and point to the need for a more basic materials approach in order to obtain improved electrodes. This work was supported by the Solar Energy Research Institute.

Abstract

Presently available semiconducting electrodes suitable for the photoelectrolysis of water are inefficient because their action spectra are not well matched to the solar spectrum. In terms of band structure, the energy gaps of the usual stable semiconducting electodes are too large. Semiconductors with smaller energy gaps, on the other hand, are usually not photoelectrochemically stable; furthermore, they have electron affinities which are too large. One solution to these contradictory requirements, the use of layer type compounds with d-energy bands, has been recently investigated Tributsch and co-workers. Another possible solution is the use of composite electrodes. In such electrodes, interface states become particularly significant. Not only are the states at the electrode-electrolyte interface important, but also those at interfaces between elements of the composite electrode. In this talk we will discuss the electronic properties of several kinds of composite electrodes in terms of energy band structure and interface states, as well as their fabrication and operation in photoelectrochemical cells. This work was supported by the Solar Energy Research Institute.

Literature Cited

1. Kohl, P. A.; Frank, S. N.; Bard, A. J., J. Electrochem. Soc., 1977, 124, 225.
2. Tomkiewicz, M. and Woodall, J. M., J. Electrochem.
   Soc., 1977, 124, 1436.
3. Yu, K. Y.; Helms, C. R.; Spicer, W. E.; Chye, P. W.,
   Phys. Rev., 1977, 15, 1629.
4. Goodenough, J. B.,"Metallic Oxides", in Progress in
   Solid State Chemistry, Editor H. Reiss, (Pergam-
   mon Press, New York) 1972, p. 145.
5. Tributsch, H., Ber. Bunsenges. Phys. Chem., 1977,
   81, 362.
6. Tributsch, H., Ber. Bunsenges. Phys. Chem., 1978,
   82, 169.
7. Tributsch, H., Ber. Bunsenges. J. Electrochem. Soc.,
   1978, 125, 1086.
8. Gobrecht, J.; Tributsch, H.; Gerischer, H., J. Electrochem. Soc., 1978, 125, 2085.
9. Hardee, K. L. and Bard, A. J., J. Electrochem. Soc.,
   1976, 123, 1024.
10. Quinn, R. K.; Nasby, R. D.; Baughman, R. J., Mater.
    Res. Bull., 1976, 11, 1011.
11. Shu, L.; Yen, R.; Hackerman, N., J. Electrochem.
    Soc., 1977, 124, 833.
12. Kennedy, J. H. and Frese, Jr., K. W., J. Electro-
    chem. Soc., 1978, 125, 709.
13. Butler, M. A.; Ginely, D. S.; Eibschutz, M., J.
    Appl. Phys., 1977, 48, 3070.
14. Rauh, R. D.; Buzby, J. M.; Reise, T. F.; Alkartis,
    S. A., J. Phys. Chem., 1977, 83, 2221.
15. Crowell, C. R. and Sze, S. M., Solid-State Electron.
    1966, 9, 1035.
16. Maruska, H. P. and Ghosh, A. K., Solar Energy Mater.
    1979, 1, 237.
17. Larach, S.; Shrader, R. E.; Stocher, C. F., Phys.
    Rev., 1957, 108, 587.
18. Augustynski, J.; Hinden, J.; Stalder, Chs., Electro-
    chem. Soc., 1977, 124, 1063.
19. Guruswamy, V. and Bockris, J. O. M., Solar Energy
    Mater., 1979, 1, 441.

RECEIVED October 17, 1980.

Photoelectrochemical Systems Involving Solid–Liquid Interfacial Layers of Chlorophylls

TSUTOMU MIYASAKA and KENICHI HONDA

Department of Synthetic Chemistry, Faculty of Engineering, University of Tokyo, Hongo, Bunkyo-ku, Tokyo 113, Japan

Photoelectrochemical conversion from visible light to electric and/or chemical energy using dye-sensitized semiconductor or metal electrodes is a promising system for the *in vitro* simulation of the plant photosynthetic conversion process, which is considered one of the fundamental subjects of modern and future photoelectrochemistry. Use of chlorophylls(Chls) and related compounds such as porphyrins in photoelectric and photoelectrochemical devices also has been of growing interest because of its close relevance to the photoacts of reaction center Chls in photosynthesis.

Although Chl, as well as most of biological pigments isolated from living organisms, is unstable under ambient conditions, the usefulness of Chl as a photoreceptor in *in vitro* studies is important for the following reasons:
(a) its capability of utilizing red incident light,
(b) strong redox reactivities in its excited states,
(c) the occurrence of highly efficient energy migration among excited molecules,
(d) the ability to form a wide variety of photoactive derivatives absorbing in far red region, such as Chl-nucleophile aggregates, and
(e) the availability of its surfactant structure which enables the ideal incorporation of the molecule into membrane structures.

Based on these characteristics, a number of investigations of the photoelectrochemical as well as the photovoltaic effects of *in vitro* Chl have appeared during the last decade.

Besides developing an efficient model system, the goal of light conversion studies using *in vitro* Chl is to obtain as much useful information about the photobehavior of Chl as possible in order to identify the *in vivo* reaction mechanisms involved. To this end, various photoelectrochemical approaches have emphasized the molecular configuration and geometry of Chl. Although studies in this field are currently progressing rapidly, this paper will review the photoelectrochemical behavior of Chl from the physical and photochemical perspectives.

0097-6156/81/0146-0231$05.25/0

1. Morphological Aspects of Chlorophyll Interfacial Layers

We are primarily concerned with several typical modes of the Chl interfacial layer relevant to photoelectric and photoelectrochemical systems. Since the Chl molecule, a surfactant pigment consisting of a partially hydrophilic porphyrin ring and a totally hydrophobic phytol chain, is insoluble in water, it is generally employed for photoelectric and photoelectrochemical measurements in the form of a solid (or quasi-solid) film deposited on an electrode surface. Such Chl films are divided into three typical modes according to molecular configuration.

1) _Amorphous Films_. An amorphous film is generally prepared by solvent evaporation of a dry organic solution of Chl on a solid substrate surface. The vacuum sublimation technique, which is widely employed for most synthetic dyes, is not applicable to Chl due to possible thermal degradation of the pigment. The red absorption peak of a dry amorphous Chl a film is around 675–680 nm. This is red-shifted from that of monomeric Chl a (660 nm) in organic solution (1,2), indicating that aggregated forms of Chl a such as dimers and oligomers (absorbing in the red at 670–680 nm (2,3,4)) are involved.

2) _Microcrystalline Chlorophyll and Its Films_. The crystallization effect is a characteristic of Chl. Jacob et $al.$(5) established evidence for the significant role of water in the formation of microcrystalline Chl. Later, Ballschmiter and Katz (6) explained this microcrystalline form of Chl a, absorbing in the red around 740 nm, in terms of (Chl a–H_2O)$_n$ adduct. A great deal of study has focused on the structural and photochemical characterization of various Chl–H_2O aggregate species (2,3,4,6–13). The relevance of these far-red absorbing aggregates as structural models to the photosynthetic reaction center P700 (14), which is believed to involve a special pair of Chl a, has also been a subject of intense study (3,4,7,9,10,11,13); for example, structual proposals for Chl a–H_2O species have suggested that the water molecule is bound to Chl a with its oxygen coordinated to a central magnesium and with its hydrogen to the C-9 keto carbonyl group (7,13) or the C-10 keto carbonyl (9,10) in the porphyrin ring. It is known that other nucleophiles can also associate to form aggregates similar to Chl–H_2O species (8). It should be noted that the formation of these hydrated Chl species as well as their light reactions may be involved more or less in the other types of Chl films prepared on electrodes, with the extent depending significantly on the method of preparing the film as well as the experimental conditions.

Chl a–H_2O microcrystals (745 nm) (5,6,10,11) and other hydrated Chl species are readily obtained by allowing a small amount of water to remain in solutions of Chl a dissolved in suitable nonpolar organic solvents. Films can be formed on the solid (electrode)surface by solvent evaporation. A Chl a-H_2O microcrystalline film can be prepared conveniently by the electrodeposition technique established by Tang and Albrecht (15); the microcrystals (positively charged) in 3-methylpentane suspension are deposited on the electrode substrate (cathode) under applied fields (about 1 kV/cm), allowing for uniform films as thick as several hundred monolayers. This microcrystalline film (740-5 nm) can easily be converted to an anhydrous and amorphous form (675 nm) by heating the film below 70°C.

   3) Monolayer Assemblies of Chlorophyll . Monolayer and multilayer films of Chls have been studied in detail since 1937 (16,17). From the biological point of view, Chl monolayer arrays are of interest because they may be important in the structure of the chloroplast-lamellae or so-called thylakoid membrane (18,19) Due to their surfactant structures, Chls are among the ideal and stable monolayer-forming pigments, and surface pressure-molecular area (Π-A) isotherms of their monolayers formed at air-water interfaces have been well characterized (20). It is believed that at the water surface Chl molecules are oriented with the porphyrin rings and phytol chains directed upward and the hydrophilic ester linkages downward. Monolayers and mixed monolayers of Chls are prepared on a neutral agueous buffer surface and can be deposited at a controlled surface concentration onto a solid substrate by the Langmuir-Blodgett technique (21). Such preparation and deposition techniques are specified in literatures (20,22,23). As a typical feature, a monolayer film of pure Chl a possesses absorption peaks at 675-680 nm and 435-440 nm with corresponding optical densities of 0.008-0.01 and 0.01-0.013 in the red and blue bands, respectively (24,25). The red-shifts observed in the absorption peaks with respect to Chl a in solution may result from dipole-dipole interaction between Chl molecules (20).
   Various inert compounds such as fatty acids, fatty alcohols, and lipids behave as two-dimensional diluents for Chl monolayers and lead to the formation of homogeneously mixed monolayers (20). These diluents have facilitated the study of Chl-Chl energy transfer within a two-dimensional plane as a function of the intermolecular Chl separation (26,27). In sufficiently dilute mixed monolayers, a majority of the Chl molecules are thought to exist in the monomeric state, with their mutual aggregations effectively suppressed within the geometrically controlled, ordered configuration. Multilayers (built-up monolayers) of Chl a have also been studied (23) and utilized for photovoltaic studies (see the next section). The molecular orientation in such Chl a multilayers has been ascertained from the observed dichroism in spectropolarization measurements with respect to absorption (23) and emission (28).
   It has been reported that Chl a in both microcrystalline films (5,10) and monolayer assemblies (23) are fairly stable over long periods of time.

was 2%. Similar differences between crystalline and amorphous films were reported earlier by Corker and Lundström (42). However, they mentioned that the photoconductive and rectifying nature of the solid Chl a layer is not attributable to the postulated organic semiconductivity of Chl a, but more likely to an oxide layer existing on the metal surface which functions as a rectifying barrier and to a large number of trapped electrons which act like dopants within the layer. Such a concept is compatible with McCree's postulation (34), although there are differences in the film configurations in their systems.

Whether or not Chl is regarded intrinsically as an organic semiconductor, the solid Chl layer in contact with a metal does display a p–type photovoltaic effect, and its efficiency depends significantly on the morphology of the Chl layer as well as the nature of the metal. The effect corresponding to a p–type photoconductor can also be expected at the junction of a metal / Chl / liquid in a photoelectrochemical system. Such a presumption is in fact compatible with the photoelectrochemical behavior observed for most of Chl-coated metal electrodes, as will be shown later.

3. Electrochemical Energetics of Chlorophyll

The ground state oxidation and reduction potentials of Chl a and b in various organic solutions have been measured by a number of investigators. These experimental data as well as other important energy parameters for Chls have been covered in a recent review by Seely (43). It is noted that the potential values reported by various investigators are not in good agreement. This might reflect the difficulty (due to junction potentials) in correlating the observed redox potentials (versus, e.g., Ag) in organic media to those in aqueous media referred to the SCE (saturated calomel electrode) or the NHE, rather than results from solvent effects. Taking the various sources of data into account, one-electron oxidation and reduction potentials for Chl a are +0.52 to +0.62 V vs. SCE and -1.08 to -1.25 V vs. SCE, respectively; those for Chl b are +0.56 to +0.72 V vs. SCE and -1.03 to -1.29 V vs. SCE, respectively. We recently measured the redox potentials for Chls by cyclic voltammetry (mainly in DMF medium), employing ferrocene as a standard redox sample to correct for the reference electrode potential. The results obtained were +0.59 V and -1.17 V vs. SCE for Chl a oxidation and reduction (one-electron), and +0.63 V and~-1.1 V vs. SCE for Chl b oxidation and reduction (44). All are within the range of previously reported values. Using singlet and triplet excitation energies reported for Chl a, 1.85 eV (45) and 1.33 eV (46), and for Chl b, 1.91 eV (45) and 1.39 eV (46), respectively, oxidation potentials of Chls in the ground and excited states are presented in Figure 1 and compared with redox potentials of other biologically important reagents.

Another topic of interest is the reaction of excited state Chl a with water. This occurs as a result of the primary photo-

2. Intrinsic Photoelectrical Behavior of Solid Chlorophyll Films

Since photoelectrochemical measurements deal mostly with
solid interfacial layers of Chl in contact with electrodes, it
is of prime importance to understand the photoelectric properties
of Chl in the solid state. Prior to any photoelectrochemical
studies, a number of investigations of the photoconductive behavior of Chl were performed based on the hypothesis (29,30) that
charge separation in photosynthetic primary reactions may result
from the photoconduction in Chl systems. Among the early studies
of Chl photophysics were those by Nelson (31) and Arnold and
Macley (32) who measured photoconductivities of solvent-evaporated
Chl films on platinum grids and of monolayer assemblies on colloi-
dal graphite grids, respectively. Therein and co-workers (33)
studied the photoconductivity of microcrystalline Chl layers sandwiched between two electrodes and characterized the solid Chl
layer as a p-type organic semiconductor. On the other hand, a
later study by McCree (34) pointed out that monolayer assemblies
of Chl are undoubtedly inefficient photoconductors. A similar
conclusion was made by Reucroft and Simpson (35) using Chl a
multilayers coated on SnO_2 or Sn electrodes. However, the photovoltaic effect of Chl a was enhanced significantly by superimposed
monolayers of an electron acceptor or doner. This effect of the
added acceptor layers has also been established by Shkuropatov
et al. (36) and Janzen and Bolton (37) using amorphous films and
monolayers of Chl a, respectively, placed on Al electrodes.

High power efficiencies of photovoltaic effects due to
metal / Chl layer contacts have recently been reported using
metal / Chl a / metal sandwich-type dry cells. Meilanov *et al.* (38)
reported quantum efficiencies as high as 10% for the charge generation in amorphous Chl a films between Al electrodes. Tang and
Albrecht (39,40), employing the electrodeposition technique, have
extensively studied the metal / microcrystalline Chl a / metal sand-
wich cells with respect to various combinations of metals with
different work functions. In their systems, microcrystalline
Chl a (the hydrated form absorbing at 745 nm) behaved as an efficient p-type semiconductor, and the highest power conversion
efficiency, 5×10^{-2}% at 745 nm (0.7% photocurrent quantum efficiency), was obtained in the cell mode, Cr / Chl a / Hg (40). They
attributed the photovoltaic effect to a potential barrier
(Schottky barrier) formed at the contact of the low work-function
metal (e.g., Cr) and the p-type organic semiconductor (crystalline
Chl a). More recently Dodelet *et al.* (41) studied Al / Chl a / Ag
photovoltaic cells (the Al / Chl a contact is photoactive) where
the photoconductivities of microcrystalline layers and amorphous
(anhydrous) layers were compared. The power conversion efficiency
in the former ($\leqslant 0.2$%) was much higher than in the latter ($\leqslant 0.04$%),
which supports the contention that the crystalline form is a
better photoconductor than the amorphous form. In this study, the
maximum photocurrent quantum efficiency for the crystalline Chl a

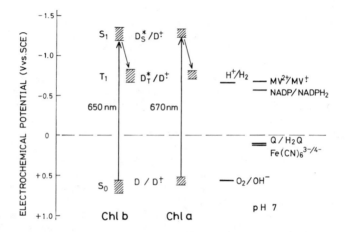

Figure 1.   Correlation of one-electron oxidation potentials for Chl a *and Chl* b *in the ground (S$_0$) and excited singlet (S$_1$) and triplet (T$_1$) states with redox potentials of some donor and acceptor species at pH 7 ((MV) methylviologen; (H$_2$Q) hydroquinone)*

synthetic processes in terms of the enzyme-assisted oxidation of water by the excitation of photosystem II reaction center Chl a (P680) and reduction of NADP (or protons in the presence of hydrogenase) by the excitation of photosystem I reaction center Chl a (P700). From Figure 1, one can see that the excited singlet state of *in vitro* Chl a, as well as Chl b, can reduce water on an energetic basis since an overpotential of about 0.7 V is available. However, it is known (47,48) that excited Chl a in solution undergoes electron transfer reactions with redox species exclusively via a long-lived triplet state (lifetime, ca. 2 ms (49)). Since the oxidation potentials of ground and triplet excited state Chls are situated close to the oxidation (+0.57 V vs. SCE) and reduction (-0.66 V vs. SCE) potentials of water, respectively, in neutral solution, the photodecomposition of water by triplet excited Chls, namely,

$$Chl^*_T + H^+ \longrightarrow Chl^+ + 1/2\ H_2 \qquad (1)$$

$$Chl^+ + OH^- \longrightarrow Chl + H^+ + 1/2\ O_2 \qquad (2)$$

may not be efficient at pH 7. On the other hand, either reaction might be promoted when the pH of solution (i.e., redox potential of water) is varied so as to favour the reaction. This conjecture is consistent with the demonstration of water splitting either in an acidic electrolyte (for the reaction 1) or in an alkaline electrolyte (for the reaction 2) as will be described in the next section. In connection with reaction 2, Watanabe and Honda (50) recently found that the rate constant for the reaction of the Chl a cation radical with water in acetonitrile medium is extremely low. However, the redox potential as well as the reaction behavior investigated in organic solutions may not be directly applicable to the energetics of a solid Chl layer-aqueous solution interface, the general condition in a photoelectrochemical system.

4. Photoelectrochemical Systems Involving Chlorophyll-Coated Semiconductor and Metal Electrodes

1) General Technique for Chlorophyll Utilization. In photoelectrochemical measurements, the methods commonly employed for preparing Chl interfacial layers on electrode substrates are solvent evaporation, electrodeposition (for crystalline Chl), and monolayer deposition techniques, as outlined previously.

Chl is purified from chloroplast extracts, usually obtained from spinach leaves, by dioxane precipitation method (51) and conventional sugar column chromatography (52). For rapid and easy preparation, the method recently developed by Omata and Murata (53) is satisfactory; synthetic (DEAE-) Sepharose is substituted for sugar on the column. Besides using spectroscopic criteria (52), the purity of Chl samples can be checked readily by means of silica-gel thin layer chromatography (54). Colorless contaminations in Chl

preparations can be determined by measuring Chl monolayer Π - A
isotherms (20). Spectral parameters for Chl a, such as extinction
coefficient, in various polar and nonpolar organic solvents have
been investigated by Seely (1).

   2) Chlorophyll-Coated Semiconductor Electrodes. Chl has
first been employed by Tributsch and Calvin (55,56) in dye
sensitization studies of semiconductor electrodes. Solvent-evaporated films of Chl a, Chl b, and bacteriochlorophyll on n-type
semiconductor ZnO electrodes (single crystal) gave anodic sensitized photocurrents under potentiostatic conditions in aqueous
electrolytes. The photocurrent action spectrum obtained for Chl a
showed the red band peak at 673 nm corresponding closely to the
amorphous and monomeric state of Chl a. The addition of super-
sensitizers (reducing agents) increased the anodic photocurrents,
and a maximum quantum efficiency of 12.5% was obtained for the
photocurrent in the presence of phenylhydrazine.
   Based on the observed strong supersensitization effect,
generation of the anodic photocurrent was explained in terms of
an electron transfer mechanism involving several possible processes. For example, electron injection from excited Chl a
(singlet and/or triplet) into the conduction band of ZnO could take
place followed by a rapid reduction of the Chl a radical cation
by a reducing agent, and/or Chl a could undergo photoreduction by
the reducing agent to produce a radical anion which subsequently
injects an electron to the conduction band. Participation of both
singlet and triplet excited states of Chl a in the above electron
transfer processes was proposed. Knowing that the singlet state
donor level of Chl a (around -1.3 V vs. SCE) is located considerably above the conduction band edge of ZnO (ca. -0.75 V vs. SCE
at pH 7 (57)) while the triplet donor level (around -0.75 V vs. SCE)
is situated very close to the ZnO level, it seems reasonable to
assume that the direct electron injection from the excited Chl a
should occur predominantly via the singlet state. Since only a short
time (ca. 10^{-10} s (58)) is necessary for electron injection from a sensi-
tizing dye to a semiconductor, electron transfer from an excited
singlet state of Chl a (lifetime ca. 6×10^{-9} s (59)) must be
efficient. In contrast, another possible process initiated by the
reduction of an excited Chl a by a reducing agent, if involved,
is considered favourable to the triplet state of Chl a , in view
of the fact that the photoionization of Chl a by the reaction with
redox species has so far been found only via triplet excitation
(43,47,48). Besides ZnO electrode, it was also found that n-type
CdS can be sensitized by Chls.
   We have studied the photoelectrochemical behavior of Chl a
and Chl b on an n-type SnO_2 (60) optically transparent electrode
(OTE; thickness of the SnO_2 layer on the glass substrate, ca. 2000 Å;
donor density, 10^{20-21} cm^{-3}). Chl monolayer assemblies, deposited
by means of the Langmuir-Blodgett technique (20,21), were employed.
The use of such monolayer assemblies as interfacial dye layers

in photoelectrochemical studies has the following advantages:
(a)   Strong attachment of dye layers to the electrode surface
      can be established via hydrophilic and/or hydrophobic inter-
      actions, without accompanying chemical reactions of the dye,
(b)   two-dimensional homogeneous dye layers can be formed on the
      electrode surface,
(c)   the surface concentration as well as the mean intermolecular
      distance of the dye can be controlled, and
(d)   the total thickness of the dye layers can be precisely con-
      trolled at the molecular level.
Besides these practical advantages, it is of biological importance
to study the photoeffects of Chl in an ordered structure, because
such a structure is a crucial factor in regulating energy migration among Chls as well as promoting electron transfer processes
in photosynthetic organisms. Photoredox reactions involving Chl
incorporated into artificial membranes have long been studied by
the use of bilayer lipid membranes (BLM) (61). BLM studies are
based on photogalvanic effects caused by excited Chl and mediated
via the solution, while our membrane-electrode system deals with
the direct capture of an electron (or hole) from photoexcited Chl
at the electrode-solution interface.

Monolayer and mixed monolayers of Chl a and/or Chl b were
deposited on a SnO_2 OTE at a constant surface pressure of 10-20
dyn/cm. This resulted in a closely packed, ordered film on the
electrode. The high transmittance of the OTE allowed for the
direct measurement of the absorbance of a monolayer at the
electrode-electrolyte interface. Quantum efficiencies for the
generation of photocurrents were calculated using this information.
Monolayers of Chl a as well as Chl b gave anodic sensitized photocurrents under potentiostatic conditions in aqueous electrolyte
containing a reducing agent (hydroquinone, H_2Q) (25). The spectral
dependence of the photocurrent matched the absorption spectrum
of Chl at a SnO_2-electrolyte interface. The spectra are shown in
Figure 2; the photoresponse of SnO_2 itself is negligible over the
visible region due to its large band gap (3.5-3.8 eV). Under open
circuit condition, Chl a-coated SnO_2 developed negative photovoltages (ca. 10^{-2} V) with the action spectrum similar to the photocurrent spectrum. The photocurrents were significantly increased
by the addition of H_2Q in the range of 0-0.03 M, and also by the
anodic polarization of the electrode.

Based on the observed supersensitization effect, generation
of the photocurrents is explained in terms of the electron transfer
mechanism as in ZnO-Chl system. Since the conduction band edge
potential of SnO_2 in neutral solution, -0.35 V vs. SCE (60), is
lower than both the singlet and triplet state donor levels of
excited Chl a and Chl b, electron injection from either of these
excited states into SnO_2 becomes thoeretically feasible.
Accordingly, one of the main sensitization processes involved is
illustrated in Figure 3, where H_2Q acts as an electron donor to
photooxidized Chl.

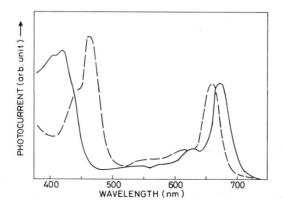

Figure 2. Photocurrent spectra for a Chl a monolayer (———) and a Chl b mono-layer (– – –) on a SnO₂ electrode. Electrolyte, aqueous phosphate buffer (0.025M, pH 6.9) containing 0.05–0.1M hydroquinone; electrode potential, 0.05 V vs. SCE. The red band maxima of both spectra are normalized.

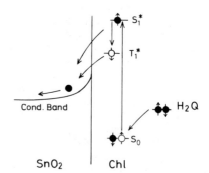

Figure 3. Scheme for the electron transfer in dye sensitization process at the inter-facial layer of Chl on SnO₂ electrode

Attempts have been made to intersperse a suitable inert surfactant diluent in the monolayer in order to control the surface concentration as well as the mean intermolecular separation of Chl. Since Chl has highly overlaping absorption and emission (fluorescence) spectra, efficient migration of excitation energy among molecules is allowed. It has been well established both in solutions (62,63) and in membranes (26,27,59) that Chl undergoes strong self-quenching of singlet excitation energy at high concentration. This concentration quenching phenomenon may arise from the formation of non-fluorescent pair traps of Chl possessing a critical separation (64) and can be suppressed effectively by the homogeneous dilution of the Chl. Among the surfactants which can act as ideal two-dimensional diluents for a Chl *a* monolayer, we chose the phospholipid, dipalmitoyllecithin (DPL) (65), which is one of the main components of biological membranes. Two-dimensional homogeneity in the Chl *a*-DPL mixed monolayer system was confirmed by the Π-A isotherms as well as by the observed enhancement of the fluorescence yield of the mixed monolayer upon decreasing the Chl *a* surface concentration (66).

It was found that the quantum efficiency of the anodic photocurrent in the Chl *a*-DPL system was increased from 3-4% in the pure Chl *a* monolayer up to about 25% (65) in highly diluted mixed monolayers. This increase in efficiency was accompanied by red shifts of the photocurrent peak positions in the red and blue bands (e.g., in the red, from 675 nm (pure Chl *a*) to 665-670 nm (Chl *a*/DPL ≤ 1/49)) and probably reflects the change in chromophore-chromophore interaction between Chl *a* molecules. The considerable efficiencies obtained apparently result from the suppression of the self-quenching of Chl *a* by separating the Chl-Chl intermolecular distance. Based on photoelectrochemical measurements, an extention of this study may elucidate the energy interactions between Chl and other photosynthetic pigments in the mixed monolayer systems.

The effect of the film thickness on the photocurrent efficiency was investigated by the use of built-up multilayers of Chl *a* (67). Since a single Chl *a* monolayer is about 14 Å thick (16), the precise control of the film thickness is possible by stacking monolayers. The magnitude of the anodic photocurrent increased with the number of monolayers until it was saturated at around 10-15 layers, while the quantum efficiency of the photocurrent monotonously decreased. The latter phenomenon may reflect enhanced excitation energy quenching among neighbouring monolayers in the three-dimensional arrey and/or an increase in total electric resistance of the multilayer with increasing number of layers. Lowering of the anodic photoeffect might also be attributed to p-type photoconductivity developing in thicker stacks of monolayers which leads to the reinforcement of the cathodic photoeffect of Chl *a*.

From the above results we conclude that a monolayer-thick film having well interspersed Chl is most efficient for the conversion of light energy using Chl-sensitized semiconductor electrodes.

The photocurrent generated by a Chl a multilayer in contact with an aqueous electrolyte without any added redox agents has been measured as a function of pH of the solution (67). Significant increase in the anodic photocurrent was observed at pHs higher than 7. Since the photooxidized Chl a (cation radical) is considered to oxidize water better at higher pH above pH 7 (see Section 3), water probably behaves as a reducing agent (supersensitizer) for Chl a in alkaline solutions.

In connection with the study using Chl-coated SnO_2 electrodes, an attempt has been reported to coat a chloroplast- or algae-immobilized polymer film onto SnO_2 OTE (68), and water oxidation at the illuminated electrode has been performed.

Chl-coated semiconductor (n-type) electrodes have thus far been studied using ZnO, CdS, and SnO_2 , all of which act as efficient photoanodes for converting visible light. Such Chl-sensitized photoanodes could be regarded as in $vitro$ models for the photosystem II (oxygen evolution) function in photosynthesis. p-type semiconductor electrodes have not been utilized successfully to produce cathodic Chl-sensitized photocurrents with satisfactory efficiencies. On the other hand, Chl-coated metal electrode systems seem to overcome this problem.

3) Chlorophyll-Coated Metal Electrodes. Photoelectrochemical reactions at Chl-coated metal electrodes have been investigated respecting the various configurational modes of Chl. Platinum electrodes have widely been employed as substrates of Chl films.

Takahashi and co-workers (69,70,71) reported both cathodic and anodic photocurrents in addition to corresponding positive and negative photovoltages at solvent-evaporated films of a Chl-oxidant mixture and a Chl-reductant mixture, respectively, on platinum electrodes. Various redox species were examined, respectively, as a donor or acceptor added in an aqueous electrolyte (69). In a typical experiment (71), NAD and $Fe(CN)_6^{4-}$, each dissolved in a neutral electrolyte solution, were employed as an acceptor for a photocathode and a donor for a photoanode, respectively, and the photoreduction of NAD at a Chl-naphthoquinone-coated cathode and the photooxidation of $Fe(CN)_6^{4-}$ at a Chl-anthrahydroquinone-coated anode were performed under either short circuit conditions or potentiostatic conditions. The reduction of NAD at the photocathode was demonstrated as a model for the photosynthetic system I. In their studies, the photoactive species was attributed to the composite of Chl-oxidant or -reductant (70). A p-type semiconductor model was proposed as the mechanism for photocurrent generation at the Chl photocathode (71).

Photoelectrochemical reactions of hydrated microcrystalline Chl a electrodeposited on platinum electrodes have been studied by Fong and co-workers (72,73,74,75). The experiments were performed in short circuit electrochemical cells. In aqueous electrolytes without added redox agents, the microcrystalline (hydrated aggregates of) Chl a-coated, platinized-platinum electrodes gave cathodic photocurrents upon illumination, and the photocurrent peaked in the red at 745 nm. The quantum efficiency of the photocurrent was increased in acidic solutions and was in the order of 10^{-3} electrons/incident photon (73). Their studies in particular focused on the *in vitro* simulation of the photosynthetic water splitting reaction by using far-red absorbing Chl a-H_2O species. These species have long been proposed as the *in vitro* structural models for photosynthetic reaction centers (refer to Section 1). Using various photoactive Chl a-H_2O species, such as Chl a-H_2O, (Chl a-H_2O)$_2$, (Chl a-$2H_2O$)$_n$, Fong *et al*. postulated, on the basis of redox titration experiments (76) and ESR observations of the reaction of excited Chl a hydrate with water (11), that the dihydrated Chl a oligomers, (Chl a-$2H_2O$)$_n$, are the most powerful redox species and are capable of both reducing and oxidizing water under far red excitation. Evidence for the water splitting reaction at an illuminated Chl a dihydrate-platinum (platinized) electrode was claimed (74) based on the mass spectrometric analysis of the photolytic products from isotopically labeled water. More recently, photoreduction of CO_2 in this system has also been undertaken (75).

In view of the fact that the Fong's experiments involved repeated platinization of the surfaces of the platinum and Chl a layers (74,75), the observed photoreactivity may arise from a photocatalytic effect of the composite of Chl a hydrates and platinum rather than Chl a hydrates themselves. In this respect, it has been reported that Pt dispersions catalize dye-sensitized photoreduction of water (77,78).

Another type of Chl interfacial layer employed on a metal electrode was a film consisting of ordered molecules. Villar (79) studied short circuit cathodic photocurrents at multilayers of Chl a and b built up on semi-transparent platinum electrodes in an electrolyte consisting of 96% glycerol and 4% KCl-saturated aqueous solution. Photocurrent quantum efficiencies of multilayers and of amorphous films prepared by solvent evaporation were compared. The highest efficiency (about 10^{-3} electrons/absorbed photon, calculated from the paper) was obtained with Chl a multilayers, and the amorphous films of Chl a proved to be less efficient than Chl b multilayers.

We have investigated the photocurrent behavior of multilayers of a Chl a-DPL (molar ratio 1/1) mixture on platinum in an aqueous electrolyte without added redox agents (80). Cathodic photocurrents with quantum efficiencies in the order of 10^{-4} were obtained with films consisting of a sufficient number of monolayers. The photocurrent was increased in acidic solutions. However, no appreciable photocurrent was observed with a single monolayer coated on platinum. The latter fact most probably results from minimal rectifying property of the metal surface and/or an efficient energy quenching of dye excited states by free electrons in

the metal (81,82). This indicates that a reasonably thick film of
Chl, which may behave like a photoconductor, is required to produce measurable photocurrents on metal electrodes.

Aizawa and Suzuki (83,84,85,86) utilized, as an ordered
system, liquid crystals in which Chl was immobilized. Electrodes
were prepared by solvent-evaporating a solution consisting of Chl
and a typical nematic liquid crystal, such as n-(p-methoxybenzyl-
idene)-p'-butylaniline, onto a platinum surface. Chl-liquid crystal
electrodes in acidic buffer solutions gave cathodic photocurrents
accompanied by the evolution of hydrogen gas (83). This was the
first demonstration of photoelectrochemical splitting of water
using *in vitro* Chl. Of particular interest in these studies is
the effect of substituting the central metal in the Chl molecule.
In contrast to Chl (Mg-pheophytin) which behaves as a photocathode, Mn-pheophytin immobilized in a liquid crystal functioned
as a photoanode, producing oxygen (oxidizing water) in alkaline
buffer solutions (84,85). More recently, Ru-pheophytin was also
examined and found to generate cathodic photocurrent with a quantum
efficiency of around 0.5% (86). The effect of the central metal
on the sign of photoresponse may reflect variations of redox potential among these metal pheophytin analogues. In these studies,
the authors suggested that a Chl-water adduct formed in the
liquid crystal system was the photoactive species (83,85).

Taking a general view of the above studies, we note that
Chl-coated metal (platinum) electrodes commonly function as photo-
cathodes in acidic solutions, although the photocurrent effcien-
cies tend to be lower compared to systems employing semiconductors.
This cathodic photoresponse may arise from a p-type photoconductive nature of a solid Chl layer and/or formation of a contact
barrier at the metal-Chl interface which contributes to light-induced carrier separation and leads to photocurrent generation.

5. Other Related Systems Involving Model Compounds

Among what have been widely employed as model compounds for
Chl, are porphyrins, phthalocyanines, and some photoactive transition metal complexes, which are more stable and easier to obtain
than Chl. Interfacial layers of these insoluble compounds are
generally prepared by means of vacuum sublimation or solvent
evaporation.

Tetraphenylporphine (TPP) and other metal porphyrine derivatives coated on platinum (87,88,89) or gold (89,90) electrodes
have been investigated in photoelectrochemical modes. Photocurrents reported are cathodic or anodic, depending on the pH as
well as the composition of the electrolyte employed. Photocurrent quantum efficiencies of 2% (89) to 7% (87) were reported
in systems using water itself or methylviologen as the redox
species in aqueous electrolyte. Photocurrent generation at
Zn-TPP-coated metal cathodes (89) was interpreted in terms of a
rectifying effect of the Schottky barrier formed at a metal-p-type

semiconductor contact, and its efficiency was found to increase with increasing work function of the metal.

Wang (91) obtained considerable negative photovoltages (1.1 V under filtered incident light from 150 W tungsten lamp) and anodic photocurrents with Zn-TPP multilayers on Al electrodes in contact with a ferro/ferri redox electrolyte. Based on the Wang's system, Kampas *et al.* (92) recently compared the quantum efficiencies of a wide variety of amorphous metal porphyrin derivatives. The highest quantum efficiency, around 10%, was obtained with Mg-porphyrin analogues having more negative oxidation potentials. The photoeffect was attributed to the formation of Schottky type contact in terms of metal (Al) / insulator (Al$_2$O$_3$) / p-type semiconductor (porphyrins) system.

Photoelectrochemical behavior of metal phthalocyanine solid films (p-type photoconductors) have been studied at both metal (93,94,95,96) and semiconductor (97,98) electrodes. Copper phthalocyanine vacuum-deposited on a SnO$_2$ OTE (97) displayed photocurrents with signs depending on the thickness of film as well as the electrode potential. Besides anodic photocurrents due to normal dye sensitization phenomenon on an n-type semiconductor, enhanced cathodic photocurrents were observed with thicker films due to a bulk effect (p-type photoconductivity) of the dye layer. Meier *et al.* (95) studied the cathodic photocurrent behavior of various metal phthalocyanines on platinum electrodes where the dye layer acted as a typical p-type organic semiconductor.

Another model compound, the tris(2,2'-bipyridine)ruthenium(II) complex, has prompted considerable interest because its water-splitting photoreactivity has been demonstrated in various types of photochemical systems (77,99,100,101). Memming and Schröppel (102) have attempted to deposit a monolayer of a surfactant Ru(II) complex on a SnO$_2$ OTE. In aqueous solution, an anodic photocurrent attributable to water oxidation by the excited triplet Ru complex was observed. A maximum quantum efficiency of 15% was obtained in alkaline solution.

Other dye-layer-coated photoelectrode systems have been studied but are not covered in this review due to space restrictions. This area of study, including the photoelectrochemistry of Chl, is still in an inchoate but rapidly developing stage.

6. Concluding Remarks: Relevance to Photosynthetic Model Systems.

Chl-coated semiconductor (n-type) electrodes and metal electrodes can act as efficient photoanodes and photocathoes, respectively, for visible light conversion. The former system functions as a dye-sensitized semiconductor electrode, while the latter is presumably driven by the photoconductive properties of a Chl solid layer and/or charge separation involving the Chl-metal contact barrier.

Morphological differences in the solid layer of Chl are crucial for determining the photoelectric properties of Chl.

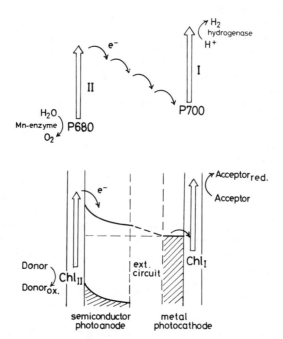

Figure 4.   Schematic for the photoelectrochemical simulation of the photosynthetic electron-pumping processes (upper sketch) by means of a Chl-semiconductor photo-anode and a Chl-metal photocathode

Accordingly, such factors as molecular configuration or surface concentration of Chl and thickness of the layer significantly affect the efficiency of photocurrent generation in electrochemical systems.

As pointed out previously, Chl-sensitized photoelectrodes are important to studies of *in vitro* simulation of the photosynthetic light conversion process. In photosynthetic primary processes, the pumping of electrons which enables the splitting of water is achieved via two photosystems in which light-induced charge separations occur efficiently coupled with one-directional electron transfers in redox systems. Though total process for the photosynthetic water splitting reaction necessarily involves enzyme-assisted dark reactions as secondary processes, the primary processes for electron-pumping could be simulated electrochemically using a couple of Chl-sensitized photoelectrodes which have inverse rectifying properties and different redox levels (i.e., work functions). An example is sketched in Figure 4; in this model, Chl-semiconductor photoanode and Chl-metal photocathode, corresponding respectively to photosystem II and photosystem I, are combined in series with an external circuit which takes the place of *in vivo* electron transfer chain across the thylakoid membrane. In order that the photocathode (system I) possesses energetically higher potentials than photoanode (system II), different molecular configurations of Chl, e.g., hydrated aggregates or monolayer assemblies, may be applicable to each photoelectrode.

Based on such an electrochemically simulated system, further modification and characterization of the Chl-containing interfacial layer on electrodes are expected to contribute to the discovery of useful information on the electron and energy transfer reactions involving Chl and other compounds of photosynthetic importance.

Acknowledgements

The authors wish to thank Drs. A. Fujishima and T. Watanabe for their useful discussions.

Literature Cited

1. Seely, G. R., Jensen, R. G., Spectrochim. Acta (1965), 21, 1835-1845.
2. Shipmann, L. L., Cotton, T. M., Norris, J. R., Katz, J. J., J. Am. Chem. Soc. (1976), 98, 8222-8230.
3. Katz, J. J., Norris, J. R., in "Current Topics in Bioenergetics", Vol. 5, Academic Press, New York, N. Y., (1973), pp. 41.
4. Fong, F. K., Koester, V. J., Biochim. Biophys. Acta (1976), 423, 52-64.
5. Jacobs, E. E., Vatter, A. E., Holt, A. S., Arch. Biochem. Biophys. (1954), 53, 228-238.

6. Ballschmiter, K., Katz, J. J., Nature (1968), 220, 1231-1233.
7. Shipmann, L. L., Cotton, T. M., Norris, J. R., Katz, J. J., Proc. Natl. Acad. Sci. USA (1976), 73, 1971-1974.
8. Cotton, T. M., Loach, P. A., Katz, J. J., Ballschmiter, K., Photochem. Photobiol. (1978) 27, 735-749.
9. Fong, F. K., Proc. Natl. Acad. Sci. USA (1974), 71, 3692-3695.
10. Brace, J. G., Fong, F. K., Karweik, D. H., Koester, V. J., Shepard, A., Winograd, N., J. Am. Chem. Soc. (1978), 100, 5203-5207.
11. Fong, F. K., Hoff, A. J., Brinkman, F. A., J. Am. Chem. Soc. (1978), 100, 619-621.
12. Chow, H. C., Serlin, R., Strouse, C. E., J. Am. Chem. Soc. (1975), 97, 7230-7237.
13. Boxer, S. G., Closs, G. L., J. Am. Chem. Soc. (1976), 98, 5406-5408.
14. Kok, B., Biochim. Biophys. Acta (1961), 48, 527-533.
15. Tang, C. W., Albrecht, A. C., Mol. Cryst. Liq. Cryst. (1974), 25, 53-62.
16. Langmuir, I., Schafer, V. J., J. Am. Chem. Soc. (1937), 59, 2075-2076.
17. Hanson, E. A., Rec. Trav. Botan. Néerl. (1939), 36, 180-266.
18. Bishop, D.G., Photochem. Photobiol. (1974), 20, 281-299.
19. Anderson, J. M., Nature (1975), 253, 536-537.
20. Ke, B., in "The Chlorophylls", Vernon, L. P., Seely, G. R, Eds., Academic Press, New York, N. Y., (1966), pp. 253.
21. Blodgett, K. B., Langmuir, I., Phys. Rev. (1937), 51, 964-982.
22. Gaines, G. L., Jr., "Insoluble Monolayers at Liquid-Gas Interfaces", Interscience, New York, N. Y., (1966).
23. Sperling, W., Ke, B., Photochem. Photobiol. (1966), 5, 857-863, 865-876.
24. Bellamy, W. D., Gaines, G. L., Jr., Tweet, A. G., J. Chem. Phys. (1963), 39, 2528-2538.
25. Miyasaka, T., Watanabe, T., Fujishima, A., Honda, K., J. Am. Chem. Soc. (1978), 100, 6657-6665.
26. Trosper, T., Park, R. B., Sauer, K., Photochem. Photobiol. (1968), 7, 451-469.
27. Costa, S. M. de B., Froines, J. R., Harris, J. M., Leblanc, R. M., Orger, B. H., Porter, G., F.R.S., Proc. Roy. Soc. (1972), A326, 503-519.
28. Leblanc, R. M., Galinier, G., Tessier, A., Lemieux, L., Can. J. Chem. (1974), 52, 3723-3727.
29. Katz, E., in "Photosynthesis in Plants", Frank, J., Loomis, W. E., Eds., Iowa State College Press, Ames, Iowa, (1949), pp. 291.
30. Bradley, D. F., Calvin, M., Proc. Natl. Acad. Sci. USA (1955), 41, 563-571.
31. Nelson, R. C., J. Chem. Phys. (1957), 27, 864-867.
32. Arnold, W., Maclay, H. K., Brookhaven Symp. Biol. (1958), 2, 1-9.
33. Terenin, A. N., Putseiko, E., Akimov, I., Disc. Faraday Soc. (1959), 27, 83-93.

34. McCree, K. J., Biochim. Biophys, Acta (1965), 102, 90-95, 96-102.

35. Reucroft, P.J., Simpson, W. H., Disc. Faraday Soc. (1971), 51, 202-211.

36. Shkuropatov, A. Ya., Kurbanov, K. B., Stolovitskii, Yu. M., Yevstigneyev, V. B., Biofizika (1977), 3, 407-412.

37. Janzen, A. F., Bolton, J., J. Am. Chem. Soc. (1979), 101, 6342-6348.

38. Meilanov, I. S., Benderskii, V. A., Blyumenfel'd, L. A., Biofizika (1970), 15, 822-827.

39. Tang, C. W., Albrecht, A. C., J. Chem. Phys. (1975), 62, 2139-2149.

40. Tang, C. W., Albrecht, A. C., Nature (1975), 254, 507-509.

41. Dodelet, J. P., Le Brech, J., Leblanc, R. M., Photochem. Photobiol. (1979), 29, 1135-1145.

42. Corker, G. A., Lundström, I., Photochem. Photobiol. (1977), 26, 139-149.

43. Seely, G. R., Photochem. Photobiol. (1978), 27, 639-654.

44. unpublished data.

45. Goedheer, J., in "The Chlorophylls", Vernon, L. P., Seely, G. R., Eds., Academic Press, New York, N. Y., (1966), pp. 147.

46. Krasnovskii, A. A., Jr., Lebedev, N. N., Litvin, F. F., Dokl. Akad, Nauk. SSSR (1974), 216, 1406-1409.

47. Kelly, J. M., Porter, G., Proc. Roy. Soc. (1970), A319, 319-329.

48. Brown, R. G., Harriman, A., Harris, L., J. C. S. Faraday II (1978), 74, 1193-1199.

49. Mau, A. W., Puza, M., Photochem. Photobiol. (1977), 32, 601-603.

50. Watanabe, T., Honda, K., J. Am. Chem. Soc. (1980), 102, 370-372.

51. Iriyama, K., Ogura, N., Takamiya, A., J. Biochem. (1974), 76, 901-904.

52. Strain, H. H., Svec, W. A., in "The Chlorophylls", Vernon, L. P., Seely, G. R., Eds., Academic Press, New York, N. Y., (1966), pp. 21.

53. Omata, T., Murata, N., Photochem. Photobiol. (1980), 31, 183-185.

54. Shiraki, M., Yoshiura, M., Iriyama, K., Chem. Lett. (1978), 103-104.

55. Tributsch, H., Calvin, M., Photochem. Photobiol. (1971), 14, 95-112.

56. Tributsch, H., Photochem, Photobiol. (1972), 16, 261-269.

57. Gleria, M., Memming,R., J. Electroanal. Chem. (1975), 65, 163-175.

58. Arden, W., Fromherz, P., J. Electrochem. Soc. (1980), 127, 370-378.

59. Beddard, G. S., Carlin, S. E., Porter, G., Chem. Phys. Lett. (1976), 43, 27-32.

60. Bolts, J. M., Wrighton, M. S., J. Phys. Chem. (1976), 80, 2641-2645.
61. Tien, H. T., "Bilayer Lipid Membranes (BLM) : Theory and Practice", Marcel Dekker, New York, N. Y. (1974).
62. Watson, W. F., Livingston, R., J. Chem. Phys. (1950), 18, 802-809.
63. Beddard, G. S., Porter, G., Nature (1976), 260, 366-367.
64. Kelly, A. R., Porter, G., Proc. Roy. Soc. (1970), A315, 149-161.
65. Miyasaka, T., Watanabe, T., Fujishima, A., Honda K., Nature (1979), 277, 638-640.
66. Miyasaka, T., Tessier, A., Leblanc, R. M., in preparation.
67. Miyasaka, T., Watanabe, T., Fujishima, A., Honda, K., Photochem. Photobiol. (1980), 32, 217-222.
68. Ochiai, H., Shibata, H., Sawa, Y., Kato, T., Proc. Natl. Acad. Sci. USA (1980), 77, 2442-2444.
69. Takahashi, F., Kikuchi, R., Biochim. Biophys. Acta (1976), 430, 490-500.
70. Takahashi, F., Kikuchi, R., Bull. Chem. Soc. Japan (1976), 49, 3394-3397.
71. Takahashi, F., Aizawa, M., Kikuchi, R., Suzuki, S., Electrochim. Acta (1977), 22, 289-293.
72. Fong, F. K., Winograd, N., J. Am. Chem. Soc. (1976), 98, 2287-2289.
73. Fong, F. K., Polles, J. S., Galloway, L., Fruge, D. R., J. Am. Chem. Soc. (1977), 99, 5802-5804.
74. Fong, F. K., Galloway, L., J. Am. Chem. Soc. (1978), 100, 3594-3598.
75. Fruge, D. R., Fong, G. D., Fong, F. K., J. Am. Chem. Soc. (1979), 101, 3694-3697.
76. Galloway, L., Roettger, J., Fruge, D. R., Fong, F. K., J. Am. Chem. Soc. (1978), 100, 4635-4638.
77. Kiwi, J., Grätzel, M., Nature (1979), 281, 657-658.
78. McLendon, G., Miller, D. S., J. C. S. Chem. Commun. (1980), 533-534.
79. Villar, J. -G., J. Bioenerg. Biomembr. (1976), 8, 199-208.
80. Miyasaka, T., Fujishima, A., Honda, K., Bull. Chem. Soc. Japan, in press.
81. Memming, R., Photochem. Photobiol. (1972), 16, 325-333.
82. Gerischer, H., Willing, F., Topics Curr. Chem. (1976), 61, 31-84.
83. Aizawa, M., Hirano, M., Suzuki, S., J. Membr. Sci. (1978), 4, 251-259.
84. Aizawa, M., Hirano, M., Suzuki, S., Electrochim. Acta (1978), 23, 1185-1190.
85. Aizawa, M., Hirano, M., Suzuki, S., Electrochim. Acta (1979), 24, 89-94.
86. Aizawa, M., Yoshitake, J., Suzuki, S., submitted.
87. Yamamura, T., Umezawa, Y., Chem. Lett. (1977), 1285-1288.

88. Umezawa, Y., Yamamura, T., J. C. S. Chem. Commun. (1978), 1106-1107.
89. Kawai, T., Tanimura, K., Sakata, T., Chem. Phys. Lett. (1978), 56, 541-545.
90. Soma, M., Chem. Phys. Lett. (1977), 50, 93-96.
91. Wang, J. H., Proc. Natl. Acad. Sci. USA (1969), 62, 653-660.
92. Kampas, F. J., Yamashita, K., Fajer, J., Nature (1980), 284, 40-42.
93. Shumov, Y. S., Komissarov, G. G., Biofizika (1974), 19, 830-834.
94. Villar, J. -G., J. Bioenerg. Biomembr. (1976), 8, 173-198.
95. Meier, H., Albrecht, W., Tschirwitz, U., Zimmerhackl, E., Geheeb, N., Ber. Bunsenges. Phys. Chem. (1977), 81, 592-598.
96. Ilatovskii, V. A., Dmitriev, I. B., Komissarov, G. G., Zh. Fiz. Khim. (1978), 52, 126-129.
97. Minami, N., Watanabe, T., Fujishima, A., Honda, K., Ber. Bunsenges. Phys. Chem. (1979), 83, 476-481.
98. Shepard, V. R., Jr., Armstrong, N. R., J. Phys. Chem. (1979), 83, 1268-1276.
99. Lehn, J. M., Sauvage, J. P., Nouv. J. Chim. (1977), 1, 449-451.
100. Brown, G., Brunschwig, B. S., Creutz, C., Endicott, J. F., Sutin, N., J. Am. Chem. Soc. (1979), 101, 1298-1300.
101. Kalyanasundaram, K., Grätzel, M., Angew. Chem. (1979), 91, 759-760.
102. Memming, R., Schröppel, F., Chem. Phys. Lett. (1979), 62, 207-210.

RECEIVED October 3, 1980.

16

Supra-Band-Edge Reactions at Semiconductor–Electrolyte Interfaces

Band-Edge Unpinning Produced by the Effects of Inversion

J. A. TURNER, J. MANASSEN[1], and A. J. NOZIK

Solar Energy Research Institute, Golden, CO 80401

I. Introduction

A recent modification (1-3) of the conventional model (4) for photoelectrochemical reactions suggests that photo-generated minority carriers may, under certain conditions, be injected into the electrolyte before they reach thermal equilibrium within the semiconductor space charge layer. This process is called "hot carrier injection. More efficient conversion of optical energy into chemical energy may be possible with hot carrier injection because a greater fraction of the incident photon energy can be deposited in the electrolyte to do chemical work.

Experiments designed to test for hot electron injection from p-type semiconductors were conducted by probing reduction reactions having redox potentials more negative than (i.e., above) the conduction band edge. The oxidized components of such redox couples should not be reduced by thermalized photo-generated electrons coming from the conducting band edge. However, hot electrons, having more negative potential, could reduce the higher lying redox couples. The energy level diagram for this process is shown in Figure 1. Analogous results should be obtained for oxidation reactions having redox potentials more positive than the valence band edge of n-type photoanodes. We call such oxidation or reduction reactions lying outside the semiconductor band gap "supra band-edge reactions".

Experiments using p-Si in non-aqueous electrolytes (5,6) indicate that the reduction of redox species with redox potentials more negative than the conduction band edge could apparently be achieved. However, further analysis showed that these results are not caused by hot electron injection, but by the unpinning of the semiconductor band edges at the semiconductor-electrolyte interface; this unpinning effect is caused by the creation of an inversion layer at the p-Si surface. This is an important effect, especially for small band gap semiconductors, that has received little attention

[1]On Sabbatical leave from the Weizmann Institute of Science, Rehovot, Israel.

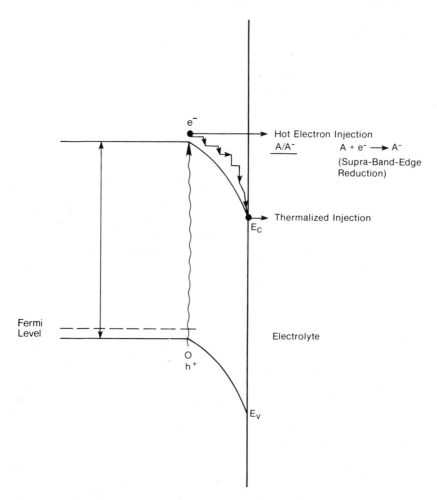

*Figure 1.   Hot electron injection from a p-type semiconductor electrode. The pho-
togenerated electron is injected into the electrolyte to reduce A to A⁻ before it is
thermalized in the space-charge layer. A thermalized electron emerging at the conduction band edge (E_c) would have insufficient energy to drive the redox couple
A/A⁻.*

in photoelectrochemistry. Unpinned band edges in photoelectrochemical systems have also recently been proposed by other workers.
Bard, et al. (7) explain the effect as the result of a large density of surface states, while Morrison (8) invokes interface states
arising from oxide layers.

The unpinned band edges can move with applied potential, making the semiconductor electrode behave similar to a metal electrode
in that changes in applied potential occur across the Helmholtz
layer rather than across the semiconductor space charge layer.
This situation, of course, drastically alters the energy relationships between the semiconductor bands and the electrolyte redox
potentials compared to the conventional model (4); in the latter,
the semiconductor bands are pinned at the surface, the potential
drop across the Helmholtz layer is constant, and potential changes
in the electrode produce corresponding changes in the band bending
within the semiconductor space charge layer. The difference between pinned and unpinned band edges is illustrated in Figure 2.
In the former case, the positions of the conduction and valence
band edges at the interface (E_C^S and E_V^S) are fixed with respect to
the electrolyte redox levels, while in the latter case the positions of E_C^S and E_V^S with respect to the electrolyte redox levels
vary with electrode potential. Thus, with unpinned band edges,
redox couples lying outside the band gap under flat band conditions can lie within the band gap under band bending conditions or
under illumination.

II. Experimental Results

The experiments were performed with single crystal (111) p-Si
electrodes with a resistivity of about 5.5 ohm cm; non-aqueous
electrolytes were used consisting of absolute methanol containing
tetramethylammonium chloride (TMAC) or acetonitrile containing
tetraethyl ammonium perchlorate (TEAP). The flat-band potentials
or p-Si in the two electrolytes were determined from Mott-Schottky
plots (in the dark) in the depletion range of the p-Si electrode,
from open-circuit photopotential measurements, and from the values
of electrode potential at which anodic photocurrent is first observed in n-type Si electrodes. These three methods all yielded
consistent flat-band potential values for p-Si of + 0.05V (vs SCE)
± .05V in both methanol and acetonitrile. These values, combined
with the doping density and the band gap of 1.12 eV for p-Si places
the conduction band edge in methanol and acetonitrile at -0.85V
(vs SCE). The supraband edge redox couples chosen for the two
electrolytes were 1,3 dimethoxy-4-nitrobenzene (E_0=-1.0V vs SCE)for methanol, and 1 nitronaphthalene (E_0=-1.08), 1,2 dichloro 4-nitrobenzene (E_0=
-0.95), and anthraquinone (E_0=-0.95) for acetonitrile. These redox
couples lie from 0.1V to 0.24V above the conduction band edge of p-Si, and
hence, in the conventional model, could not be photoreduced by p-Si.

Photocurrent-voltage data showed that all the supra-band-edge
redox species listed above are reduced upon illumination of the
p-Si electrode; no appreciable reduction was achieved in the dark.
Typical results for p-Si with methanol and acetonitrile are shown

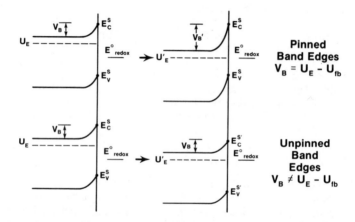

Figure 2. Pinned vs. unpinned band edges. In the former case the position of E_c^s and E_v^s remain fixed with respect to the electrolyte redox couples as the electrode potential is varied. In the latter case, E_c^s and E_v^s change with applied potential.

in Figure 3. The light intensity dependence of the photocurrent-voltage behavior for anthraquinone in acetonitrile is shown in Figure 4. At high light intensity (442 mW/cm^2, white light from xenon lamp) only a cathodic reduction photocurrent is observed. As the light intensity is reduced, an oxidation wave grows in that develops on the anodic sweep. This oxidation wave is believed to represent the re-oxidation of reduced anthraquinone at the p-Si electrode. In Figure 5, it is seen that the oxidation wave can be generated if the high light intensity is suddenly reduced to zero on the anodic return sweep after reduction has occurred at the high light intensity.

Although the results in Figures 3 are those that would be expected for hot electron injection effects, additional experiments and analyses indicate that hot electron processes are not occurring here. The first type of experimental result that was inconsistent with hot electron injection involved the wavelength dependence of the reduction current for the supra-band-edge redox species. This is shown in Figure 6 for 1,3 dimethoxy-4-nitrobenzene in methanol. The relative quantum efficiency of the photoreduction current was found to be independent of wavelength over the range 400 nm to 800 nm. Above 800 n, the quantum efficiency drops and reaches zero at about 1100 nm; this drop-off is consistent with the 1.12 eV band gap of Si.

A second set of experiments involved capacitance measurements as a function of electrode potential, ac signal frequency, and light intensity. Typical results for p-Si in acetonitrile and methanol are shown in Figures 7 and 8. In acetonitrile (Figure 5), the dark capacitance first decreases with increased band bending and then at about -0.3 V (vs SCE) remains at a flat minimum at ac signal frequencies above 100 Hz. Under illumination, the capacitance increases sharply with increasing light intensity at all potentials; the capacitance-voltage curve shows a minimum at about -0.1 V (vs SCE) and increases rapidly at higher negative potentials. At frequencies below 100 Hz, the capacitance also increases in the dark with increasing negative potential beyond -0.3 (vs SCE). Similar results are obtained with methanol (Figure 6) except that a sharp decrease in capacitance is also generally observed at all frequencies at potentials more negative than -0.3 to -0.5V (vs SCE). This drop is associated with reduction of trace amounts of water in the electrolyte which leads to large dark current flow.

III. Discussion

For the particluar p-Si electrode used in these experiments (doping density of 2 x 10^{15} cm^{-3}) the predicted wavelength dependence of the photo-reduction current for hot carrier processes should show no photocurrent for wavelengths longer than 440 nm. This is because the depletion width of the p-Si electrode with 1 volt band bending is 870 nm, and photogenerated electrons arriving at the surface from the field-free region (after crossing the space-charge layer) would be thermalized. That is, the depletion

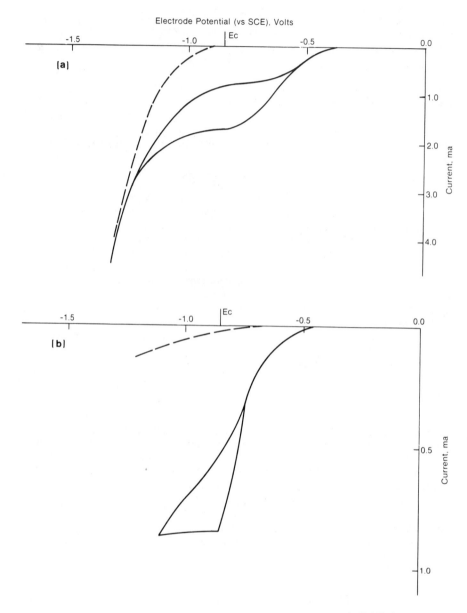

Applied Physics Letters

Figure 3. Photoreduction on p-Si *or redox couples with redox potentials lying above the conduction band edge as determined by dark flat-band potential measurements. (a) Photoreduction of 1,3 dimethoxy-4-nitrobenzene in methanol (*E_o = −1.0 V *vs. SCE); (b) photoreduction of anthraquinone in acetonitrile (*E_o = −0.95 V *vs. SCE);* E_c *for* p-Si *in both cases in the dark is* −0.85 V *vs. SCE; (– – –) dark; (———) light (5).*

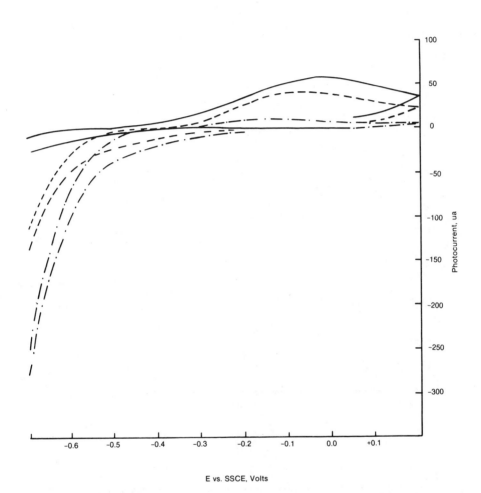

Figure 4.   Light intensity dependence of the photoreduction current of anthra-quinone in acetonitrile. An anodic wave grows in at about 0.0 V (vs. SSCE) with decreasing light intensity. ((——) 0.9 mw/cm²; (– – –) 50 mw/cm²; (· – · –) 442 mw/cm²; anthraquinone (1mM); 0.5M TEAP, Acetonitrile, 50 mV/s scan)

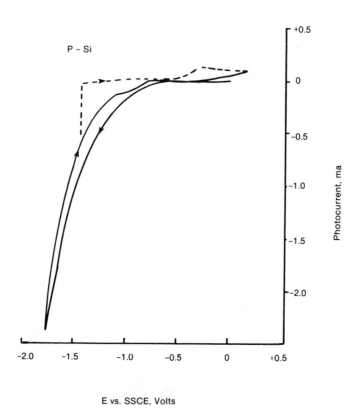

Figure 5. Current–voltage curve for anthraquinone (400 mw/cm²) in acetonitrile at high light intensities: (– – –) the current response after the light is shut off during the anodic sweep. An oxidation wave at about 0.0 V (vs. SSCE) develops if the light is shut off.

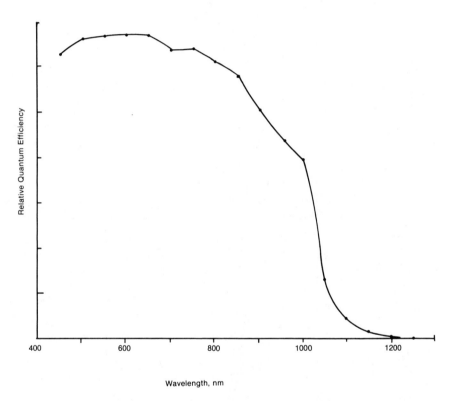

*Figure 6.   Wavelength dependence of the photoreduction current of 1,3 dimethoxy-
4-nitrobenzene (20mM) in methanol (1M TMAC)*

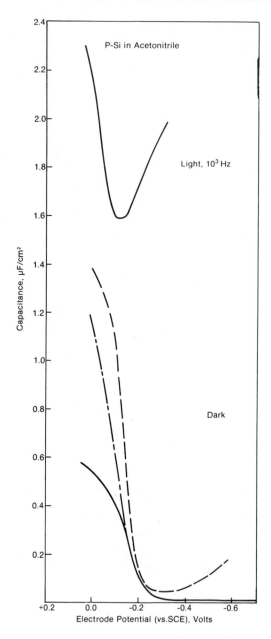

Figure 7.   Capacitance vs. p-Si electrode potential in acetonitrile as a function of frequency and light intensity (5) ((– – –) 10^2 Hz; (——) 10^3 Hz; (— - —) 5×10^3 Hz)

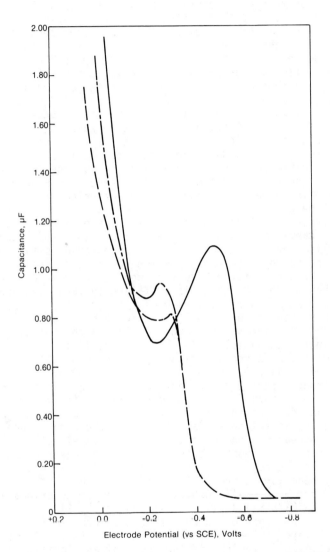

Applied Physics Letters

Figure 8. Dark capacitance vs. p-Si electrode potential in methanol as a function of frequency (5) ((———) 10^2, 5×10^1 Hz; (— —) 5×10^3 Hz; (– – –) 10^4 Hz)

width is wide enough to allow sufficient electron-phonon colli-
sions to dissipate the band bending potential energy as phonon
excitation (heat).

In order to create hot electrons at the surface of the sample,
it would be necessary to illuminate the p-Si electrode with photons
of energy greater than about 2.8 eV (wavelengths less than 440 nm).
These photons have absorption coefficients greater than 2×10^4
cm^{-1} and would be absorbed well within the space charge layer;
they would also have energies sufficiently greater than the band
gap such that photoelectrons would be created at the Si surface
with energies at least 0.3 eV above the conduction band edge. This
excess energy would be sufficient to drive the supra-band-edge re-
ductions under investigation; photons with wavelengths longer than
440 nm would be absorbed deeper into the crystal and could not
produce electrons that would arrive at the surface with sufficient
energy to drive the reduction reaction. Thus, the fact that the
wavelength dependence of the reduction current in Figure 6 follows
the absorption edge of Si means that hot carrier effects cannot be
occurring here.

It is instructive to compare the data in Figures 7 and 8 with
the ideal curves obtained for a metal-insulator-semiconductor (MIS)
device (9). For MIS devices, the space-charge density and the
capacitance decrease in the depletion region and reach a minimum
value at the onset of the inversion region. The inversion region
occurs when the band bending is sufficiently large such that the
Fermi level at the surface lies closer to the minority carrier
band rather than to the majority carrier band. As the inversion
region develops, the space-charge density in the semiconductor
surface increases rapidly. The effect of this increased charge
density on the measured capacitance depends upon the frequency of
the applied ac signal. At high frequency (>100 Hz for Si/SiO$_2$)
the electron concentration cannot follow the ac signal, and the
measured capacitance is flat and minimized in the inversion region.
However, at low frequency (<100 Hz) the electrons in the space-
charge layer can follow the ac signal and the capacitance increases
with the increasing degree of inversion. The effect of illumination in these experiments is to increase the measured capacitance
in the inversion region at the high frequencies such that the low
frequency behavior is produced. The capacitance data presented in
Figures 7 and 8 generally exhibit the behavior described above for
MIS (metal/SiO$_2$/p-Si) devices. Although the p-Si sample is etched
in HF solution before each run, it is also exposed to air and a
thin oxide layer ($\sim$10-20 Å) exists on the surface. The effect of
the oxide layer is to inhibit current flow and facilitate the main-
tenance of a large charge density in the semiconductor space charge
layer.

The creation of a large space-charge density in the semiconductor (and a corresponding increase in capacitance) can cause the
bands to become unpinned since changes in applied potential can
now occur across the Helmholtz layer.

As seen in Figure 7, the effect of light on the system is to further increase the capacitance in the inversion layer. This, of course, enhances the unpinning effect as the capacitance of the space-charge-layer approaches or exceeds that of the Helmholtz layer.

To check these effects further, detailed studies were made (10) of the capacitance-voltage behavior of p-Si electrodes in acetonitrile as a function of oxide thickness and light intensity. Oxide layers with thicknesses varying from 25 Å to 667 Å were grown under controlled conditions(11,12) and measured with an ellipsometer. The capacitance-voltage data for these samples were measured as a function of light intensity, and the data unequivocally showed the characteristic behavior expected for MIS devices (9). These studies confirmed that inversion could be readily induced in p-Si electrodes, resulting in band edge unpinning effects.

Finally, the results in Figures 4 and 5 are consistent with band edge unpinning due to illumination in the inversion region. The energy level diagram presented in Figure 9 explains the behavior in Figures 4 and 5. In the dark, the supraband-edge couple A/A⁻ lies above the conduction band edge of the p-type electrode. In inversion and under moderate illumination, the bands move up such that the redox couple lies near the conduction band edge. The photocurrent-voltage characteristic here would be semi-reversible in that oxidation could follow reduction, but the peaks would be split wide apart. With stronger illumination, the bands move further upward such that the redox couple lies well within the band gap. This produces irreversible photoreduction behavior. A sudden shut-off of the illumination under these conditions would move the bands downward, such that oxidation of the reduced A⁻ species could occur on the anodic sweep. This behavior is what is observed in Figures 4 and 5.

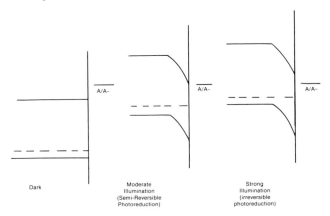

Figure 9. Energy level diagrams showing movement of semiconductor band edges with respect to the redox potential of the electrolyte as a function of illumination intensity

Acknowledgement

The work reported here was supported by the U.S. Department of Energy, Office of Basic Energy Sciences, Division of Chemical Sciences.

Literature Cited

1. Boudreaux, D.S.; Williams, F.; Nozik, A.J. J. Appl. Phys., 1980, 51, 2158.
2. Nozik, A.J.; Boudreaux, D.S.; Chance, R.R.; Williams, F. Advances in Chemistry Series, 1980, 184, 155 (Amer. Chem. Soc. Washington).
3. Williams, F.; Nozik, A.J. Nature, 1978, 271, 137.
4. Gerischer, H. Topics in Applied Physics, Vol. 31: Solar Energy Conversion, B.O. Seraphin, ed., Springer-Verlag, Berlin, 1979.
5. Turner, J.; Manassen, J.; Nozik, A.J. Applied Physics Letts., 1980, 37, 488.
6. Bocarsly, A.B.; Bookbinder, D.C.; Dominey, R.N.; Lewis, N.S.; Wrighton, M.S. J. Amer. Chem. Soc., 1980, 102, 3683.
7. Bard, A.J.; Bocarsly, A.B.; Fen, F.; Walton, E.G.; Wrighton, M.S. J. Amer. Chem. Soc., 1980, 102, 3671.
8. Morrison, S.R., private communication.
9. Sze, S.M. Physics of Semiconductor Devices, Wiley-Interscience, N.Y., Chap. 9, 1969.
10. Turner, J.; Klausner, M.; Nozik, A.J., to be published.
11. Deal, B.E. J. Electrochem. Soc., 1963, 110, 527.
12. Archer, R.J. J. Electrochem. Soc., 1957, 104, 444.

RECEIVED October 3, 1980.

Study of the Potential Distribution at the Semiconductor–Electrolyte Interface in Regenerative Photoelectrochemical Solar Cells

MICHA TOMKIEWICZ, JOSEPH K. LYDEN, R. P. SILBERSTEIN, and FRED H. POLLAK

Department of Physics, Brooklyn College of CUNY, Brooklyn, NY 11210

Liquid junction solar cells (1) offer the possibility of high efficiency energy conversion using low cost electrode materials. In addition, an expensive junction fabrication process is not needed since the junction is formed at the semiconductor-electrolyte interface simply by immersing the semiconductor in the appropriate electrolyte. In this paper we discuss some of the problems and techniques in evaluating the structure of the space charge region in such cells under operating conditions. Specifically we discuss several recent results on regenerative solar cells with a single crystal CdSe photoanode, a metal counter-electrode, and a NaOH/S$^=$/S 1:1:1M electrolyte solution. (2) The combined energy diagram of such a cell is shown in Fig. 1 with a few of the charge accumulation modes that can determine the potential distribution at the junction. This is a highly simplified energy diagram. We are assuming that the only potential drop is at the space charge layer of the semiconductor, that there is a homogeneous distribution of dopants, and that the potential distribution is one-dimensional. Effects of surface roughness and lateral inhomogeneity in dopant distributions are neglected. All these effects will have to be included for a more complete characterization of the junction. Within the one-dimensional model which is presented in Fig. 1, it is important to characterize the dark charge carrier distribution before its role in the light-induced charge transfer process is considered. Therefore, our emphasis is on dark "equilibrium" measurements.

We will illustrate the difficulties and the opportunities which are associated with two complementary measuring techniques: Relaxation Spectrum Analysis and Electrolyte Electroreflectance. Both techniques provide information on the potential distribution at the junction of a "real" semiconductor. Due to the individual characteristics of each system, care must be taken before directly applying the results which were obtained on our samples to other, similarly prepared crystals.

0097-6156/81/0146-0267$05.00/0

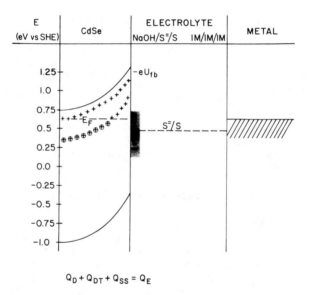

$$Q_D + Q_{DT} + Q_{SS} = Q_E$$

Figure 1.   Combined energy diagram for a regenerative photoelectrochemical cell with n-CdSe as the anode, metallic cathode and polysulfide as the electrolyte. The diagram indicates some of the charge accumulation modes that might contribute to the potential distribution at the interface. ((Q_D) ionized donors; (Q_{DT}) deep traps; (Q_{SS}) surface state; (Q_E) compensating charge in the electrolyte)

Relaxation Spectrum Analysis

    The details of this method for analyzing the impedance of photoelectrochemical cells were published elsewhere. (3) We assume that the various modes of charge accumulation are additive and that they are compensated by the charge accumulation in the Helmholtz layer. This additivity of charges will make the differential capacitance of the various modes appear in parallel with each other, each with its own relaxation time. The equivalent circuit is presented in Fig. 2. The basic premise of this technique is to use the frequency dispersion to construct the equivalent circuit in terms of a set of completely passive elements. Once this is accomplished, one tries to determine the physical origin of these elements by monitoring their values as a function of such parameters as potential, pH, doping level, temperature, etc. This technique was used to study a non-regenerative solar cell with TiO_2 as the photoanode, (4) and parameters such as the doping distribution, and the pH dependence of the flatband potential and capacitance of the Helmholtz layer were determined. We present here preliminary results on regenerative cells with CdSe as the photoanode and a polysulfide solution as the electrolyte.
    An alternative approach that was used in the past was to treat the photoelectrochemical cell as a single RC element and to interpret the frequency dispersion of the "capacitance" as indicative of a frequency dispersion of the dielectric constant. (5) In its simplest form the frequency dispersion obeys the Debye equation. (6) It can be shown that in this simple form the two approaches are formally equivalent (7) and the difference resides in the physical interpretation of modes of charge accumulation, their relaxation time, and the mechanism for dielectric relaxations. This ambiguity is not unique to liquid junction cells but extends to solid junctions where microscopic mechanisms for the dielectric relaxation such as the presence of deep traps were assumed.
    In our experience with CdSe regenerative cells we have found that unlike TiO_2 (4) the complete frequency range of the impedance measurements extends to the limits of our experimental capability, which is 10 MHz. At these high frequencies, the inductance of various components in the measuring apparatus seems to play an important role. While attempts were made to reduce its contributions to the impedance, we were unable to eliminate it and thus we included it in the equivalent circuit in Fig. 2. The high frequency values of the real and imaginary parts are used to calculate L, C_{SC} and R_S, and these values are used to calculate B_{SS} as previously described. (3) Fig. 3 shows the relaxation spectrum of single crystal CdSe in 1M $NaOH/S^=/S$ electrolyte, with a potential U= +0.8 V vs $NaOH/S^=/S$ 1:1:1M electrode which was separately measured to be at -0.72 V vs SCE. The sharp dip in the imaginary part is fully consistent with the presence of the inductive element in series as shown in Fig. 2

EQUIVALENT CIRCUIT

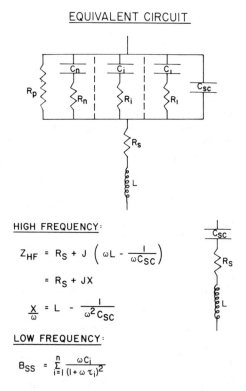

HIGH FREQUENCY:

$$Z_{HF} = R_S + J\left(\omega L - \frac{1}{\omega C_{SC}}\right)$$

$$= R_S + JX$$

$$\frac{X}{\omega} = L - \frac{1}{\omega^2 C_{SC}}$$

LOW FREQUENCY:

$$B_{SS} = \sum_{i=1}^{n} \frac{\omega C_i}{(1 + \omega \tau_i)^2}$$

Figure 2. Assumed generalized equivalent circuit of the semiconductor–electrolyte interface. Reduced equivalent circuit at high frequencies and the expression for the impedance at low and high frequencies.

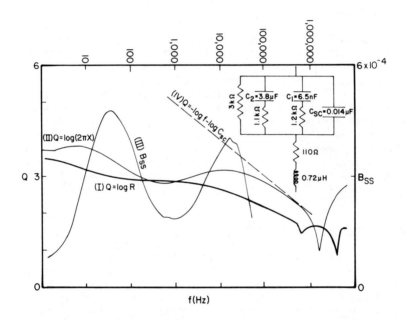

Figure 3.   Impedance response curves for n-CdSe single crystal in polysulfide (CdSe in NaOH/$S^=$/S/1:1:1; potential +0.8 V, area = 0.15 cm²)

while the two smaller dips in the real part are not consistent
with this presentation. They could result from the presence of
inductive elements parallel to R_S. The possible role of stray
capacitance in the system is also a matter for further investigation.

Noting these uncertainties, we have evaluated the equivalent circuit and present the results in Fig. 3. The potential
dependence of the three capacitive elements is shown in Figs.
4 and 5. As shown in Fig. 4 the agreement of the C_{SC} data with
the Mott–Schottky relation is good and indicates that C_{SC} orig-
inates from the majority carriers in the space charge region.
The normalized slope indicates a doping level of 8 x 10^{16}/cm^3,
and the intercept is at -0.96 V vs NaOH/$S^=$/S 1:1:1M electrode.
The potential dependence of the other two capacitive elements
are given in Fig. 5. These elements do not obey the Mott-
Schottky relation and their origin is still being investigated.
Preliminary measurements have been made on Au Schottky barriers
on CdSe single crystals. Neither Relaxation Spectrum Analysis
nor Deep Level Transient Spectroscopy (10) has provided evidence for the slower relaxing states. These null results and
the potential dependence of C_1 and C_2 indicate that these
slower states might be associated with the crystal–liquid
interface. However, a word of caution is indicated as the
crystals used for the Schottky barrier work had a much lower
doping level (5 x 10^{13} / cm^3 as determined from their Mott-
Schottky data) than the crystals used in the photoelectrochemical cells.

Electrolyte Electroreflectance

Electrolyte Electroreflectance (EER) is a sensitive optical technique in which an applied electric field at the surface
of a semiconductor modulates the reflectivity, and the detected
signals are analyzed using a lock-in amplifier. EER is a
powerful method for studying the optical properties of semiconductors, and considerable experimental detail is available
in the literature. (11, 12, 13, 14, 15) The EER spectrum is
automatically normalized with respect to field-independent
optical properties of surface films (for example, sulfides),
electrolytes, and other experimental particulars. Significantly, the EER spectrum may contain features which are sensitive to both the AC and the DC applied electric fields, and
can be used to monitor in situ the potential distribution at
the liquid junction interface. (14, 15, 16, 17, 18)

We have measured the EER spectra of single crystal CdSe
in a polysulfide electrolyte using the same configuration for
which the relaxation spectrum analysis was applied. Preliminary results are shown in Fig. 6.

Two principal spectral features can be seen in Fig. 6
near 1.75 and 2.16 eV, corresponding to the E_o^{AB} (direct gap
at $\vec{k}$=0) and E_o^C (spin-orbit split component) peaks previously

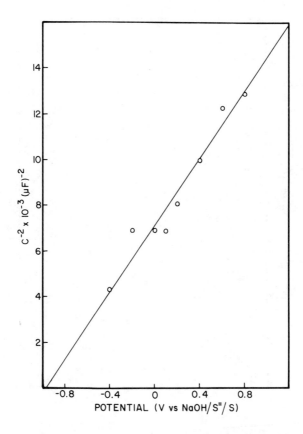

Figure 4.   Mott–Schottky plot of C_{sc} from Figure 3

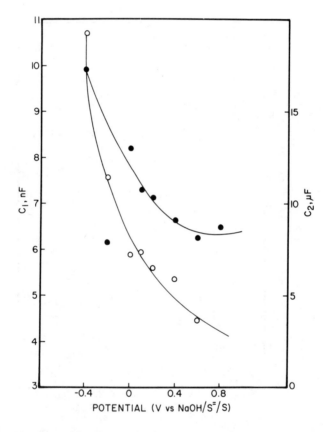

Figure 5.   Variation of C_1 (●) and C_2 (○) from Figure 3 with the electrode potential

observed. (11) As the potential is reduced from +0.5 V to
-1.0 V the spectral lineshape changes significantly. Additional measurements carried out with zero DC applied voltage have
shown that the intensity of the spectrum scales linearly with
the amplitude of the modulation voltage from V_{ac} = 0.05 V to
V_{ac} = 1.0 V peak to peak. This indicates that the EER spectra
are "low-field" spectra. (14, 15, 16, 17) However, EER spectra in the low-field regime are expected to be independent of
the DC electric field at the surface, hence, independent of
potential. The variation in the EER lineshape as a function
of potential may be a result of the mismatch between the electric field profile near the surface of the semiconductor and
the penetration depth of the light. (13, 14, 15) In this case,
the EER spectra would be independent of potential for values
corresponding to deep depletion, where the field over the penetration depth of the light is relatively constant. Additional
data show that this relation is approximately maintained for
potentials in the range +0.5 V to +1.0 V. For low-field
electroreflectance, the signal observed at the fundamental
frequency is expected to pass through a minimum as we approach
the flatband potential. (13, 18) Despite the complications
in the lineshape due to field nonuniformity within the penetration depth of the light, the EER spectra of Fig. 6 do show that
the signal intensity goes through a minimum at U = -0.8 V vs
$NaOH/S^{=}/S$ electrode. This value for the flatband potential is
in approximate agreement with that from the capacitance data.
As the flatband potential is crossed, the energy bands at the
surface go from depletion to accumulation and the EER signal
is expected to reverse in sign, (13, 18) as can be observed in
Fig. 6 for U = -1.0 V. By measuring the amplitude and phase
of the EER signal at a spectral feature for which the photon
energy is independent of the electrode potential, a very accurate value of the flatband potential can be obtained simply by
scanning through the potential range. The appropriate conditions are approximately satisfied for $h\nu$ = 1.74 eV, the energy
gap of CdSe, which is near the average position of the main
electroreflectance peak. Fig. 7 shows the EER signal for
$h\nu$ = 1.74 eV as a function of potential under conditions identical with those for the data in Fig. 6. For comparison, the
variation in the photocurrent due to chopped white light, as a
function of the electrode potential, is also shown. We can see
that the turn-on potential and the inversion of the EER signal
agree, indicating that the liquid junction cell resembles a
"well behaved" diode. For the uniform field case the EER
method should be superior to both the capacitance technique
which relies upon the separate evaluation of the various charge
accumulation modes, and the turn-on potential technique which
is dependent on kinetic factors.

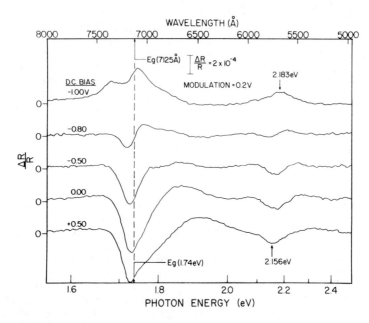

Figure 6.   Variation with potential of the electrolyte electroreflectance spectra of single-crystal n-CdSe in polysulfide solution (CdSe T = 300 K in NaOH/S=/S 1:1:1; modulation = 0.2 V)

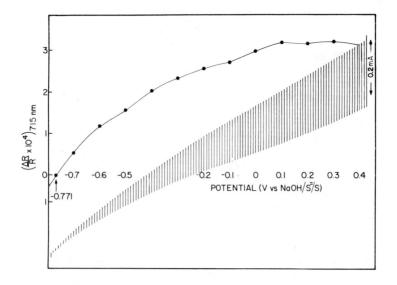

Figure 7.   Potential sweep of the electroreflectance signal at 1.74 eV superimposed on potential sweep of chopped light photoresponse

Conclusions

We have extended the technique of Relaxation Spectrum Analysis to cover the seven orders of magnitude of the experimentally available frequency range. This frequency range is required for a complete description of the equivalent circuit for our CdSe–polysulfide electrolyte cells. The fastest relaxing capacitive element is due to the fully ionized donor states. On the basis of their potential dependence exhibited in the cell data and their indicated absence in the preliminary measurements of the Au Schottky barriers on CdSe single crystals, the slower relaxing capacitive elements are tentatively associated with charge accumulation at the solid–liquid interface.

The close agreement among the flatband potentials determined by the capacitance measurements, electrolyte electroreflectance, threshold for light-induced charge separation, indicates that these cells are "well behaved." This close agreement also suggests that the open circuit voltage should approach the theoretical value — the difference between the polysulfide redox potential and the flatband potential of the photoelectrode—once the polarization at the counter electrode and the dark current are minimized. The dependence of the electroreflectance lineshape on the electrode potential offers the promise of direct measurement of the electric field at the junction and of the penetration of the light relative to the electric field. To fully realize this promise a full analysis of the lineshape is required.

ABSTRACT

The Relaxation Spectrum Analysis was carried out for a cell consisting of n–CdSe in a liquid junction configuration with $NaOH/S^=/S$ 1:1:1M as the electrolyte. Three parallel RC elements were identified for the equivalent circuit of this cell, and the fastest relaxing capacitive element obeys the Mott–Schottky relation.

Electroreflectance spectra were measured for n–CdSe in the liquid junction configuration, and variations of the lineshape as a function of potential were observed. As the potential was reduced below the flatband potential, the electroreflectance signals changed sign. The potential at which this change occurs correlates well with the turn-on potential for light-induced photocurrent and with the intercept of the Mott–Schottky plot.

Research supported in part by the Solar Energy Research Institute under contract XS-9-832-1 and in part by the Office of Naval Research under contract N00014-78-C-0718.

Literature Cited

1. Heller, A., ed.; "Semiconductor Liquid Junction Solar Cells" The Electrochemical Society, Inc. Princeton, N.J., 1977.
2.a Hodes, G., Manassen, J., and Cahen, D., Nature 261, 406 (1976).
 b Ellis, A.B., Kaiser, S.W. and Wrighton, M.S.; J. Am. Chem. Soc. 98, 1635 (1976).
3. Tomkiewicz, M.; J. Electrochem. Soc. 126, 2220 (1979).
4. Tomkiewicz, M.; J. Electrochem. Soc. 126, 1505 (1979).
5. Dutoit, E.C., Van Meirhaeghe, R.L., Cardon, F. and Gomes, W.P.; Berichte der Bunsen-Gesellshaft 79, 1206 (1975).
6. Debye, P.; Phys. Z 35, 101 (1934).
7. Cole, K.S. and Cole, R.H.; J. Chem. Phys. 9, 341 (1941).
8. Losee, D.L.; Appl. Phys. Lett. 21, 54 (1972).
9. Crowell, C.R. and Nakano, K.; Solid State Electron. 15, 605 (1972).
10. Lang, D.V.; J. Appl. Phys. 45, 3014, 3023 (1974).
11. Cardona, M., Shaklee, K.L. and Pollak, F.H.; Phys. Rev. 154, 696 (1967).
12. Cardona, M.; "Modulation Spectroscopy" (Academic Press, New York 1969).
13. Seraphin, B.O.; "Semiconductors and Semimetals," Vol. 9, eds. Willardson, R.L. and Beer, A.C. (Academic Press, New York 1972) P. 1.
14. Hamakawa, Y. and Nishino, T.; "Optical Properties of Solids: New Developments," ed. Seraphin, B.O. (North-Holland, Amsterdam 1976) P. 225.
15. Aspnes, D.E.; "Handbook on Semiconductors," Vol. 2, ed. Balkanski, M. (North-Holland, New York 1980) P. 109.
16. Aspnes, D.E.; Surf. Sci. 37, 418 (1973).
17. Aspnes, D.E.; Phys. Rev. B10, 4228 (1974).
18. Pond, S.F. and Handler, P.; Phys. Rev. B6, 2248 (1972); B8, 2869 (1973); Pond, S.F.; Surf. Sci. 37, 596 (1973).

RECEIVED October 3, 1980.

Luminescence and Photoelectrochemistry of Surfactant Metalloporphyrin Assemblies on Solid Supports

JOHN E. BULKOWSKI, RANDY A. BULL, and STEVEN R. SAUERBRUNN

Department of Chemistry, University of Delaware, Newark, DE 19711

Photoelectrochemical cells based on charge separation at semiconductor-liquid junctions offer promise as a practical means for converting solar energy into electricity or storable chemical fuels (1). One approach to achieving such devices involves the sensitization of wide-gap semiconductors by modification of their surfaces with visible-light absorbing dyes. In this well-established spectral sensitization process (2,3,4), the light-excited dye molecules transfer majority carriers into the semiconductor, and the interfacial energetics are advantageously employed to inhibit back reactions. For instance, if the energy levels are suitably matched, dyes bound to n-type semiconductor electrodes inject electrons into the conduction band upon illumination. The electrons are swept away from the surface into the bulk of the semiconductor as a result of the space-charge layer developed beneath the semiconductor surface. In the electrochemical cell, the photooxidized dyes trapped at the surface are reduced by appropriate redox species in the electrolyte. Coupling of the photoelectrode via an external circuit to a counter electrode also immersed in the electrolyte completes the cell.

For dyes absorbing in the visible region, wide-gap semiconductors are chosen to maximize electron transfer from the excited dye molecules on the surface to the solid (5,6). However, a major difficulty associated with this approach is the restriction of overall quantum conversion efficiencies due to low light absorption by thin dye layers. Thick dye layers, although they may absorb all of the light, do not result in significantly greater conversion efficiencies, since they suffer from increased quenching probabilities and large resistances. These properties of thick dye films are discouraging with regard to fabricating practical photoelectrochemical cells. Consequently, we are employing a molecular design approach to systematically examine photoinduced charge and energy transfer in highly ordered dye assemblies of a controllable architecture. The aim is to understand the fundamental sensitization processes, both within

0097-6156/81/0146-0279$05.00/0

the assemblies and at the interfaces, in order to ultimately realize useful sensitization materials.

The objective of this molecular design approach is the unique three-dimensional array schematically represented in Figure 1. In this idealized structure, coordination compounds are organized in distinct monomolecular layers. Each layer is connected to adjacent ones by bifunctional molecular linkages coordinated to the axial sites of the metal centers. The dye molecules we are using in our efforts to synthesize such a structure are metalloporphyrins, whose coordination chemistry and structural characteristics appear to be well suited to accomplishing the synthetic goals. The interlayer linkages to be used are bifunctional bases such as pyrazine, which is shown in the illustration. Their functions are two-fold: to provide binding sites for positioning successive layers in the array, and to provide an adjustable inner-sphere pathway for photoinduced charge transfer between adjacent layers. Although it is not our intent to develop the detailed rationale behind our design here, it should be noted that construction of such an array in a step-wise manner would provide a convenient means for tuning the optical and electrical properties of the film's architecture. For example, formation of a multi-layered assembly with successive layers comprised of metalloporphyrins with decreasing redox potentials could result in the generation of an asymmetric film.

The synthetic strategy we are following to develop these ordered films involves a combination of monolayer techniques and coordination chemistry substitution reactions. This synthetic approach is depicted in Figure 2 for a two layered system in which the layers are interconnected by pyrazine. The first step is formation of a template layer on a solid support. This is accomplished by transferring an ordered layer of surfactant metalloporphyrins from an air-water interface onto a hydrophobically treated support surface by passing the support down through the oriented film (Step 1). It is imperative that this monolayer is homogeneous and that the molecular planes of each porphyrin molecule are oriented parallel to the support surface, leaving the metal centers exposed at the hydrophilic interface. The second and third steps involve successive additions of the pyrazine and non-surfactant water soluble metalloporphyrin complexes to the template submerged in the aqueous phase (Steps 2 and 3, respectively). The two-layered assembly is then removed from the water through a protective monolayer of stearic acid to prevent disruption of the film structure (Step 4). If instead of removing the assembly in Step 4, Steps 2 and 3 were successively repeated, multi-layered assemblies having the essential features described in Figure 1 might be realized.

Since achievement of template formation is crucial to the success of this approach, much of our effort to date has been directed at elucidating the monolayer properties of various por-

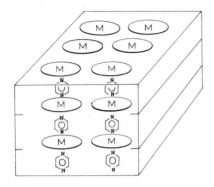

Figure 1.   Schematic of the proposed dye assembly where M is coordinated in the plane of a macrocyclic ligand (e.g., a porphyrin) and pyrazine is axially coordinated between two metals in adjacent layers

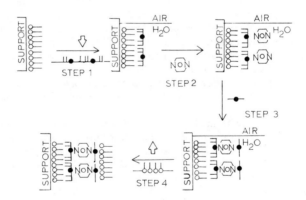

Figure 2.   Representation of the dye assembly construction procedure: (O–) stearic acid; (–●–) surfactant metalloporphyrin (MTOAPP); and (–●–) water soluble porphyrin

phyrin compounds. This has led to our work with surfactant
porphyrin derivatives of the type schematically represented in
Figure 3. The results of our monolayer and luminescence studies
with these modified porphyrins do in fact indicate that they form
well-ordered monolayers at the air–water interface with characteristics suitable for template formation. This is in agreement
with reported results for similar surfactant porphyrin derivatives
(7,8). The purpose of this article is to present the results
of our luminescence and photoelectric studies with these porphyrin
derivatives, since they demonstrate that these compounds provide
a good opportunity for examining electron and energy transfer
phenomena at a uniquely defined dye sensitized semiconductor-
electrolyte interface. The results are also discussed with regard
to the molecular design of the novel photosensitive films.

Experimental

Preparation and Purification of Materials.

The tetraoctadecylamide derivative (H_2TOAPP) of meso-tetra
($\alpha,\alpha,\alpha,\alpha$-o-aminophenyl)porphyrin was prepared by modification
of the "picket-fence" porphyrin procedure of Collman and coworkers
(9). Purification was by column chromatography on silica gel
using gradient elution with ether-petroleum ether solvent combinations. The chromatographic procedure was repeated until the
porphyrins were determined to be greater than 99% isomerically
pure. Purity was checked using high-performance liquid chromatography by comparing the pure product to a standard mixture of the
four possible atropisomers. Refluxing the free-base surfactant
porphyrin with zinc acetate in dimethylformamide and nickel
chloride in chloroform-ethanol gave the zinc and nickel metalloporphyrin derivatives, respectively.

Octadecane (Eastman) and stearic acid (Aldrich) were recrystallized from acetone and acetonitrile, respectively. All
other organic solvents and inorganic compounds were reagent
grade or better and used as received. Triply distilled water
(from an all glass system having permanganate and sulfuric acid
stages) was used for the monolayer subphases and the electrolyte
solutions.

Solid supports were either 25 x 75 mm glass microscope slides
(Fisherbrand) or 25 x 75 mm Sb doped SnO_2 coated glass slides
(Practical Products Co.). The SnO_2 coating was 3000Å thick and
had a resistance of 100 $\Omega/\square$. Optical transmittance was 80–85% in
the visible region above 400 nm. The glass microscope slides
were treated with hot organic solvents, hot HNO_3, dilute NH_4OH,
and then were rinsed several times with triply distilled water.
They were air dried in an inverted position. The SnO_2 optically
transparent electrodes (OTE's) were treated with hot $CHCl_3$ and
CH_3OH, etched with H_2SO_4, and then washed several times with
triply distilled water. A conductive silver coating (Practical

Products Co.) was painted as a contact on one end of the OTE, which was then cured in an oven at 160°C. After cleaning and drying, the glass slides and OTE's were stored in sealed individual glass containers until use.

Monolayer Techniques.

The general methods used for forming and manipulating monolayers are described by Kuhn and coworkers (10). Pressure-area curves were determined on a paraffin coated Cenco Langmuir balance calibrated by suspending weights from an attached side arm of known length. The calibration was checked periodically by measuring the pressure-area curve for stearic acid (SA). The films at the air-water interface were formed by delivering 100-200 µl of approximately 10^{-4} M solutions of the required surfactant mixtures in chloroform onto the aqueous surface using a calibrated 200 µl syringe. The surface-pressure area isotherms were measured by compressing the film in successive steps at one minute intervals. The aqueous subphase, unless noted otherwise, was buffered at pH 6.5 with 5 x 10^{-5} M NaHCO$_3$ and contained 3 x 10^{-4} M CdCl$_2$.

The precleaned glass slides for the spectroscopic studies were precoated on one side with three layers of cadmium stearate (formed by addition of SA to the cadmium-containing subphase) by passing two slides positioned back-to-back through the monolayer three times at a rate of 1.0 cm/min. The surface pressure was held constant at 30 dyne/cm by a weight and pulley transfer apparatus with a motor driven lift (11). Deposition of the surfactant porphyrin mixtures was by passing the stearate coated slides down through the appropriate film at a constant pressure of 20 dyne/cm. If a single porphyrin layer was required on the support, the porphyrin monolayer was then removed from the subphase surface and a stearic acid film was formed through which the slide containing the porphyrin monolayer was removed. For formation of two porphyrin dye layers face-to-face on the support, the slide was simply dipped and then removed through the porphyrin film on the subphase. Deposition of surfactant porphyrins on the SnO$_2$ OTE's for the photoelectric measurements was by vertical removal of the OTE (two slides at a time with the glass surfaces facing each other) through the porphyrin mixture on the aqueous subphase. Surfactant porphyrin (MTOAPP): stearic acid:octadecane monolayers in the ratio of 1:4:3 were held at a constant pressure of 25 dyne/cm for these depositions. In this support-film arrangement, the porphyrin rings were positioned next to the SnO$_2$ surface. Coated slides were stored in separate glass containers in the dark before use. Deposition ratios were routinely measured and used as a criterion to assess the quality of the coatings.

Luminescence Measurements.

Corrected luminescence spectra were determined with a Perkin-Elmer MPF-44B spectrophotofluorimeter equipped with a R-928 photomultiplier and a DCSU-2 microprocessor. Calibration was with a Perkin-Elmer standard tungsten lamp unit. Excitation and emission spectra of monolayers on the glass slides were from the front side with the slide positioned in a sample holder with a non-fluorescing black background. The exciting light was at an angle of 30° to the normal of the sample slide with detection of luminescence at right angles to the incident light through a 565 nm cut-off filter (Turner, Sharp Cut #22).

Photoelectrochemical Measurements.

The electrochemical cell consisted of the porphyrin treated SnO_2 OTE mounted as a window on a machined Teflon block which contained a compartment for the electrolyte solution. The semiconductor–porphyrin face contacted the electrolyte with an area of 1.0 cm^2. Electrical contact to the SnO_2 electrode was made via a conductive silver coating on one end. A platinum wire and a saturated calomel electrode (SCE) served as the counter and reference electrodes, respectively. The SCE contacted the electrolyte via a salt bridge. The cell compartment was provided with an inlet and outlet for introduction of the electrolyte. The supporting electrolyte was 0.1 M KCl and the pH was maintained at 7.0 using a phosphate buffer. The electrolyte was rigorously deaerated with a N_2 purge. Current versus potential measurements were made using circuitry similar to that described by Honda and coworkers (12). The potentiostat and signal programmer were of conventional design and constructed by us from commercially available components. Photocurrents were measured by a Keithley electrometer (Model 610C) and recorded on an X-Y recorder. The cell mounts and all electrical components were connected to a common ground. Typical cell internal resistances were 20–30 kΩ.

The light source for current versus potential and current versus time measurements was a 300-watt ELH lamp (General Electric) operated at an integrated irradiance of ca. 75 mw/cm^2 above 350 nm. A 350 nm cut-off filter was positioned between the electrochemical cell and the light source. The photocurrent backgrounds of SnO_2 electrodes having only a cadmium stearate layer were 10^{-2} less than those having the porphyrin dye monolayers. Determination of action spectra was accomplished by placing the electrochemical cell in the sample chamber of the Perkin-Elmer MPF-44B spectrometer and using the 150-watt Xe light source and excitation optics for irradiating the electrode.

Results and Discussion

Monolayer Properties.

Previous studies of dye sensitization with porphyrin materials

(<u>13</u>–<u>19</u>) have usually involved formation of solid films by techniques such as solution evaporation or vapor deposition.   Little
is known about the structures of such films.   As a result, we have
been studying methods for generating well-defined porphyrin assemblies using monolayer methods at the air-water interface (<u>10</u>).
Our attempts to generate monolayers of metallotetraphenylporphyrins by spreading solutions of them on an aqueous subphase did
not result in the generation of monomolecular layers. As expected,
a high degree of aggregation and microcrystallinity was readily
apparent.   Compression of these films results in curves atypical
of monolayers with extrapolated area/molecule values less than
15 $Å^2$/molecule at observable surface pressures.   Addition of
surfactant compounds known to enhance monolayer-forming properties (e.g., stearic acid) in the spreading process did not
alleviate the aggregation problems.   Codeposition of mixtures
of various surfactant pyridine bases (e.g., an octadecylamide
derivative of 4-aminopyridine) and metallotetraphenylporphyrins
resulted in no visible aggregation.   Pressure-area isotherms
characteristic of monomolecular layers were observed; however,
the porphyrin area/molecule values ranged from 80–120 $Å^2$/molecule
depending on the nature of the metallotetraphenylporphyrin.   This
suggests that the porphyrins are either aggregated, oriented
perpendicularly to the aqueous surface, or tucked up inside the
surfactant layer.   Although these systems are unsuitable for
template formation in our assembly procedure, they may be useful
for examining dye sensitization orientational effects at electrode
surfaces.

To achieve porphyrin monolayers suitable for use in our proposed assembly scheme, we examined the properties of surfactant
porphyrins of the general type shown in Figure 3.   A typical surface pressure-area isotherm for H_2TOAPP on a pH 6.5 buffered
$NaHCO_3$ aqueous subphase is given in Figure 4.   This curve is characteristic of formation of monomolecular layers at the air-water
interface (<u>20</u>).   Extrapolation of the condensed phase region
to zero pressure gives a value of ca. 160 $Å^2$/molecule for the
porphyrin, which is in agreement with previous observations for
this type of compound (<u>8</u>).   Similar values are found for both
the Zn and NiTOAPP derivatives.   This area/molecule value is
consistent with that expected for the porphyrin ring oriented
nearly parallel to the aqueous surface (values of ca. 165 $Å^2$/
molecule are calculated from crystal structures of tetraphenyl-
porphyrin).   The decrease in area to ca. 130 $Å^2$/molecule upon
compression to 20 dyne/cm probably results from tilting of the
porphyrin heads as the loosely packed chains (total area of ca.
80 $Å^2$/molecule) are squeezed into a tighter arrangement.   This
could result in some overlap of the porphyrin π orbital system.
The porphyrin orientation indicated by this result is encouraging
with regard to using these compounds for template formation to
build up the proposed porphyrin-containing arrays described
earlier.

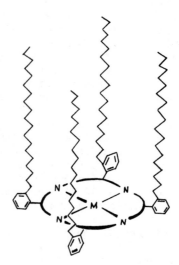

Figure 3.   Schematic of a surfactant metalloporphyrin, MTOAPP

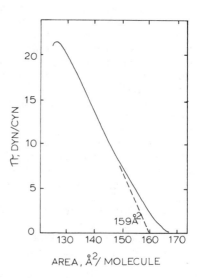

Figure 4.   Surface pressure–area isotherm of H₂TOAPP (aqueous phase, 5 × 10⁻⁵M NaHCO₃; pH 6.5; air atmosphere at 25°C)

Mixed monolayer properties of the MTOAPP derivatives with various combinations of SA and octadecane were determined to investigate ways for controlling the surface concentration using photoinert spacers. The objective was to characterize their stability and homogeneity upon dilution. Surface pressure-area isotherms for differing mixing ratios of ZnTOAPP and SA are given in Figure 5. Two features of this family of curves indicate homogeneity of mixing (20). They are: 1) increasing breakdown pressure as a function of mixing ratio, and 2) the calculated curve, Figure 5c', which is based on additivity of the areas of pure ZnTOAPP and SA (a 1:4, ZnTOAPP:SA mixture) does not agree with the experimentally determined curve, Figure 5c. Results obtained for a ZnTOAPP-octadecane mixture indicated that at low mixing ratios (less than 1:10, ZnTOAPP:octadecane) layers are stabilized relative to pure ZnTOAPP. Maximum film stability was observed for the 1:3 mixture of ZnTOAPP:octadecane. It appears that the octadecane molecules can be accommodated by vacancies in the loosely packed hydrophobic region. Studies of mixtures of porphyrin-stearic acid-octadecane indicated that a 1:4:3 ratio of ZnTOAPP:SA:octadecane gives good, stable monolayers so this combination was chosen for the photoelectrochemical measurements.

Using established methods (10), the ZnTOAPP monolayers were readily transferred to solid supports, such as suitably treated glass slides or SnO_2 optically transparent electrodes.

Luminescence Studies.

Luminescence spectroscopy was used as a sensitive technique to characterize the ZnTOAPP layers at the air-support interface. Figure 6 compares the emission (λ_{ex}, 435 nm) and excitation (λ_{em}, 660 nm) spectra of a monolayer of ZnTOAPP and SA (1:4) on a hydrophobic glass support (arrangement of layers is glass/SA/ZnTOAPP/SA) with those of a 10^{-6} M ZnTOAPP chloroform solution. There is a slight bathochromic shift (ca. 10 nm) of the monolayer peaks and a broadening of the Soret band which qualitatively agrees with results reported for vapor deposited solid films of ZnTPP on quartz (21,22). These spectra show that our compounds are quite pure (e.g., no additional peaks at 630 and 690 nm due to chlorin impurities (22)) and that they are not aggregated. Even 10^{-4} M porphyrin solution excitation spectra of these compounds show several additional bands in the Soret region due to intermolecular interactions. Also, the broadening and peak shifts are rather insensitive to dilution of the monolayers with SA, although fluorescence intensities do significantly decrease. Figure 7 shows the self-quenching of the 660 nm fluorescence emission (normalized to relative intensity per chromophore using concentration values based on our aqueous subphase monolayer data) as a function of dilution of the ZnTOAPP by SA. A sharp decrease in the quenching occurs at ZnTOAPP:SA ratios of 1:4, and a con-

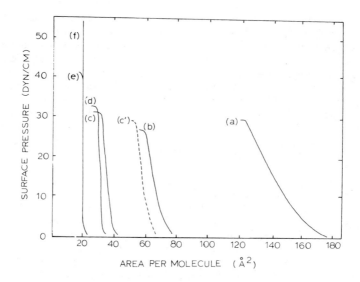

*Figure 5. Surface pressure–area isotherms for ZnTOAPP:SA mixing ratios—
(a) pure ZnTOAPP; (b) 1:2; (c) 1:4; (c′) 1:4 calculated; (d) 1:10; (e) 1:500; and
(f) pure SA. Aqueous phase contains 5 × 10⁻⁵M NaHCO₃ buffer, pH 6.5, and
3 × 10⁻⁴M CdCl₂.*

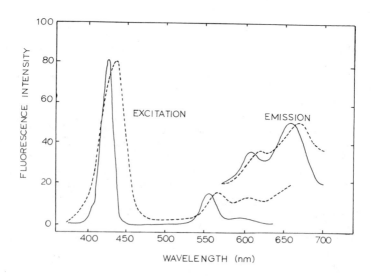

*Figure 6. Excitation and emission spectra of a 10⁻⁶M solution of ZnTOAPP in
chloroform (——) and a monolayer of ZnTOAPP on a hydrophobically treated,
Cd–SA glass support (– – –). Relative intensities are arbitrarily adjusted.*

stant limiting value is reached at a ratio of ca. 1:10. This is consistent with the formulation of a homogeneous, tightly packed monolayer on the support surface. A heterogeneous surface of islands of porphyrin and spacers would be expected to show little effect upon SA dilution. Also, this luminescence data indicates that energy transfer associated with self-quenching takes place over very short ranges (i.e. less than 1 Å spacings between porphyrin ring peripheries). This may even require contact of orbitals on adjacent chromophores as might be expected for porphyrins arranged in a tilted configuration on the surface. By slight separation using SA, the quenching effect per molecule is decreased by a factor of 3.

Inclusion of various ratios of non-fluorescing NiTOAPP traps into the ZnTOAPP monolayer (MTOAPP:SA, 1:4) gives the dependency indicated in Figure 8. This result indicates significant energy transfer from the Zn centers to the traps, even at concentrations of one trap per one hundred dye centers. The curve shape agrees well with one calculated for a dipolar-dipolar coupling mechanism (23) between porphyrins. Preliminary evidence with a two-layer system in which the ZnTOAPP monolayer (ZnTOAPP:SA, 1:4) concentration was held constant, but which was in face-to-face contact with a second NiTOAPP-SA monolayer of varying Ni concentration, indicates a similar intensity dependence on mixing ratio as was found for the single layer system. This result suggests that interlayer quenching also occurs via a dipolar-dipolar mechanism. However, the critical distance for quenching (I/Io = 0.5) was found to be less than that measured for the single layer assembly. This indicates that there is less effective coupling between weakly interacting chromophores in adjacent layers than between molecules tightly packed in individual layers. We are currently studying these effects in greater detail.

Photoelectrochemical Studies.

Our initial photoelectrochemical studies have been conducted with monolayers of ZnTOAPP:SA:octadecane mixtures in the ratio of 1:4:3. They are deposited directly on the SnO_2 OTE's with the surfactant porphyrin head groups in contact with the electrode surface. The electrolyte contained 0.1 M KCl and was maintained at pH 7.0 with a phosphate buffer. The electrolyte was deoxygenated by a N_2 gas purge. An anodic photocurrent was generated under short-circuit conditions which increased with applied potential. The anodic photocurrent is consistent with electron injection toward the SnO_2. Additionally, an open circuit voltage was measured upon irradiation by the ELH lamp source. The observed photovoltage (ca. 10 mV) was negative and consistent with the generation of an anodic photocurrent. Negligible photoeffects were observed for a SnO_2 OTE coated with just a cadmium stearate layer under identical conditions.

This behavior is consistent with an energy level scheme for

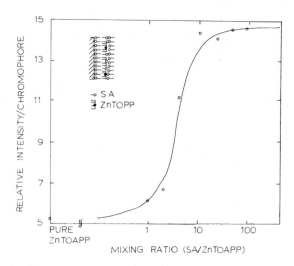

Figure 7.   Dependence of fluorescence intensity (λ_{em}, 660 nm) of ZnTOAPP on SA:ZnTOAPP mixing ratios. Mixtures are deposited on a hydrophobically treated, Cd–SA glass slide and have a Cd–SA outer layer as represented in the insert.

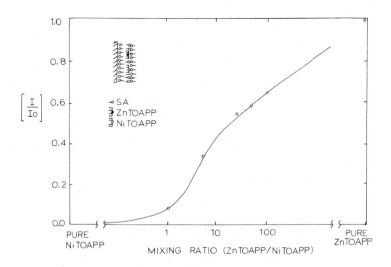

Figure 8.   Dependence of fluorescence quenching on ZnTOAPP:NiTOAPP mixing ratios (MTOAPP:SA, 1:4) in a monolayer on Cd–SA–treated glass with a Cd–SA outer layer as represented in the insert

electron transfer in which the electron donor is situated at a more negative potential than the conduction band edge of the semiconductor. The flatband potential for SnO_2 at pH 7.0 is -0.45 V vs. SCE (24). The energy level of the excited donor is approximated from its oxidation redox potential and excitation energy (4). The oxidation potential for ZnTOAPP is taken as +0.70 V, based on reported values for substituted ZnTPP's (25); the singlet excitation energy is 1.88 eV, based on the 660 nm emission. This results in a calculated oxidation redox potential for the excited singlet of -1.18 V vs. SCE. The excited triplet potential would be expected to be no greater than 0.4-0.6 V less than the singlet, based on known Zn porphyrin triplet energies (26). Since these excited state levels are energetically higher than the SnO_2 conduction band, electron transfer to the semiconductor should be possible via either pathway to produce anodic photocurrents with these metalloporphyrin sensitizers.

A typical time response for a short-circuited photocurrent in the presence of hydroquinone (H_2Q) as an added solution redox species is shown in Figure 9. These photocurrents were stable for several hours. In the absence of H_2Q in the electrolyte, the photocurrent also increased rapidly upon the onset of illumination, but subsequently decayed exponentially to 70% of its initial value in a half-decay time of ca. 25 s. This behavior is similar to that observed for chlorophyll monolayers deposited on SnO_2 (12). Photocurrents under potentially-controlled conditions were also stable upon illumination, but exhibited slower decay characteristics when the light was turned off. This effect is unusual and is currently under further investigation.

A typical photocurrent action spectrum is illustrated in Figure 10 together with the excitation spectrum of the ZnTOAPP monolayer. The good correspondence between the two curves indicates that the dye is, in fact, the source of the photocurrents observed.

We are currently exploiting the unique ability to control the architecture in these monolayers to further investigate their photoelectrochemical properties with respect to such factors as light intensity, solution and film redox components, assembly structure, dye orientation, etc. We are particularly interested in using the monolayer structural information we now have to correlate film electronic properties with charge transfer effects at both the dye-solid and dye-liquid interfaces.

Acknowledgement

This work was supported by the University of Delaware Research Foundation and by the U.S. Department of Energy (Grant No. DE-FG02-79ER-10533).

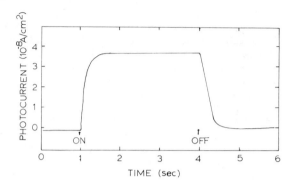

Figure 9.   Short-circuited photocurrent vs. time for a monolayer of ZnTOAPP (ZnTOAPP:SA, 1:4) directly on a SnO₂ OTE (0.1M KCl, pH 7.0, 0.05M H₂Q, N₂ purged)

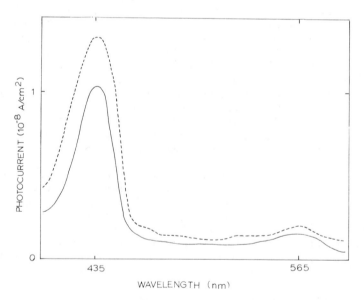

Figure 10.   Action spectrum of a monolayer of ZnTOAPP (ZnTOAPP:SA:octa-decane, 1:4:3) directly on a SnO₂ OTE: 0.1M KCl, pH 7.0, electrode potential is +0.3 V vs. SCE, N₂ purged; (– – –) the excitation spectrum (λ_{em}, 660 nm) of ZnTOAPP monolayer at SnO₂–electrolyte interface

Literature Cited

1. Wrighton, M.S. Acc. Chem. Res., 1979, 9, 303.
2. Gerischer, H. in "Physical Chemistry: An Advanced
    Treatise," Eyring, H.; Henderson, D.; Jost, W., Eds.;
    Academic Press, New York, 1970; Vol. 9A, Chapter 5.
3. Memming, R. in "Electroanalytical Chemistry," A.J. Bard, Ed.;
    Marcel Dekker, New York, Vol. 11; 1978.
4. Gerischer, H. Topics in Current Chemistry, 1976, 61, 31.
5. Memming, R. Photochem. Photobiol., 1972, 16, 325.
6. Spitler, M.T.; Calvin, M. J. Chem. Phys., 1977, 66, 4294.
7. Whitten, D.G.; Eaker, D.W.; Horsey, B.E.; Schmehl, R.H.;
    Worsham, P.R. Ber. Bunsenges. Phys. Chem., 1978, 82, 858.
8. Mercer-Smith, J.A.; Whitten, D.G. J. Am. Chem. Soc., 1979,
    101, 6620.
9. Collman, J.P.; Gagne, R.R.; Reed, C.A.; Halbert, T.R.; Lang,
    G.; Robinson, W.T. J. Am. Chem. Soc., 1975, 97, 1427.
10. Kuhn, H.; Mobius, D.; Bucher, H. "Physical Methods of
    Chemistry," Vol. 1; Part 3B; Weissburger, A.; Rossiter,
    B., Eds.; Wiley, New York, 1972; p. 588.
11. Bucher, H.; Elsner, O.; Mobius, D.; Tillmann, P.; Wiegand, J.
    Z. Phys. Chem., 1969, 65, 152.
12. Miyasaka, T.; Watanabe, T.; Fujishima, A.; Honda, K.
    J. Am. Chem. Soc., 1978, 100, 6657.
13. Meier, H. Topics in Current Chemistry, 1976, 61, 85.
14. Wang, J.H. Proc. Natl. Acad. Sci., U.S.A., 1969, 62, 653.
15. Adler, A.D. J. Polymer Sci. Part C, 1970, 29, 73.
16. Adler, A.D.; Varadi, V.; Wilson, N. Ann. N.Y. Acad. Sci.,
    1975, 244, 685.
17. Umezawa, Y.; Yamamura, T. J. Chem. Soc. Chem. Comm., 1978,
    1106.
18. Tang, C.W. U.S. Patent No. 4,164,431, August, 1979.
19. Umezawa, Y.; Yamamura, T. J. Electroanal. Chem., 1979, 95,
    113.
20. Gaines, G.L., Jr. "Insoluble Monolayers at Liquid-Gas
    Interfaces"; Wiley Interscience, New York, 1966.
21. Tanimura, K.; Kawai, T.; Sakata, T. J. Phys. Chem., 1979,
    83, 2639.
22. Tanimura, K.; Kawai, T.; Sakata, T. J. Phys. Chem., 1980,
    84, 751.
23. Kuhn, H. J. Photochem., 1979, 10, 111.
24. Mollers, F.; Memming, R. Ber. Bunsenges. Phys. Chem., 1975,
    76, 469.
25. Wolberg, A. Isr. J. Chem., 1975, 12, 1031.
26. Hopf, F.R.; Whitten, D.G. in "Porphyrins and Metalloporphy-
    rins," Smith, K.M., Ed.; Elsevier, New York, 1975; Chapter
    16.

RECEIVED October 3, 1980.

Effects of Temperature on Excited-State Descriptions of Luminescent Photoelectrochemical Cells Employing Tellurium-Doped Cadmium Sulfide Electrodes

ARTHUR B. ELLIS and BRADLEY R. KARAS

Department of Chemistry, University of Wisconsin—Madison, Madison, WI 53706

The need for alternate energy sources has led to the rapid development of photoelectrochemical cells (PECs). A PEC consisting of an n-type semiconductor, a counterelectrode, and a suitably chosen electrolyte can convert optical energy directly into chemical fuels and/or electricity (1,2,3,4). We recently reported that tellurium-doped CdS (CdS:Te) mimics undoped CdS in its ability to sustain the conversion of monochromatic ultraband gap light ($\geq$2.4 eV; $\lambda \lesssim 500$ nm (5)) into electricity at $\sim$7% efficiency in PECs employing aqueous polychalcogenide electrolytes (6,7,8,9). A novel feature of the CdS:Te photoanodes is that they emit ($\lambda_{max} \sim$600 nm for 100 ppm CdS:Te) with $\sim$0.1% efficiency while effecting the oxidation of polychalcogenide species.

Luminescence results from the introduction of intraband gap states by the substitution of Te for S in the CdS lattice. Because of its lower electron affinity, Te sites trap holes which can then coulombically bind an electron in or near the conduction band to form an exciton. Subsequent radiative collapse of this exciton leads to emission (10,11,12,13). In the context of the PEC, emission thus serves as a probe of electron-hole ($e^- - h^+$) pair recombination which competes with $e^- - h^+$ pair separation leading to photocurrent. Except for intensity, the emitted spectral distribution is found to be independent of the presence and/or composition of polychalcogenide electrolyte, excitation wavelength (Ar ion laser lines, 457.9–514.5 nm) and intensity ($\lesssim$30 mW/cm^2), and applied potential (-0.3V vs. SCE to open circuit) (6,7,8,9).

Optical penetration depth plays a significant role in the PEC properties we observe. The absorptivity of 100 ppm CdS:Te for 514.5 nm light is $\sim$10^3 cm^{-1} (5,10,11,12,13). Since the depletion region in which $e^- - h^+$ pairs are efficiently separated by band bending to yield photocurrent is $\sim$10^{-4}–10^{-5} cm thick (14), a significant fraction of 514.5 nm light is absorbed beyond this region. We therefore expect and observe greater emission intensity and smaller photocurrent with 514.5 nm excitation than with ultraband gap wavelengths (8) for which the CdS:Te

0097-6156/81/0146-0295$05.00/0

absorptivity is $\sim 10^4$-10^5 cm^{-1} (5,10,11,12,13). Additionally, whereas there is little potential dependence of emission intensity with 514.5 nm excitation, a strong dependence is observed with ultraband gap irradiation: in passing from -0.3V vs. SCE to open circuit in aqueous polychalcogenide electrolytes, increases in emission intensity of ~ 15-1400% obtain (6,7,8,9). This is in accord with the premise that variations of potential correspond to alterations in the degree of band bending (14).

Temperature is another PEC parameter which can potentially modify the efficiencies of photocurrent and luminescence. Among the materials whose temperature dependent PEC properties have been studied are SnO$_2$ (15), TiO$_2$ (16) and CuInS$_2$ (17). Undoped CdS has a known optical band gap temperature dependence of -5.2x10^{-4} eV/°K between 90 and 400 K (18). Owing to the general similarity of CdS:Te to CdS, we anticipated a comparable red shift in the onset of absorption. In this paper we summarize the results of temperature studies on CdS:Te-based PECs employing aqueous polyselenide (9) and sulfide electrolytes.

Experimental

Single-crystal plates of vapor-grown, 100 ppm CdS:Te were purchased from Cleveland Crystals, Inc., Cleveland, Ohio. Emissive spectral features were consistent with those previously reported for CdS:Te (6-13) and confirmed (Roessler's correlation, (12)) that the Te concentration was ≤ 100 ppm. The ~ 5x5x1 mm samples had resistivities of ~ 2 ohm-cm (four point probe method) and were oriented with the 5x5 mm face perpendicular to the c-axis. Samples were first etched with 1:10 (v/v) Br$_2$/MeOH and then placed in an ultrasonic cleaner to remove residual Br$_2$. The electrolyte was either sulfide, 1M OH$^-$/1M S^{2-}, or polyselenide, typically 5M OH$^-$/0.1M Se^{2-}/0.001M Se$_2^{2-}$; short optical pathlengths (≤ 0.1 cm) were used to make the latter essentially transparent for $\lambda \gtrsim 500$ nm. Electrode and electrolyte preparation as well as electrochemical and optical instrumentation employed have been described previously (8). Electrolytes were magnetically stirred and blanketed under N$_2$ during use.

Assembling the PEC inside the sample compartment of a spectrophotofluorometer permitted measurement of emission spectral data (200-800 nm; ~ 5 nm bandwidth). Front-surface emissive properties were recorded by inclining the photoelectrode at $\sim 45°$ to both the incident Coherent Radiation CR-12 Ar ion laser beam (501.7 or 514.5 nm) and the emission detection optics. In all experiments the ~ 3 mm dia. beam was 10X expanded and masked to fill the electrode surface; incident intensities were generally ≤ 10 mW/cm^2. Temperature of the PEC was adjusted as previously described (9).

Results and Discussion

Emissive and Photocurrent Properties. Three characteristics
of emissive PECs which might display thermal effects are the
emissive spectral distribution, emission intensity, and photocurrent. We observed the emission spectrum of 100 ppm CdS:Te in
both polyselenide and sulfide electrolytes from 20-100°C. Low-
resolution spectra of the various samples revealed red-shifts of
λ_{max} with increasing temperature of at most 5-10 nm; the displacement of λ_{max} for many samples, however, was within the bandwidth
of the spectrometer (∿5 nm). These results are in the range of
an extrapolation of Roessler's data (12) from which we would
predict a red-shift in λ_{max} of ∿7-11 nm between 20 and 100°C.
Parallel shifts have also been reported for CdS:Te absorption and
excitation spectra over subambient temperature ranges (13). These
spectra incorporate a low-energy band which distinguishes CdS:Te
from CdS and, in fact, masks the exact position of the CdS:Te band
gap (7,8,11,12,13). Several other features of the CdS:Te
emission spectrum are particularly relevant in the context of PEC
experiments. For a given temperature the spectral distribution
is independent of whether 501.7 or 514.5 nm excitation is used and
of electrode potential between +0.7 V vs. Ag (pseudoreference
electrode, PRE) and the onset of cathodic current. Such an
insensitivity to potential indicates that the energies of the
intraband gap Te states are affected by potential in the same
manner as the conduction and valence band energies.
    A very profound temperature effect was observed for the
emission intensity. Figure 1 presents an emission-temperature
profile at open circuit in sulfide electrolyte; the relative
invariance of the sample's spectral distribution with temperature
allowed us to monitor emission intensity at the band maximum.
Emission intensity was matched for 501.7 and 514.5 nm excitation
at 20°C using ∿17 times as much 501.7 nm intensity. Over the
20-100°C excursion emission intensity is seen to drop by factors
of ∿8 and 30 for 501.7 and 514.5 nm excitation, respectively.
These factors are consistent with the 10- to 20-fold decline
observed in polyselenide electrolyte; in those experiments there
also appeared to be little potential dependence of the results
(9). Similar thermal quenching data has been reported for dry
CdS:Te samples irradiated with UV light (10,12,13), electron
beams (11), and α particles (19). The temperature dependence of
the decline in emission intensity has been linked to the
ionization energy of the Te-bound hole, ∿0.2 eV (10,11,12,13,19).
    Given the inherently competitive nature of emission and
photocurrent, it should not be surprising that photocurrent was
generally observed to increase with temperature. Figure 2 is
a photocurrent-temperature profile obtained at +0.7 V vs. Ag (PRE)
in sulfide electrolyte; the light intensity and sulfide concentration employed ensured that photocurrents were limited by
excitation rate and not by mass transport, i.e., photocurrents

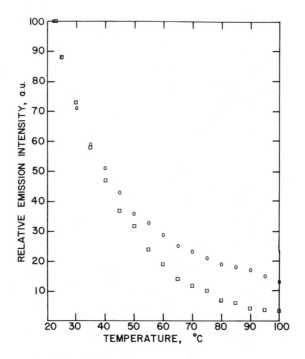

Figure 1.   Relative emission intensity monitored at 600 nm vs. temperature in 1M OH⁻/1M S²⁻ electrolyte of CdS:Te (100 ppm) excited at open circuit with 514.5 (□) and 501.7 nm (○) light in identical geometries. The excitation intensity at 501.7 nm is ∼ 17X that at 514.5 nm in order to match approximately room temperature emission intensities.

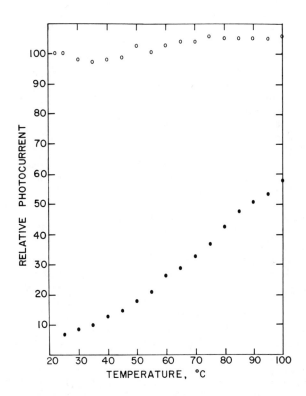

Figure 2. Relative photocurrent vs. temperature for the CdS:Te (100 ppm) photo-electrode of Figure 1 in 1M OH⁻/1M S²⁻ electrolyte excited in identical geometries with equivalent intensities (ein/s) of 514.5 nm (●) and 501.7 nm (○) light at +0.7 V vs. Ag (PRE). The scale is such that the 25°C, 501.7 nm photocurrent has been arbitrarily set to 100 and corresponds to a current density of ~ 0.38 mA/ cm² and a photocurrent quantum yield of ~ 0.66.

were not saturated with respect to light intensity and were not
affected by stirring. At 20°C for matching intensities (ein/sec)
∿15 times the photocurrent is observed with ultraband gap
501.7 nm excitation as with band gap edge 514.5 nm light. As
the temperature is increased to 100°C, a modest increase in
501.7 nm photocurrent is observed, while about an order of
magnitude growth is obtained for 514.5 nm excitation. In fact,
at 100°C the 514.5 nm photocurrent has reached ∿60% of the
ultraband gap photocurrent. Similar effects were observed in
polyselenide electrolyte for both CdS:Te and undoped CdS (9).

The photocurrent enhancement for 514.5 nm excitation is a
predictable consequence of an absorption edge which red shifts
with temperature: As the absorptivity for 514.5 nm light
increases with temperature, progressively larger fractions of
light will be absorbed in the depletion region. In this sense
the accelerated decline in emission intensity observed for
514.5 nm relative to 501.7 nm excitation (Figure 1) may include
a contribution reflecting decreasing optical penetration depth;
weaker emission would be expected as 514.5 nm light acquires
status as an ultraband gap wavelength, since near—surface
nonradiative recombination sites could play a more significant
role in excited-state deactivation (7,8). Although the rate
at which 514.5 nm photocurrent increases with temperature is
in reasonable accord with the CdS optical band gap temperature
coefficient, the effect of potential-dependent absorptivity must
be considered.

Electroabsorption measurements have been made on undoped CdS
at room temperature. They reveal that for electric fields of
∿10^5 V/cm, the change in absorptivity, $\Delta\alpha$, is ∿$+1\times10^3$ and -4×10^3
cm^{-1} at 515 and 500 nm, respectively (20). While this effect
helps to blur the discrepancy in 514.5 and 501.7 nm optical
penetration depths, the resultant absorptivities are still
sufficiently disparate relative to the width of the depletion
region to yield very different photocurrents at 295 K. To our
knowledge electroabsorption data are not presently available
for CdS:Te. A crude attempt to gauge the magnitude of this
effect at 295 K for 514.5 nm light suggests it is small (8). In
the absence of electroabsorption data over the 20-100°C range,
however, the electric field and thermal effects are not completely
decoupled and the results presented here should be so treated.

We should point out that the temperature effects on emission
intensity and photocurrent are completely reversible. Although
this result suggests that electrode stability obtains over the
duration of the experiments, the properties measured may not be
very sensitive to variations in surface or near-surface
composition. There is now considerable evidence, in fact, that
surface reorganization processes do occur in CdS- and CdSe- based
PECs in polychalcogenide electrolytes (17, 21-26). In particular,
the occurrence of such an exchange reaction for CdS:Te in polyselenide electrolyte would yield CdSe to whose lower band gap

(1.7 eV (27)) some of the observed properties could be attributed. The reversibility of the temperature effects as well as the similar behavior seen in sulfide electrolyte argue against such an explanation. Additionally, we have employed low light intensities ($\lesssim 10$ mW/cm^2) and current densities ($\lesssim 2$ mA/cm^2 with a total charge of generally $\lesssim 1$ C/cm^2) to further minimize exchange processes. But we do recognize that our techniques by no means rule out the possibility of surface reorganization processes at some level.

Simultaneous measurement of current, luminescence and voltage can be presented in iLV curves which summarize much of our data. We find that the ratio of open-circuit to in-circuit luminescence intensity (ϕ_{r_0}/ϕ_r) is a useful expression of the emission's potential dependence, with the in-circuit value taken at a potential where the photocurrent is saturated. Low values of photocurrent quantum yield, $\phi_x \lesssim 0.1$, occur with band gap edge excitation and yield ϕ_{r_0}/ϕ_r values close to unity. Ultraband gap excitation generally yields $0.5 \lesssim \phi_x \lesssim 1.0$. Pulsing the electrode between open circuit and the potential corresponding to saturated photocurrent easily demonstrates the non-unity value of ϕ_{r_0}/ϕ_r. Since photocurrent quantum yield increases markedly with temperature for band gap edge illumination, we predict that ϕ_{r_0}/ϕ_r will exceed unity at elevated temperatures. Figure 3 presents full iLV curves for a CdS:Te-based PEC employing polyselenide electrolyte. Equivalent intensities (ein/sec) of 501.7 and 514.5 nm light were employed at both room and elevated temperatures (49°C for 501.7 nm; 86°C for 514.5 nm). These iLV curves may be summarized as follows: Photocurrent at 23°C is ~ 18 times greater for 501.7 nm excitation (curve A vs. curve B) and open-circuit emission intensity is ~ 5 times smaller (curve A' vs. curve B') than for 514.5 nm light. The ratio of ϕ_{r_0}/ϕ_r is 1.0 for 514.5 nm (curve B') and 3.5 (curve A') for 501.7 nm illumination. Increasing the temperature to 49°C increases the photocurrent from 501.7 nm light by $\sim 15\%$ (curve C) and diminishes the luminescence intensity by a factor of 2. However, a similar ϕ_{r_0}/ϕ_r ratio of 3.4 obtains (curve C'). At 86°C the 514.5 nm photocurrent increases by a factor of almost 8 (curve D). Despite its approximately 10-fold drop in intensity, emission from 514.5 nm excitation now exhibits a potential dependence with a ϕ_{r_0}/ϕ_r value of 1.27 (curve D' - note 10X scale expansion). Similar non-unity ϕ_{r_0}/ϕ_r values were observed in sulfide electrolytes at temperatures exceeding ~ 80°C.

<u>Interrelationships of Excited-State Decay Routes.</u> The iLV curves conveniently display the competitive nature of photocurrent and luminescence intensity as excited-state deactivation pathways. Our analysis is limited in the sense that we have obtained absolute numbers for ϕ_x but have had to content ourselves with relative ϕ_r measurements. We lack measures of nonradiative recombination efficiency (ϕ_{nr}), although they now appear to be

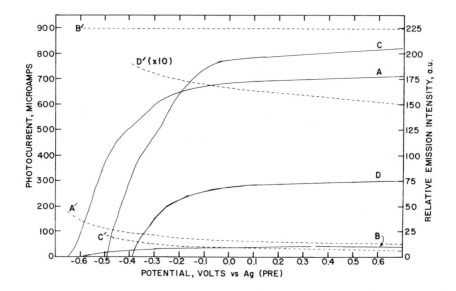

Journal of the Electrochemical Society

Figure 3. Current–luminescence–voltage (iLV) curves for a 100-ppm CdS:Te electrode in polyselenide electrolyte.

Unprimed, solid-line curves are photocurrent (left-hand scale) and primed, dotted-line curves are emission intensity (right-hand scale) monitored at $\lambda_{max} \sim 600$ nm. Curves A and A' result from excitation at 501.7-nm, 23°C; Curves B and B' from 514.5-nm, 23°C; Curves C and C', 49°C and 501.7-nm excitation; Curves D and D', 86°C, 514.5-nm irradiation. Note that the ordinate of Curve D' has been expanded by a factor of 10. Equivalent numbers of 501.7- and 514.5-nm photons were used to excite the photoelectrode in identical geometric configurations. The exposed electrode area is ~ 0.41 cm^2, corresponding to an estimated ϕ_x for 501.7-nm excitation at 23°C and +0.7 V vs. Ag (PRE) of ~ 0.50, uncorrected for solution absorbance and reflectance losses (9).

obtainable by the technique of photothermal spectroscopy ($\underline{28},\underline{29}$).
Treating these as the only possible decay routes yields eq. 1

$$\phi_x + \phi_r + \phi_{nr} = 1 \tag{1}$$

At open circuit $\phi_x = 0$ and photogenerated $e^- - h^+$ pairs are forced
to recombine. The ratio of ϕ_r to ϕ_{nr} is shown to be wavelength
dependent by comparing the open-circuit emission intensities of
curve A' and curve B' in Figure 3. This ratio is also temperature
dependent, as shown in Figure 1.
     Correlation of ϕ_x, ϕ_r, and ϕ_{nr} involves knowing the relative
extent to which nonradiatively and radiatively recombining $e^- - h^+$
pairs are prevented from recombining, i.e., their relative
contribution to photocurrent. Three possible schemes are:
photocurrent, ϕ_x, interconverts (1) exclusively with ϕ_r; (2)
exclusively with ϕ_{nr}; (3) with both ϕ_r and ϕ_{nr} such that
$\phi_{nr} = k\phi_r$ for any ϕ_x.
     Scheme 1 is unlikely because of the relative magnitudes of
ϕ_r and ϕ_x. Measured values of ϕ_r and ϕ_x are 10^{-3} and 10^{-1},
respectively, for band gap edge excitation ($\underline{8}$). In passing from
+0.7 V vs. Ag (PRE) to open circuit, a $\sim$100-fold increase in
emission intensity is predicted. This is inconsistent with the
insensitivity displayed in curve B', Figure 3. Scheme 2 argues
that ϕ_r will be independent of potential. While it appears that
curve B' (Figure 3) agrees with this, it is contradicted by
curves A',C', and D'. Although not perfect, Scheme 3 is most
compatible with our data. This assumption when combined with
eq. 1 leads to a simple relationship for monochromatic excitation
between ϕ_x and ϕ_{r_o}/ϕ_r:

$$\frac{\phi_{r_o}}{\phi_r} = \frac{1}{1 - \phi_x} \tag{2}$$

Table I consists of a compilation of ϕ_{r_o}/ϕ_r ratios as a function
of ϕ_x. Our results and those presented for p-GaP and n-ZnO are
in rough agreement with this simple model ($\underline{8},\underline{9},\underline{30},\underline{31},\underline{32}$).
Construction of a more refined model awaits incorporation of
other data (nonexponential lifetimes, electroabsorption,
carrier properties, intensity effects, quantitative evaluation
of ϕ_{nr} by photothermal spectroscopy, e.g.) and examination of
other systems.

Acknowledgment

     We are grateful to the Office of Naval Research for support
of this work. BRK acknowledges the support of the Electrochemical
Society through a Joseph W. Richard Summer Fellowship. David J.
Morano, Daniel K. Bilich, and Holger H. Streckert are thanked for
their assistance with some of the measurements.

Table I. Relationship Between ϕ_x and ϕ_{r_o}/ϕ_r [a] (9)

ϕ_x	ϕ_{r_o}/ϕ_r
0.001	1.00
0.01	1.01
0.05	1.05
0.10	1.11
0.20	1.25
0.30	1.43
0.40	1.67
0.50	2.00
0.60	2.50
0.70	3.33
0.80	5.00
0.90	10.00
1.00	∞

[a]Calculated from eq.2 where ϕ_x is the photocurrent quantum yield, and ϕ_{r_o}/ϕ_r is the ratio of emission quantum yields between open circuit and the potential where ϕ_x is measured.

Journal of the Electrochemical Society

Abstract

The effect of temperature on excited-state deactivation processes in a single-crystal, n-type, 100 ppm CdS:Te-based photoelectrochemical cell (PEC) employing aqueous polychalcogenide electrolytes is discussed. While serving as electrodes these materials emit ($\lambda_{max} \sim 600$ nm). Photocurrent (quantum yield ϕ_x) from ultraband gap ($\gtrsim 2.4$ eV; $\lambda \lesssim 500$ nm) 501.7 nm excitation increases modestly by $\lesssim 20\%$ between 20° and 100°C; photocurrent from band gap edge 514.5 nm excitation increases by about an order of magnitude, reaching ~ 50–100% of the room temperature 501.7 nm value. Highlighting the competitive nature of emission and photocurrent as excited-state decay processes, luminescence (quantum yield ϕ_r) declines over the same temperature regime by factors of between 10 and 30. At most, modest red shifts of λ_{max} (<10 nm) are observed in the spectral distribution of emission with temperature. These effects are discussed in terms of optical penetration depth, band bending, and the known red shift of the CdS absorption edge with temperature. Correlations involving ϕ_x and ϕ_r suggested by the data are discussed.

Literature Cited

1. Bard, A.J. Science, 1980, 207, 139.
2. Memming, R. Electrochim. Acta, 1980, 25, 77.
3. Wrighton, M.S. Acc. Chem. Res., 1979, 12, 303.
4. Nozik, A.J. Ann. Rev. Phys. Chem., 1978, 29, 189.
5. Dutton, D. Phys. Rev., 1958, 112, 785.

6. Ellis, A.B.; Karas, B.R. J. Am. Chem. Soc.,1979, 101, 236.
7. Ellis, A.B.; Karas, B.R. Adv. Chem. Ser., 1980, 184, 185.
8. Karas, B.R.; Ellis, A.B. J. Am. Chem. Soc., 1980, 102, 968.
9. Karas, B.R.; Morano, D.J.; Bilich, D.K.; Ellis, A.B.
    J. Electrochem. Soc., 1980, 127, 1144.
10. Aten, A.C.; Haanstra, J.H.; deVries, H. Philips Res. Rep.,
    1965, 20, 395.
11. Cuthbert, J.D.; Thomas, D.G. J. Appl. Phys., 1968, 39, 1573.
12. Roessler, D.M. J. Appl. Phys. 1970, 41, 4589.
13. Moulton, P.F., Ph.D. Dissertation, Massachusetts Institute
    of Technology, 1975.
14. Gerischer, H. J. Electroanal. Chem., 1975, 58, 263.
15. Wrighton, M.S.; Morse, D.L.; Ellis, A.B.; Ginley, D.S.;
    Abrahamson, H.B. J. Am. Chem. Soc., 1976, 98, 44.
16. Butler, M.A.; Ginley, D.S. Nature, 1978, 273, 524.
17. Heller, A.; Miller, B. Adv. Chem. Ser., 1980, 184, 215.
18. Bube, R.H. Phys. Rev., 1955, 98, 431.
19. Bateman, J.E.; Ozsan, F.E.; Woods, J.; Cutter, J.R. J. Phys.
    D. Appl. Phys., 1974, 7, 1316.
20. Blossey, D.F.; Handler, P. in "Semiconductors and Semimetals";
    Willardson, R.K., Beer, A.C.,Eds.; Academic Press, New York,
    1972; Vol. 9, Chapter 3 and references therein.
21. Noufi, R.N.; Kohl, P.A.; Rodgers, J.W. Jr.; White, J.M.;
    Bard, A.J. J. Electrochem. Soc., 1979, 126, 949.
22. Heller, A.; Schwartz, G.P.; Vadimsky, R.G.; Menezes, S.;
    Miller, B. ibid., 1978, 125, 1156.
23. Cahen, D.; Hodes, G.; Manassen, J. ibid., 1978, 125, 1623.
24. Gerischer, H.; Gobrecht, J. Ber. Bunsenges Phys. Chem.,
    1978, 82, 520.
25. Hodes, G.; Manassen, J.; Cahen, D. Nature (London), 1976,
    261, 403.
26. DeSilva, K.T.L.; Haneman, D. J. Electrochem. Soc., 1980,
    127, 1554.
27. Wheeler, R.G.; Dimmock, J.O. Phys. Rev., 1962, 125, 1805.
28. Fujishima, A.; Brilmyer, G.H.; Bard, A.J. in "Semiconductor
    Liquid Junction Solar Cells", Heller, A., Ed.; Proc. Vol.
    77-3: Electrochemical Society, Inc.: Princeton, N.J. 1977,
    p. 172.
29. Fujishima, A.; Maeda, Y.; Honda, K.; Brilmyer, G.H.; Bard,
    A.J. J. Electrochem. Soc., 1980, 127, 840.
30. Ellis, A.B.; Karas, B.R.; Streckert, H.H. Faraday Discuss.
    Chem. Soc., in press.
31. Beckmann, K.H.; Memming, R. J. Electrochem. Soc., 1969, 116,
    368.
32. Petermann, G.; Tributsch, H.; Bogomolni, R. J. Chem. Phys.,
    1972, 57, 1026.

RECEIVED October 3, 1980.

20

The Role of Ionic Product Desorption Rates in Photoassisted Electrochemical Reactions

HAROLD E. HAGER

Department of Chemical Engineering, University of Washington, Seattle, WA 98195

A rapidly growing interest in the photoelectrolysis of water has been evident in recent published research. The goal of this work is the development of a system which can directly convert solar energy to a storable chemical fuel (H_2) by driving an endo-ergic reaction ($H_2O \rightarrow H_2 + \frac{1}{2}O_2$).

Practical utilization of photo-assisted electrolysis systems is hampered by poor overall conversion efficiencies. Essentially, this problem results from poor quantum efficiencies at low bias.

The transient current response of photo-electrodes to stepped-illumination changes suggests itself as a method of mechanisti-cally interpreting this quantum efficiency problem. Though such transients have been studied for p-type GaP ([1]) and a number of n-type transition metal compounds ([2], [3], [4], [5], [6]), published explanations of the observed behavior are not well developed.

Below stepped-illumination experiments are presented for the photo-assisted electrolysis of water using n-type TiO_2 or SnO_2 photoanode/dark Pt cathode systems. An analysis of these results will be performed, focusing on the influence of the anodic half-cell reaction products upon the electronic state of the semi-conductor/electrolyte interface.

Experiment

The photo-assisted electrolysis current vs. time scans were obtained with the following experimental set-up:
three electrode cell, including:
a. Pt black reference electrode
b. smooth Pt counter electrode
c. n-type TiO_2 or SnO_2 working electrode
A 2½ inch diameter by 1/8" thick quartz plate was attached to the bottom of the cell, forming a UV window. A pen-ray quartz UV lamp and power supply were employed for the illumination source. The lamp was housed in a minibox fixed with a shutter (ILEX No. 3 Universal).

The potential difference between working and reference

0097-6156/81/0146-0307$07.50/0
© 1981 American Chemical Society

electrodes was controlled via a pontiostat (PAR Model 173) employed in conjunction with a universal programmer (PAR Model 175). Electrochemical current was determined by amplification (Keithly Model 160B digital multimeter) of the instantaneous voltage drop across a reference resistor. The multimeter output went to a differential amplifier module (Tektronix 5A20N), employed as the y-input for the current vs. time oscilloscope scans (Tektronix 5103N). The horizontal rate of sweep was controlled by a time base/amp module (Tektronic 5B10N).

The bias between the working and reference electrode was monitored by a second multimeter, or by direct scope measurement using a differential amplifier module.

The electrolyte solutions were made up from de-ionized water which was obtained by passing distilled water through an ion exchanger (Barnstead purifier with #D0809 cartridge). All acids used were reagent grade.

Two methods of preparing TiO_2 were employed. Technique I utilized the flame deposition of small TiO_2 particles, producing a uniform TiO_2 film on a Ti substrate. The TiO_2 was formed by passing $TiCl_4$ in an O_2 stream through the flame of a hydrogen/oxygen torch.

Alternatively, TiO_2 coated plates were produced by bringing a sheet of Ti foil into direct contact with a bunsen flame (7). This is designated technique II.

Following deposition, the TiO_2 coated plates were subjected to a reduction treatment under H_2 at 600°C for 2 hours. After verification of a good electrical contact at the back surface of the Ti sheet, a copper plate with an electrode lead was attached to the electrode back with silver paint. The electrode back and edges were covered with silicone rubber adhesive (GE RTV 108).

A chemical vapor deposition method was employed for the SnO_2 electrode preparation. A solution of 5 weight % $SnCl_4$ in ethanol was sprayed on a glass plate held at 450°C. The electrode contact was obtained by coating with silver paint along one edge of the SnO_2 film and then putting down a strip of copper foil of equal width. The contact was insulated by applying a coating of epoxy (Devcon 5-minute).

Experimental Results

Two types of experimental current scans were performed. After introducing the photoanode into solution, successive current vs. potential (i vs. ϕ) scans were performed at slow sweep rates (2mV/S). All of the photoanode materials studied exhibited an initial drift of the i vs. ϕ scans to more anodic bias. This occurred for both uninterrupted scans as well as intermittent scanning. The origin of this aging phenomena is not understood. However, it was observed that after approximately thirty minutes the i vs. ϕ scans achieved a profile which was reproduced during further cyclic sweeps. This result was taken as a criterion for proper electrode pre-conditioning.

Our emphasis in this paper will be to describe and analyze current vs. time scans determined after making stepped-changes in the illumination intensity. These scans were performed after proper photoanode pre-conditioning was indicated.

Typical experimental results are shown in Figure 1 for TiO_2 electrodes made by technique I. Figure 2 presents the results for a similar study with an SnO_2 photoanode. These scans were made in $1.0N$ H_2SO_4. Current vs. time profiles obtained with technique I TiO_2 photoanodes in electrolyte made up by introducing various additions of HCl to 120 ml of $1.0N$ H_2SO_4 exhibited generally similar behavior (Figure 3).

Overshoot of the steady state level is an intriguing feature of many of the transient current responses. This behavior occurs following both opening and closing the lightbox shutter. The following observations characterize this "overshoot" phenomena:

1.   The magnitude of the overshoot current, relative to the difference between the photo-assisted and dark current steady state levels, decreases with increasing bias.

2.   The timescale for decay of the overshoot current is dependent upon the illumination intensity, ψ, the electrolyte, and bias: (a)   the decay time decreases with increasing ψ.   (b)   For solutions containing H_2SO_4 and no HCl the decay time decreases with increasing bias.   (The decay time for solutions containing both H_2SO_4 and HCl pass through a minimum with increasing bias.) (c)   At constant ψ and bias, the decay time increases with increasing HCl addition.

Photoanode Response under Stepped-Illumination. In the following paragraphs, arguments are developed which ascribe the experimental observations to ionic adsorption at the photoanode/ electrolyte interface. The form of analysis was chosen to clearly demonstrate the role of ionic products upon the associated half-cell reaction charge transfer processes.

The anodic half-cell reaction occurring at the photoanode/ electrolyte interface may be written:

$$2H_2O + 4h^+ \rightleftarrows 4H^+ + O_2 \qquad\qquad [1]$$

Occurrence of the photo-assisted electrolysis reaction, from left to right, produces an increased H^+ concentration at the photoanode surface.

We have previously analyzed the photo assisted corrosion of n-type materials during photoanode applications (8). This work suggested that the H^+ ions produced during the electrolysis reaction underwent desorption from the photoanode surface much less readily than would be suspected for fully hydrated H^+ ions, and is summarized in Appendix I.

The following H^+ ion adsorption balance is proposed:

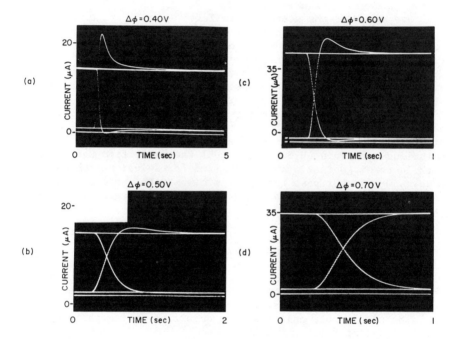

Figure 1.   Oscilloscope traces of current vs. time response following opening and closing of the lightbox shutter, Type I TiO₂ photoanode, 1.0N H₂SO₄ ((a) 0.4 V; (b) 0.5 V; (c) 0.6 V; (d) 0.7 V)

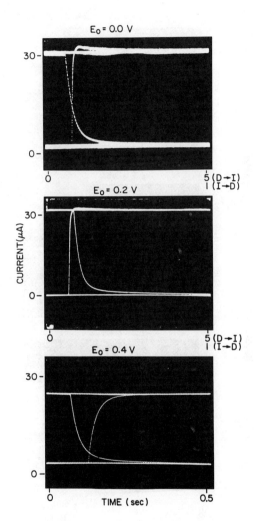

Figure 2. Oscilloscope traces of current vs. time response following opening and closing of the tlightbox shutter, SnO_2 photoanode, 1.0N H_2SO_4 ((top) 0.0 V; (middle) 0.2 V; (bottom) 0.4 V)

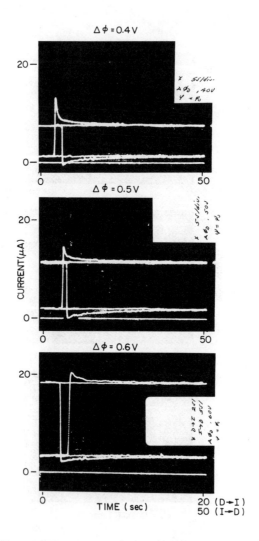

Figure 3. Oscilloscope traces of current vs. time response following opening and closing of the lightbox shutter, Type I TiO₂ photoanode, 120 mL 1.0N H₂SO₄ + 15 mL 1.0N HCl (top) 0.4 V; (middle) 0.5 V; (bottom) 0.6 V)

$$H^+ \text{ accumulation} = H^+ \text{ generation} - H^+ \text{ desorption} \qquad [2]$$

or

$$\frac{dC_{H^+_s}}{dt} = \nu_{H^+} - k_{des}(C_{H^+_s} - C^o_{H^+_s}) \qquad [3]$$

where $\nu_{H^+} = H^+$ generation rate = photo-assisted electrolysis reaction rate and $C^{o+}_{H_s}$ is the surface H^+_s concentration at $t = 0$ for a photoelectrode initially under dark steady state conditions.

Solving eqn. [3] with the initial condition

$$C_{H^+_s} = C^o_{H^+_s} \quad \text{at } t = 0 \qquad [4]$$

we obtain

$$C_{H^+_s} = C^o_{H^+_s} + \frac{\nu_{H^+}}{k_{des}} (1 - \exp[-k_{des}t]) \qquad [5]$$

or

$$\theta_{H^+} = \theta^o_{H^+} + \frac{\nu_{H^+}}{\Gamma k_{des}} (1 - \exp[-k_{des}t]) \qquad [6]$$

where Γ is the number of sites/cm^2.

Consider the electronic energy diagram for an n-type semiconductor, as shown in Figure 4. The total potential energy in the semiconductor conduction band at x, relative to the corresponding band energy deep in the bulk may be expressed as:

$$\Phi_{PE_{Tot}}(x) = \frac{qN_D}{\epsilon} (H - x)^2 - \frac{q^2}{4\epsilon x} + \Phi_B(x) \qquad [7]$$

where the first term on the right represents the conduction band energy at x, relative to the bulk level, as given from the solution of Poisson's equation. The second term on the right accounts for the image potential energy lowering. This contribution arises from the interaction between the electron, at position x, and its associated image charge, induced at -x, and is applicable only for very small ratios of semiconductor dopant concentrations to electrolyte concentrations, such that H is very much larger than the electrical double layer thickness in solution, d_s.

In practice, this limitation is not seriously restrictive. Moderate electrolyte concentrations ($\gtrsim$ 0.5N) are required by solution IR drop concerns, yielding d_s on the order of a few angstroms. Optimal semiconductor doping levels are dictated by developing H with the approximate value of α^{-1}, where α is the semiconductor absorption coefficient, and H is the depletion width. As a consequence, a typical order of magnitude for H is 1 micron (10^4 Å). Thus the ratio H/d_s

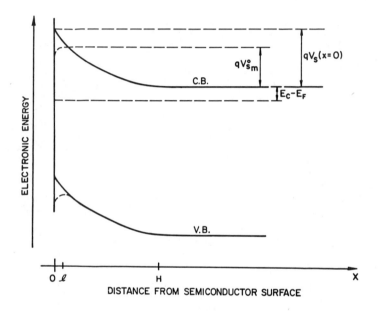

Figure 4.   Energy diagram of a Schottky barrier formed at an n-type semiconductor–electrolyte interface

has the order of magnitude 10^4, and the charge density distribution at the semiconductor/electrolyte interface closely resembles that for the semiconductor/metal interface of a Schottky barrier.

Finally, the term $\Phi_B(x)$ represents the potential energy of interaction between the electron at x and the net charge adsorbed at the electrode/electrolyte interface. In the following treatment we will assume that under the illumination conditions employed in the experiments reviewed above, only the interfacial H^+ concentration is appreciably altered by changes in the illumination level, ψ. Thus we neglect the photo-adsorption/desorption of other ionic species.

Appendix II delineates our approach for determining the adsorption potential energy, Φ_B. In essence, this was accomplished by breaking Φ_B into two terms, with one accounting for direct interaction with the central H^+ ion. The second term accounts for direct interaction between the electron and successive nearest neighbor H^+ ions. The final expression for the adsorption potential energy at the conduction band energy maximum, $x = \ell$, is given by:

$$\Phi_B(\ell) = \frac{-q^2}{(w\frac{\varepsilon}{\varepsilon'} + \ell)\varepsilon} - \frac{4q^2\theta_{H^+}^{\frac{1}{2}}}{\varepsilon r_{H^+}} \sum_{n=1}^{\infty} n^{-1} \left\{ 1.57 - \tan^{-1} \left[\frac{\ell + w\frac{\varepsilon}{\varepsilon'}}{nr_{H^+}/\sqrt{\theta_{H^+}}} \right] \right\} \quad [8]$$

where
ε = semiconductor dielectric constant
ε' = solution dielectric constant
w = location of adsorbed H^+
r_{H^+} = H^+ ion radius
θ_{H^+} = fractional H^+ ion coverage

The discrete charge model of equation [8] is suggested by combined capacitance-photocurrent vs. bias studies (9, 10). These results indicate that the adsorbed charge is localized, yielding a local potential distribution which can not be simply described by an averaged charge density approach, such as is obtained from solution of Poisson's equation.

The value of ℓ is determined from the condition:

$$\frac{d\Phi_{PE_{Tot}}}{dx} = 0 \quad [9]$$

Using relations [7] and [8] in eqn. [9] we obtain:

$$\frac{d\Phi_{PE_{Tot}}}{dx}\bigg|_{x=\ell} = \frac{q^2}{4\varepsilon\ell^2} + \frac{q^2}{(w\frac{\varepsilon'}{\varepsilon} + \ell)^2\varepsilon}$$

$$+ \frac{4q^2\theta_{H^+}^{\frac{1}{2}}}{\varepsilon r_{H^+}} \sum_{n=1}^{\infty} \left\{ n^{-1}(nr_{H^+}\sqrt{\theta_{H^+}})^{-1} \left[1 + \left[\frac{\ell + w\frac{\varepsilon'}{\varepsilon}}{nr_{H^+}\sqrt{\theta_{H^+}}} \right]^2 \right]^{-1} \right\} - \frac{2qN_D}{\varepsilon}(w - \ell) = 0 \qquad [10]$$

Evaluation of Photocurrent

Electrochemical processes which pass through a surface intermediate species will exhibit a pseudo-capacitive current response
(11). This form of behavior is observable in Figure 1(d).
Indeed, at high bias (Figure 1(d)) the transient current response
appears to be solely determined by this pseudo-capacitance, for
the capacitance values determined from these scans ($\sim 10^{-3}$ F) are
two orders of magnitude larger than typical semiconductor space
charge capacitances (12).

These observations, and application of Kirchoff's law, suggest that the time dependent electrochemical current density,
Je(t), may be written as the product of the fractional intermediate surface coverage, $\theta_{int}(t)$, and the difference between the
photo-generation and recombination current densities, $J_L(t)$ and
$J_r(t)$, respectively. Thus:

$$Je(t) = \theta_{int}(t)[J_L(t) - J_r(t)] \qquad [11]$$

where $J_L(t)$ is the total electron/hole pair generation current
flux minus the bulk recombination current, and $J_r(t)$ is the interfacial recombination current flux, normalized per unit area of
intermediate coverage.

The recombination current density, J_r, can be treated effectively as a Schottky barrier diode current density. Including
both thermionic emission and diffusion charge transport mechanisms
(13) J_r can be written as

$$J_r = \frac{qN_D\nu_r}{1 + \frac{\nu_r}{\nu_D}} \exp\left(-\frac{qV_{S_m}^o - (E_C-E_F)}{kT} \right) \left(\exp\frac{qV_{app}}{kT} - 1 \right) \qquad [12]$$

where ν_r is the recombination velocity. The maximum potential
difference between the conduction band in the space-charge region
and the bulk conduction band is expressed as $V_{S_m}^o$, and (E_C-E_F) is
the energy difference between the conduction and Fermi levels
in the bulk (see Figure 4). The applied voltage is designated by
V_{app}. Finally, ν_D is the effective diffusion velocity associated
with electron transport from the edge of the depletion region to
the top of the conduction band barrier, as defined (13) by

$$\nu_D = \left\{ \int_\ell^H \frac{q}{\mu kT} \exp\left[\Phi_{PE_{Tot}}(x) / kT \right] dx \right\}^{-1} \qquad [13]$$

where μ is the electron mobility.

In summary, we have developed eqns. [8], [10], and [13] for the evaluation of $\Phi_B(\ell)$, ℓ, and ν_D, respectively.

Two computer programs for determination of ℓ, ν_D, and $\Phi_B(\ell)$ are listed in Appendices II and III of reference [9]. The former accounts for adsorption potential energy effects. The latter neglects these contributions.

Table I presents the results of such computations for a broad range of semiconductor material parameters. The system parameters ν_{H^+} and k_{des} are taken from Figures 1(a), (b), and (c) corresponding to applied potentials of 0.4, 0.5, and 0.6V, respectively.

The potential energy barrier is lowered by accounting for the adsorption potential energy contribution. Note that though the absolute magnitude of the barrier lowering increases with increasing applied potential, the relative barrier lowering, $(\Delta PE)/PE_o$, decreases with higher bias (Table II). The latter result is in qualitative agreement with our experimental observation 1.

<u>Transient Response.</u> The current analysis performed above demonstrates that H^+ ion adsorption at the photoanode/electrolyte interface decreases the electronic energy barrier for electron transfer from the bulk conduction band to the electrode surface. We explore below the hypothesis that the observed overshoot current vs. time behavior arises from this effect. In essence, the time dependent change in H^+ produces an associated change in the barrier height for electron transfer to the semiconductor surface, altering the surface electron/hole recombination rate.

Determination of $\theta_{int}(t)$ is made with the following assumptions:

rate of intermediate formation = $K_f[1 - \theta_{int}(t)]$ $\qquad [14i]$

rate of intermediate destruction = $K_d \, \theta_{int}(t)$ $\qquad [14ii]$

A transient balance (Appendix III) on the intermediate yields:

$$\theta_{int}(t) = \frac{1 - e^{-\frac{K_{form}+K_{dis}}{\Gamma} t}}{1 + K_{dis}/K_{form}} \qquad [15]$$

The surface recombination current density, J_r, has the form:

$$J_r \propto e^{-\Phi_B/kT} \qquad [16]$$

TABLE I.   The Position of Maximum Potential Energy, ℓ, the Maximum Potential Energy Barrier, PE, and the Electron Diffusion Velocity, VD. Subscript o indicates H^+ Adsorption has been Neglected.

$\Delta\phi = 0.4V$

RN(1/cm³)	DC	ℓ_o(cm)	VD_o(cm/s)	PE_o(EV)	ℓ(cm)	VD(cm/s)	PE (EV)
5.0×10^{16}	10	7.6×10^{-7}	2.4×10^{3}	.278	1.9×10^{-6}	2.9×10^{3}	.139
5.0×10^{17}	10	4.4×10^{-7}	9.6×10^{3}	.187			
5.0×10^{18}	10	2.7×10^{-7}	3.6×10^{4}	.036			
5.0×10^{16}	50	4.9×10^{-7}	9.4×10^{2}	.363	1.2×10^{-6}	9.8×10^{2}	.319
5.0×10^{17}	50	2.8×10^{-7}	3.1×10^{3}	.335	6.8×10^{-7}	3.4×10^{3}	.257
5.0×10^{18}	50	1.6×10^{-7}	1.0×10^{4}	.285	4.0×10^{-7}	1.3×10^{4}	.151
5.0×10^{16}	100	4.1×10^{-7}	6.5×10^{2}	.378	9.8×10^{-7}	6.7×10^{2}	.352
5.0×10^{17}	100	2.3×10^{-7}	2.1×10^{3}	.361	5.5×10^{-7}	2.2×10^{3}	.315
5.0×10^{18}	100	1.3×10^{-7}	6.9×10^{3}	.331	3.1×10^{-7}	7.7×10^{3}	.250

$\Delta\phi = 0.5V$

RN(1/cm³)	DC	ℓ_o(cm)	VD_o(cm/s)	PE_o(EV)	ℓ(cm)	VD(cm/s)	PE (EV)
5.0×10^{16}	10	7.1×10^{-7}	1.9×10^{3}	.371	1.9×10^{-6}	2.4×10^{3}	.221
5.0×10^{17}	10	4.1×10^{-7}	6.8×10^{3}	.274	1.2×10^{-6}	1.0×10^{4}	.013
5.0×10^{18}	10	2.5×10^{-7}	3.8×10^{4}	.109			

TABLE I (Cont.)

$\Delta\phi = 0.5V$

RN(1/cm³)	DC	ℓ_0(cm)	VD_0(cm/s)	PE_0(EV)	ℓ(cm)	VD(cm/s)	PE(EV)
5.0×10^{16}	50	4.6×10^{-7}	7.7×10^{2}	.461	1.2×10^{-6}	8.2×10^{2}	.414
5.0×10^{17}	50	2.6×10^{-7}	2.5×10^{3}	.431	6.6×10^{-7}	2.8×10^{3}	.348
5.0×10^{18}	50	1.5×10^{-7}	8.8×10^{3}	.378	3.8×10^{-7}	1.0×10^{4}	.234
5.0×10^{16}	100	3.9×10^{-7}	5.4×10^{2}	.477	9.6×10^{-7}	5.5×10^{2}	.449
5.0×10^{17}	100	2.2×10^{-7}	1.7×10^{3}	.459	5.4×10^{-7}	1.8×10^{3}	.410
5.0×10^{18}	100	1.2×10^{-7}	5.3×10^{3}	.427	3.0×10^{-7}	6.2×10^{3}	.341

$\Delta\phi = 0.6V$

RN(1/cm³)	DC	ℓ_0(cm)	VD_0(cm/s)	PE_0(EV)	ℓ(cm)	VD(cm/s)	PE(EV)
5.0×10^{16}	10	6.7×10^{-7}	1.6×10^{3}	.465	1.9×10^{-6}	1.9×10^{3}	.304
5.0×10^{17}	10	3.9×10^{-7}	6.5×10^{3}	.362	1.1×10^{-6}	9.2×10^{3}	.077
5.0×10^{18}	10	2.3×10^{-7}	2.3×10^{4}	.188			
5.0×10^{16}	50	4.4×10^{-7}	6.5×10^{2}	.559	1.2×10^{-6}	6.8×10^{2}	.510
5.0×10^{17}	50	2.5×10^{-7}	2.1×10^{3}	.528	6.6×10^{-7}	2.3×10^{3}	.440
5.0×10^{18}	50	1.4×10^{-7}	7.0×10^{3}	.472	3.7×10^{-7}	8.7×10^{3}	.318
5.0×10^{16}	100	3.7×10^{-7}	4.5×10^{2}	.576	9.6×10^{-7}	4.6×10^{2}	.546
5.0×10^{17}	100	2.1×10^{-7}	1.4×10^{3}	.557	5.4×10^{-7}	1.5×10^{3}	.504
5.0×10^{18}	100	1.2×10^{-7}	4.7×10^{3}	.523	3.0×10^{-7}	5.3×10^{3}	.432

TABLE II. Relative Barrier Lowering for the
Conditions of Table I, with
$RN = 5.0 \times 10^{16}/cm^3$.

ε	bias	Δ PE	$\dfrac{\Delta \text{ PE}}{\text{bias}}$	$\dfrac{\Delta \text{ PE}}{\text{PE}_o}$
10	0.4	.139	.348	.50
	0.5	.15	.300	.40
	0.6	.161	.268	.35
50	0.4	.044	.11	.121
	0.5	.047	.094	.102
	0.6	.049	.082	.088
100	0.4	.026	.065	.069
	0.5	.028	.056	.059
	0.6	.030	.050	.052

Using eqn. [8], and the approximation $\tan^{-1}x = 57.29\,x$ for $x \ll 1$, we can write:

$$J_r \propto e^{b\theta_{H^+}/kT} \qquad [17]$$

for $\theta_{H^+} \ll 1$. Here b is a constant.

Under the conditions of the experiments reviewed above, b is of order 1eV. At room temperature $kT = 0.0256$. Thus for $\theta_{H^+} \ll 0.0256$

$$e^{b\theta_{H^+}/kT} \approx 1 + \frac{b\theta_{H^+}}{kT}$$

and eqn. [17] may be rewritten as:

$$(J_r - J_{r_o}) \propto \theta_{H^+} \qquad [18]$$

where J_{r_o} is the value of J_r in the limit $\theta_{H^+} \to 0$. Let us assume that $\theta_{H^+}(t=0) = 0$. Noting that J_{r_o} is independent of t, application of eqn. [6] gives:

$$J_r(t) = J_{r_{ss}} - \Delta J_{r_{ss}} e^{-k_{des}t} \qquad [19]$$

where $\Delta J_{r_{ss}}$ is the difference between the steady state recombination current density, $J_{r_{ss}}$, and the recombination rate at zero coverage, J_{r_o}.

Substituting relations [15] and [19] in eqn. [11],

$$J_e(t) = \left\{\frac{1 - e^{-Kt}}{1+K'}\right\}\left[J_L - (J_{r_{ss}} - \Delta J_r e^{-k_{des}t})\right] \qquad [20]$$

where $K = K_{form} + K_{dis}$

and $K' = K_{dis}/K_{form}$

Collecting terms in eqn. [20] we obtain:

$$J_e(t) = \frac{1}{1+K'}\left\{J_L - J_{r_{ss}} + e^{-Kt}\left(J_{r_{ss}} - J_L + \Delta J_r e^{-k_{des}t}\right) + \Delta J_r e^{-k_{des}t}\right\} \qquad [21]$$

or

$$\frac{J_e(t)}{\beta / (1+K')} = \left\{1 - e^{-Kt}\left(1 + \frac{\delta}{\beta} e^{-k_{des}t}\right) + \frac{\delta}{\beta} e^{-k_{des}t}\right\} \qquad [22]$$

where $\delta = \Delta J_r/J_{r_{ss}}$ and $J_L - J_{r_{ss}} = \beta$.

Figure 5 shows computer generated plots of $\dfrac{J_e(t)}{\beta \,/\, (1+K')}$ vs.

t for the bias conditions of Figures 1(a),(b), and (c) respectively. The parameter k_{des} was taken from the appropriate figure, leaving K and the ratio δ/β as two unknown variables. Values for these parameters were chosen to give the best agreement between the form of the computed scan and the associated experimental plot.

  Figure 5 demonstrates that the shape of the transient current response following opening of the shutter can be reasonably reproduced by the model developed above, except for the very short time limit. This discrepancy may arise from failure to account for capacitance effects in the current monitoring circuit.

  Further efforts to analyze the experimental steady state J_e levels confront a series of problems. Essentially, these arise from the need to assign values to a number of semiconductor/ solution properties (e.g., μ, ε, ε', and θ_{H^+}) with insufficient available information to make these assignments. Methods of experimentally determining these unknown parameters are being explored.

Relation to Other Work

  Bokris and Uosaki (1) have studied transient photo–assisted electrolysis current for systems including a p-type semiconductor photocathode and dark Pt anode. A set of current vs. time scans taken with a ZnTe photocathode system is shown in Figure 6.

  The results of Figure 6 exhibit the same general behavior as our experimental scans reported above: the magnitude of the overshoot current relative to the difference between the steady state dark and illuminated current levels decreases with increasing bias (or decreasing photocathode potential).

  We suggest that these results arise from adsorption effects which are the cathodic complements to the anodic phenomena outlined above. The electrolysis half-cell reaction at the photocathode:

$$2H_2O + 2e^- \rightleftharpoons H_2 + 2OH^- \qquad\qquad [23]$$

increases the local OH^- concentration at the photocathode/electrolyte interface, just as the anodic half-cell reaction produces an increase in the local H^+ concentration at the photoanode/electrolyte interface.

  As recombination current in a photocathode was treated as an electron current to the photoanode surface, the photocathode recombination current may be viewed as a hole current. Correspondingly, OH^- ions at the photocathode/electrolyte interface lower the energy barrier for hole transport to the photocathode surface.

  Figure 7 displays computer generated plots of $J_e(t)/\beta(1 + K')$ vs. t for comparison with the transient current responses reported

by Bockris and Uosaki (1). The normalized current expression
employed is the cathodic equivalent to the anodic case outlined
above.

Summary

    We have performed an experimental study of photo-assisted
electrolysis for illuminated n-type TiO_2 photoanode/dark Pt
cathode systems. Analysis of these results indicates that the
electronic state of the semiconductor/electrolyte interface is
influenced by the electrolysis reaction products, in a manner not
previously accounted for.
    Specifically, the oxidation half-cell reaction:

$$2H_2O + 4h^+ \rightleftarrows O_2 + 4H^+$$

alters the local H^+ ion concentration at the photoanode/electro-
lyte interface. A generalized model has been presented demonstrating that the interfacial ion concentration strongly affects transport processes associated with the space charge recombination
current. Our results show that the H^+ ion coverage decreases the
energy barrier for electron transport to the photoanode surface,
increasing the recombination current.
    A review of photo-assisted electrolysis studies performed
with p-type semiconductor photocathode/dark Pt anode systems suggests that a complementary phenomena arising from the presence of
OH^- ions produced during the reduction half-cell reaction,

$$2H_2O + 2e^- \rightleftarrows H_2 + 2OH^-$$

augments the hole current to the electrode/electrolyte interface,
again increasing the surface recombination current.
    These conclusions are significant and of practical concern.
Ionic products formed during both the anodic and cathodic electrolysis half-cell reactions alter the local net charge at the
electrode/electrolyte interface so as to increase the recombination currents. Hence for both photo-assisted half-cell reactions
the electrochemical current is reduced by this phenomena. The
relative size of this effect is larger at lower bias, indicating
that this phenomena presents a serious dilemma to workers designing
photo-assisted conversion systems with high overall energy conversion efficiencies.

Acknowledgements

    The support provided by the Department of Energy and the
National Science Foundation is gratefully acknowledged.

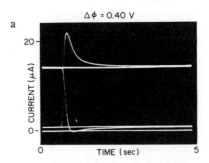

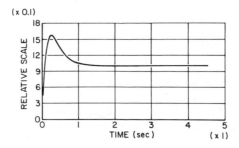

Figure 5. *Comparison between the experimental photocurrent vs. time profile and Equation 22: (a) 0.4 V; (b—facing page) 0.5 V; (c—facing page) 0.6 V*

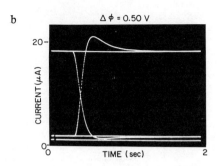

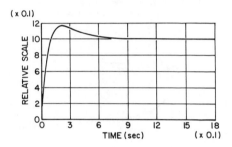

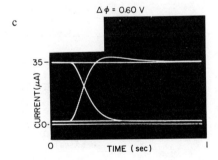

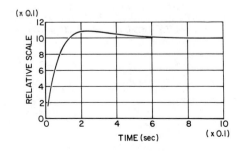

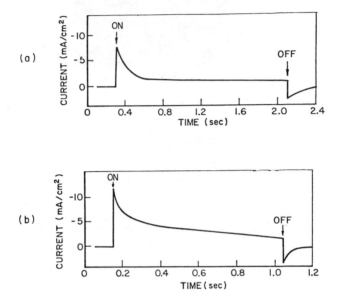

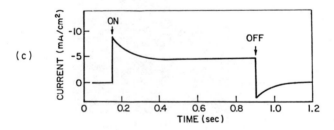

Journal of the Electrochemical Society

Figure 6.   Traces of the current vs. time response following opening and closing of the lightbox shutter, ZnTe photocathode, 1.0N NaOH (1): (a) −0.95 V (NHE); (b) −1.15 V (NHE); (c) −1.36 V (NHE)

Appendix I

Consider a photo-assisted water electrolysis cell, incorporating a photoanode and dark metal cathode. Illumination of the n-type semiconductor photoanode with a depletion space charge region results in a net flow of positive vacancies, or holes, to the semiconductor/electrolyte interface. Here the hole (h^+) may be accepted by the reduced form of the oxygen redox couple.

Alternatively, redox levels of the semiconductor material may lie at energies such that the hole, directly or through a multi-step process described below, accepts an electron from a semiconductor surface atom, producing concommitant surface oxidation.

A full description of photo-assisted semiconductor corrosion is not available. We have undertaken an analysis of this phenomena, relying on aqueous solution electrochemical equilibria considerations (16).

It has been experimentally observed that GaP is stable in the dark in weakly acidic solutions (15). This stability has been ascribed (15) to the formation of Ga_2O_3 at the semiconductor surface, with this oxide film effectively passivating the surface.

Electrochemical equilibria analysis (16) indicates that Ga_2O_3 is stable in aqueous electrolytes with $3.0 < pH < 11.2$ up to potentials appreciably more anodic than the potential associated with the oxidation of H_2O to O_2 (Figure 8). Yet upon illumination, n-type GaP employed as a photoanode in an aqueous solution at pH = 4.7 is unstable (17). Thus, under illumination, semiconductor photoelectrodes may corrode even though the bulk pH is insufficiently acidic/basic to drive such a lattice dissolution process.

We propose that this observed photo-assisted corrosion occurs via the following mechanism:

[1] Photo-produced holes, h^+, arrive at the electrode surface, oxidizing H_2O to O_2:

$$2H_2O + 4h^+ \rightarrow O_2 + 4H^+$$

[2] The production of two H^+ ions, for every molecule of water oxidized, promotes a surface H^+ coverage which is substantially higher than the dark value.

[3] For H^+ surface coverages larger than that obtained in the dark at a pH of 3.0, Ga_2O_3 is no longer stable, undergoing dissolution via the following reaction:

$$Ga_2O_3 + 6H^+ \rightarrow 2Ga^{+++}(sol'n) + 3H_2O$$

Indeed, our analysis suggests that the H^+ ions produced during the electrolysis reaction undergo removal from the photoanode surface region much less readily than would be expected for simple diffusion. A possible, more complete, explanation may be that the H^+ ion is produced on the surface with an incomplete hydration sheath, facilitating a direct H^+ ion/electrode

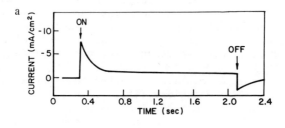

-0.95 V (NHE)

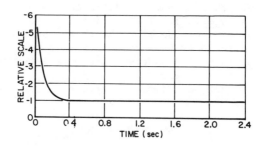

Figure 7.   *Comparison between the experimental photocurrent vs. time profile and the cathodic equivalent of Equation 22: (a)* −0.95 *V (NHE); (b*—facing page) −1.15 *V (NHE); (c*—facing page) −1.36 *V (NHE)*

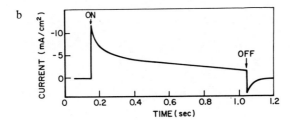

−1.16 V (NHE)

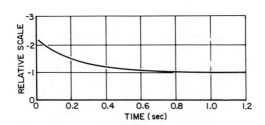

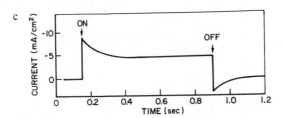

−1.36 V (NHE)

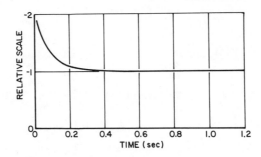

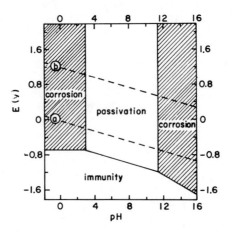

Figure 8.   Theoretical conditions of corrosion, immunity, and passivation of gallium, at 25°C, assuming passivation by a film of α-Ga₂O₃ (16)

interaction. Desorption of the H^+ ion can then occur only by sufficiently overcoming the electrode/H^+ ion interaction energy to draw the H^+ ion away from the electrode surface. This permits insertion of (a) water molecule(s) between the H^+ ion and electrode, completing the hydration sheath, prior to desorption of the fully hydrated ion.

The oxide redox energy levels for all elements except gold are cathodic to the redox level of the H_2O/O_2 couple. Gold, however, is an impractical component for compound semiconductors. All other compound semiconductors employed as electrolysis photoanodes will undergo surface oxidation in aqueous electrolytes to produce a surface oxide film which normally constitutes the stable surface of the photoelectrode. Our proposed mechanism indicates that the proton induced oxide dissolution reaction arises from product H^+ ion interactions with the oxide anion ($O^=$).

The arguments just presented suggest that all n-type semiconductor photoanodes which resist corrosion under conditions of H_2O to O_2 oxidation must have (a) stable elemental component(s) at low pH. Moreover, since the oxide dissolution reaction is associated with essentially the same $H^+/O^=$ interactions, the low pH stability limit will be at approximately the same value ($pH_{stable} \gtrsim 3$).

Pourbaix (16) has prepared theoretical stability diagrams of potential vs. pH for many common metals and nonmetalloids. A review of these results indicates that semiconductor compounds of Au, Ir, Pt, Rd, Ru, Z_r, Si, Pd, Fe, Sn, W, Ta, Nb, or Ti should serve as relatively acid-stable photoanodes for the electrolysis of water. Indeed, all of the stable photo-assisted anode materials reported in the literature, as of March, 1980 (see Table III) contain at least one element from this stability list, with the exception of CdO. Kung and co-workers (18) observed that the CdO photoanode was stable at a bulk pH of 13.3. The Pourbaix diagram for Cd (16) shows that an oxide film passivates Cd over the concentration range 10.0 < pH < 13.5. Hence the desorption of the product H^+ ion for the particular case of CdO must be exceptionally facile; without producing an effective surface pH lower than 10.0. This anamolous behavior for CdO is not well understood.

Appendix II

Determination of Adsorption Potential Energy

$$\Phi_B = \Phi_B \begin{pmatrix} \text{direct interaction} \\ \text{with central } H^+ \text{ ion} \end{pmatrix} + \Phi_B \begin{pmatrix} \text{direct interaction with} \\ \text{successive nearest neighbors} \end{pmatrix}$$

$$\Phi_B = \int_{-\infty}^{-\ell} [- \frac{q^2}{(w \frac{\epsilon'}{\epsilon} - x)^2 \epsilon} - \frac{4q^2}{\epsilon} \sum_{\substack{n=\sqrt{j^2+m^2} \\ j=1 \\ m=0}}^{\substack{m=\infty \\ j=\infty}} \{(w \frac{\epsilon'}{\epsilon} - x)^2 + n^2 R^2)^{-1}\}] \, dx \tag{1}$$

TABLE III. Stable Water Electrolysis Photoanodes

Material	Band gap energy (eV)	References
TiO_2	~3.0[*]	18–23
Fe_2O_3	2.2	20,23
WO_3	~2.6[**]	18,20
$SrTiO_3$	3.2	18,24
$BaTiO_3$	3.3	18,25
$CaTiO_3$	3.4	26
Fe_2TiO_5	2.2	27
$KTaO_3$	3.5	28,29
$Hg_2Ta_2O_7$	1.8	18
$Hg_2Nb_2O_7$	1.8	18
$KTa_{0.77}Nb_{0.23}O_3$	3.2	29
$PbTi_{1.5}W_{0.5}O_{6.5}$	2.4	18
Nb_2O_5	3.4	30
$YFeO_3$	2.6	31
$PbFe_{12}O_9$	2.3	18
$CdFe_2O_4$	2.3	18
ZrO_2	5.0	18
CdO	2.1	18
SnO_2	~3.6[***]	29,32

[*]   reported E_g 2.9 – 3.2
[**]   reported E_g 2.4 – 2.3
[***]   reported E_g 3.5 – 3.3

where
$$\begin{aligned}
\epsilon &= \text{semiconductor dielectric constant} \\
\epsilon' &= \text{solution dielectric constant} \\
w &= \text{location of adsorbed } H^+ \\
R &= \text{average separation between adsorbed } H^+ \text{ ions} \\
-\ell &= \text{distance to conduction band energy maximum}
\end{aligned}$$

Integrating (1)

$$\Phi_B = \frac{-q^2}{(w\frac{\epsilon'}{\epsilon}+\ell)\epsilon} - \int_{-\infty}^{-\ell} \frac{4q^2}{\epsilon} [\Sigma_n \{(w\frac{\epsilon'}{\epsilon})^2 + n^2R^2 - 2w\frac{\epsilon'}{\epsilon}x + x^2\}]^{-1} dx \tag{2}$$

From ref. 14, equation 109

$$\int (c+bx+ax^2)^{-1} dx = \frac{2}{\sqrt{g}} \tan^{-1}(\frac{2cx+b}{\sqrt{g}})$$

where $g = 4ac - b^2$.
    Thus (2) may be reduced to:

$$\Phi_B = \frac{-q^2}{(w\frac{\epsilon'}{\epsilon}+\ell)\epsilon} - \frac{4q^2}{\epsilon} \sum_n^{n=\infty} \frac{2}{\sqrt{g}} \tan^{-1}(\frac{2x-2w\frac{\epsilon'}{\epsilon}}{\sqrt{g}}) \Big|_{-\infty}^{-\ell} \tag{3}$$

where

$$g = 4[(w\frac{\epsilon'}{\epsilon})^2 + n^2R^2] - 4(w\frac{\epsilon'}{\epsilon})^2 = 4n^2R^2$$

or

$$\Phi_B = \frac{-q^2}{(w\frac{\epsilon'}{\epsilon}+\ell)\epsilon} - \frac{4q^2}{\epsilon}\frac{1}{R}\sum_{n=1}^{\infty} n^{-1}\left\{\tan^{-1}(\frac{-2\ell-2w\frac{\epsilon'}{\epsilon}}{2nR}) - \tan^{-1}(-\infty)\right\} \tag{4}$$

Now $-\tan^{-1}(-\infty) = \tan^{-1}(\infty) = \frac{90°}{360°} * 2\Pi = 1.57$

Thus we can write

$$\Phi_B = \frac{-q^2}{(w\frac{\epsilon'}{\epsilon}+\ell)\epsilon} - \frac{4q^2}{\epsilon R}\sum_{n=1}^{\infty} n^{-1}\left\{1.57+\tan^{-1}(\frac{-\ell-w\frac{\epsilon'}{\epsilon}}{nR})\right\} \tag{5}$$

Assuming a square adsorption pattern

$$R = r_{H^+}\Theta_{H^+}^{-\frac{1}{2}}$$

where r_{H^+} = H^+ ion radius

$$\Theta_{H^+} = \text{fractional } H^+ \text{ ion coverage}$$

employing these relations in (5) we obtain

$$\Phi_B = \frac{-q^2}{(w\frac{\varepsilon'}{\varepsilon}+\ell)\varepsilon} - \frac{4q^2\;\Theta_{H^+}^{\frac{1}{2}}}{\varepsilon\;r_{H^+}}\;\sum_{n=1}^{\infty}\;n^{-1}\;1.57+\tan^{-1}\;(\frac{-\ell-\;w\frac{\varepsilon'}{\varepsilon}}{nr_{H^+}/\sqrt{\Theta_{H^+}}}) \tag{6}$$

Appendix III

Time Dependence of Fractional Intermediate Surface Coverage

$$J_e(t) \;=\; \Theta_{intermediate}\;(t)\;*\;(J_L(t)\;-\;J_r(t))$$

rate of adsorption = $K_{form}\;(1 - \Theta_{interm.})$

rate of desorption = $K_{dis}\;\Theta_{interm.}$

Balance:

Accumulation = adsorption - desorption

$$\Gamma\;\frac{d\;\Theta_{interm}}{dt}\;=\;K_{form}\;(1-\Theta_{interm.})\;-\;K_{dis}\;\Theta_{interm.}$$

$$=\;K_{form}\;-\;(K_{form}\;+\;K_{dis})\;\Theta_{interm.}$$

$$\Rightarrow\;\frac{-\Gamma}{K_{form}+K_{dis}}\;\ell n\;[K_{form}\;-\;(K_{form}+K_{dis})\;\Theta_{interm}]\;=\;t\;+\;c$$

Initial Condition $\Theta = 0$ at $t = 0$

thus $C = \dfrac{-\Gamma}{K_{form}+K_{dis}}\;\ell n\,(K_{form})$

$$\Rightarrow\;\frac{-\Gamma}{K_{form}+K_{dis}}\;\ell n\;[\frac{K_{form}-(K_{form}+K_{dis})\;\Theta_{interm}}{K_{form}}]\;=\;t$$

or $1 - (1 + K)\Theta_{interm.}\;=\;e^{-(\frac{K_{form}\;+\;K_{dis}}{\Gamma})t}$

Thus finally: $\Theta_{interm.} = \dfrac{1 - e^{-(\frac{K_{form} + K_{dis}}{\Gamma})t}}{1 + K'}$

where $K' = \dfrac{K_{dis}}{K_{form}}$

Abstract

A growing awareness that photo-assisted electrochemical systems may provide the best device technology for some direct solar energy conversion applications is manifested by the rapidly expanding literature in this area. Nevertheless, poor device performance remains a major problem in this research. This is particularly true for many of the photogenerative systems which have been studied. The most widely investigated photogenerative reaction has been the photo-electrolysis of water. Growing evidence suggests that unfavorable photo-assisted water electrolysis cell performances arise, at least in part, from product ion adsorbate effects. The latter appears to contribute both to materials stability and photon utilization efficiency vs. bias problems. This product ion phenomena should also be of importance, in general, for any system incorporating photoelectrodes which utilize depletion space charge fields.

"Literature Cited"

1. Bockris, J. O'M.; Uosaki, K. J. Electrochem. Soc., 1977, 124, 1348.
2. Hardee, K. S.; Bard, A. F. J. Electrochem. Soc., 1977, 123, 1025.
3. Mavroides, J. G. "The Electrochemical Society Softbound Symposium Series", PV 77-3; Princeton, N.J., 1977; p. 84.
4. Pavlov, D.; Zanova, S.; Papazov, G. J. Electrochem. Soc., 1977, 124, 1523.
5. Jeh, L.-C. R.; Hackerman, N. J. Electrochem. Soc., 1977, 124, 833.
6. Curran, J. S.; Gissler, W.; J. Electrochem. Soc., 1979, 126, 56.
7. Fujishima, A.; Kohayakawa, K.; Honda, K. J. Electrochem. Soc., 1975, 122, 1488.
8. Hager, H. E.; Ollis, D. F. Abstract 247, The Electrochemical Society Extended Abstracts, Fall Meeting, Pittsburgh, Pennsylvania, Oct. 15-20, 1978.
9. Hager, H. E., Ph.D. Thesis, Princeton University, 1979.
10. Wilson, R. H. Personal communication.
11. Bockris, J. O'M. "Modern Electrochemistry" V2; Plenum Press: New York, 1970.

12. Myamlin, V.; Pleskov, Y. "Electrochemistry of Semiconductors"; Plenum Press: New York, 1967.
13. Crowell, C. R.; Sze, S. M. Solid State Electronics, 1966, 9, 695.
14. Weast, R. C., Ed. "Handbook of Chemistry and Physics"; 54th Edition, CRC Press: Cleveland, Ohio, 1973.
15. Memming, R.; Schwandt, G. Electrochim. Acta, 1968, 13, 1299.
16. Pourbaix, M. "Atlas of Electrochemical Equilibria in Aqueous Solutions"; Pergamon Press: New York, 1966.
17. Nakato, Y.; Abe, K.; Tsubomura, H. Ber. Bunsen-Gesellschaft, 1976, 80, 1002.
18. Kung, H.; Jarrett, H.; Sleight, A.; Ferretti, A. J. Appl. Phys. 1977, 48, 2463.
19. Mavroides, J.; Tcherner, D.; Kafalos, J.; Kolesar, D. Mat. Res. Bull., 1975, 10, 1023.
20. Hardee, K.; Bard, A. J. Electrochem. Soc., 1977, 124, 215.
21. Tiyishima, A.; Honda, K. Nature, 1972, 238, 37.
22. Wrighton, M.; Ginley, D.; Wolczanski, P.; Ellis, A.; Morse, D.; Linz, A. Proc. Nat. Acad. Sci. USA, 1975, 72, 1518.
23. Hardee, K.; Bard, A. J. Electrochem. Soc., 1976, 123, 1024.
24. Wrighton, M.; Ellis, A.; Wolczanski, P.; Morse, D.; Abrahamson, H.; Ginley, D. J. Am. Chem. Soc., 1976, 98, 2774.
25. Kennedy, J.; Ftese, K. J. Electrochem. Soc., 1976, 123, 1683.
26. Tchernov, D. I. Abstract C3,"Int. Conf. Photochem. Conv. Storage Solar Energy", London, Ontario, 1976.
27. Ginley, D.; Butler, M. J. Appl. Phys., 1977, 48, 2019.
28. Ellis, A.; Kaiser, S.; Wrighton, M. J. Phys. Chem., 1976, 80, 1325.
29. Bolta, J.; Wrighton, M. J. Phys. Chem., 1976, 80, 2641.
30. Clechet, P.; Martin, J.; Oliver, R.; Vallouy, O. C. R. Acad. Sci., 1976, C282, 887.
31. Butler, M.; Ginley, D.; Eibschutz, M. J. Appl. Phys., 1977, 48, 3070.
32. Kim, H.; Laitenen, H. J. Electrochem. Soc., 1975, 122, 53.

RECEIVED October 3, 1980.

Carbanion Photooxidation at Semiconductor Surfaces

MARYE ANNE FOX and ROBERT C. OWEN

Department of Chemistry, University of Texas at Austin, Austin, TX 78712

Within the last decade, many inventive procedures for the quantum utilization of visible light by organic or inorganic absorbers have been developed. The most efficient of these often involve electron exchange reactions. Here, a vexing problem persists: rapid thermal recombination of the electron-hole pairs can regenerate the ground state and effectively waste the energy of the absorbed photon. We have reasoned that this back reaction could be inhibited, or at least dramatically slowed, if an excited anion M^- (eqn 1) were used as the donor rather than a neutral molecule M (eqn 2). The reversal of equation (1), governed only by the typically low electron affinity of the photoproduced

$$M^- \xrightarrow{\ h\nu\ } M\cdot + e^- \qquad\qquad (1)$$

$$M \xrightarrow{\ h\nu\ } M^+_\cdot + e^- \qquad\qquad (2)$$

radical, should be less favored than the back reaction in equation (2), where an electrostatic attraction within the photoproduct pair favors reversion to the ground state. Indeed, in this analysis, recapture of a photo-ejected electron by the oxidized primary photoproduct formed from a dianion should be even less favorable. This approach to the inhibition of electron recapture is therefore complementary to the use of acceptors which form metastable one electron reduction products upon photoexcitation of an electron donor. In fact, the most effective acceptors (e.g., methyl viologen) are usually those which operate on this same electrostatic principle, i.e., where electrostatic attraction of the oxidized and reduced primary photoproducts is minimized. In the cases considered here, one might reasonably expect comparable stabilization of the primary photoproducts in equations (1) or (2) by the presence of neutral electron acceptors. Consequently, the relative inhibition of back electron transfer in simple processes like equations (1) and (2) should find direct parallel in the presence of neutral electron acceptors.

0097-6156/81/0146-0337$05.00/0

Further inhibition would also be anticipated if the environment of the electron–hole pair interferes with recombination. Separation of such pairs has been effectively achieved in the electric field formed at the interface between a semiconductor and an electrolyte solution (1). Accordingly, we have examined a series of excited organic anions, both in homogeneous solution and at a semiconductor electrode in a photoelectrochemical cell. As expected, redox photochemistry occurring at the semiconductor surface differs significantly from that found under homogeneous conditions. We have used this altered reactivity as a synthetic technique for controlled oxidative coupling reactions and as an investigative tool for establishing mechanism in visible light photolysis of highly absorptive anions.

Our experimental procedure parallels that described earlier in the construction of an anionic photogalvanic cell (2). A very thin layer (~1 mm) of anhydrous solution containing the absorptive anion is sandwiched between an optical flat and a poised semiconductor electrode. Upon irradiation of this stirred solution, photocurrents between the illuminated electrode and a dark platinum counterelectrode can be monitored. (See reference 2 for experimental detail.) By choosing an appropriate wavelength region from the irradiation source, we may excite preferentially either the dissolved (or adsorbed) anion or the semiconductor itself. Redox reactions occurring in the stirred solution may be followed in situ by cyclic voltammetry or by withdrawing an aliquot for chemical analysis by standard spectroscopic and/or chromatographic techniques. Preparative electrolyses were conducted in the same cell employing a PAR (Princeton Applied Research) potentiostat and digital coulometer. Solution phase photolyses were conducted by irradiating sealed, degassed, pyrex ampoules containing the reactive anions and were analyzed as above.

By exciting the red–orange cyclooctatetraene dianion $\underset{\sim}{1}$ in the presence of cyclooctatetraene in our photoelectrochemical cell (n–TiO_2/NH_3/Pt), we were able to observe photocurrents without detectable decomposition of the anionic absorber (2). Presumably, a rapid dismutation of the photooxidized product inhibited electron recombination, producing a stable hydrocarbon whose cathodic reduction at the counter electrode regenerates the original mixture essentially quantitatively (eqn 3).

We hope to confirm the principle of this mechanism by demonstrating that characterizable chemical reactions might ensue if the initial photooxidation (unlike eqn 3) were irreversible.

Previous work in our laboratory (3) and in others (4) has established that the primary photoprocess in a variety of excited carbanions involves electron ejection. This photooxidation will generate a reactive free radical if recapture of the electron is inhibited. Parallel generation of these same carbon radicals by electrochemical oxidation reveals an irreversible anodic wave, consistent with rapid chemical reaction by the oxidized organic species (5). Little chemical characterization of the products has been attempted, however (6).

A typical cyclic voltammetric trace for the anodic oxidation of the fluorenyl anion 2 at platinum is shown in Figure 1. The oxidation potential for this and several other resonance stabilized carbanions lies conveniently within the band gap of n-type TiO_2 in the non-aqueous solvents, and hence in a range susceptible to photoinduced charge transfer. Furthermore, dimeric products (e.g., bifluorenyl) can be isolated in good yield (55-80%) after a one Faraday/mole controlled potential (+1.0 eV vs Ag quasi-reference) oxidation at platinum.

If an electron acceptor is available in homogeneous solution, photochemical reaction can be observed. For example, when 2 is excited ($\lambda \geq 350$ nm) in anhydrous dimethylsulfoxide (DMSO), methylation occurs, ultimately giving rise to 9,9-dimethyl-fluorene in >80% yield. By analogy with Tolbert's mechanism for photomethylation in DMSO (4), such a process may be initiated by electron transfer to DMSO to form a caged radical-radical anion pair from which subsequent C-S cleavage occurs (eqn 4).

If no acceptor is present, recapture of the photoejected electron will be rapid and the photon energy will be lost. In accord with this prediction, after overnight photolysis of a tetrahydrofuran solution of 2, under conditions identical to those described above, 2 can be recovered essentially unchanged. Exactly parallel results have been obtained with tetraphenyl-cyclopentadienide (3).

That a semiconductor electrode can prevent back-electron transfer follows from detection of dimer (dihydrooctaphenyl-

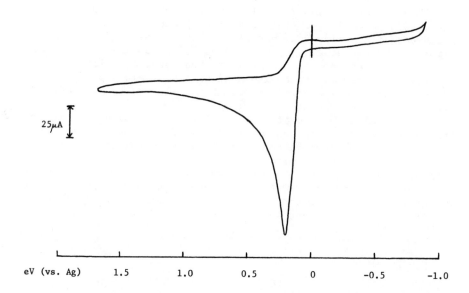

*Figure 1.   Electrochemical oxidation of fluorenyl lithium (DMSO, LiClO$_4$ (0.1M),
room temperature, scan rate: 50 mV s^{-1})*

fulvalene) (7) by HPLC and NMR, when lithium tetraphenylcyclo-pentadienide is excited at the surface of doped n-type TiO_2 crystal held at 0.0 eV (vs Ag) and connected with a platinum electrode in DMSO containing $LiClO_4$ (8) as inert electrolyte (eqn 5).

$$(5)$$

Photooxidation occurs at potentials well negative of the anodic wave. As dimer formation occurs photo-currents (0.1-3.9 μA) can be detected, implying photoinduced change transfer.

It is also significant to note that dimeric products are formed whether the anion or the semiconductor is excited. It is possible to selectively excited the anion by filtering out light of wavelengths shorter than the onset of band gap irradiation in $n-TiO_2$, i.e.,by cut off filters. For very thin layers of irradiated solution in our cell, incomplete light absorption by the solution/adsorbate occurs when unfiltered light is used as the excitation source. Under these conditions, where significant excitation of the semiconductor occurs simultaneously with anion excitation, greatly enhanced photocurrents attributable to band gap irradiation in the presence of donors are observed. Thus oxidation is achieved respectively either if an electron is injected into the conduction band from the anionic excited state or if an electron is transferred to the photogenerated hole in the valence band. As we have shown previously, sensitization by adsorbed or chemically attached dyes (9) should therefore extend the wavelength response of large band gap semiconductors in a process parallel to that observed here.

Analogous results are being obtained for the fluorenyl anion and other resonance stabilized anions. Although the mechanistic details are still under investigation and will be discussed elsewhere, these experiments demonstrate that the semiconductor interface is effective in inhibiting electron recapture in anion photolysis, in establishing mechanism in possible charge transfer photoreactions, and in acting as a catalytic surface in useful oxidative photocoupling reactions.

Acknowledgement. We are grateful to the U.S. Department of Energy for Financial support.

References and Notes

1. (a) Bard, A.J., J. Photochem. 1980, 10, 59; (b) Nozik, A.J.,
   Ann. Rev. Phys. Chem. 1978, 29, 189; (c) Memming, R. in
   "Electroanalytical Chemistry", A.J. Bard, ed., M. Dekker, N.Y.
   Vol. II, 1978; (d) Archer, M.D., J. Appl. Electrochem. 1975,
   5, 17; (e) Gerischer, H. in "Physical Chemistry – An Advanced
   Treatise", H. Eyring, D. Henderson, and W. Jost, eds.,
   Academic Press, N.Y., 463 (1970).

2. Fox, M.A.; Kabir-ud-Din, J. Phys. Chem. 1979, 83, 1800.

3. Fox, M.A.; Singletary, N.J., Solar Energy, 1980, in press.

4. For examples, see Tolbert, L.M., J. Am. Chem. Soc. 1978, 100,
   3952; Dvorak, V.; Michl, J., ibid. 1976, 98, 1080; and Fox,
   M.A., Chem. Rev. 1979, 79, 259.

5. Lochert, P.; Federlin, P., Tetrahedron Lett. 1973, 1109.

6. For exceptions to this generalization, see Schafer, H.;
   Azrak, A.A., Chem. Ber. 1972, 105, 2398; Borhani, K.J.;
   Hawley, M.D., J. Electroanal. Chem., 1979, 101, 407.

7. Pauson, P.L.; Williams, B.J., J. Chem. Soc., 1961, 4158.

8. A control experiment established that $LiClO_4$ has no effect on
   the homogeneous photolysis of the lithium salt of 2. At con-
   versions of anion greater that ~20%, the efficiency of dimer
   formation decreases. Whether this can be attributed to further oxidation of the anion derived from dimer or to cathodic
   cleavage of dimer is under investigation.

9. Fox, M.A.; Nobs, F.J.; Voynick, T.A., J. Amer. Chem. Soc.
   1980, 102, 4029, 4036.

RECEIVED October 15, 1980.

Fundamental Aspects of Photoeffects at the
_n_-Gallium Arsenide–Molten-Salt Interphase

R. J. GALE, P. SMITH, P. SINGH, K. RAJESHWAR, and J. DUBOW

Department of Electrical Engineering, Colorado State University,
Fort Collins, CO 80523

Detailed studies of the semiconductor/electrolyte interphase boundary are necessary in order to understand fully the operation of photoelectrochemical cells. It is important to be able to establish the potential distribution throughout the interphasial regions because the surface potential barrier governs the separation of photogenerated electrons and holes and thus the energy conversion efficiency. Additionally, it is important to identify those factors that control or modify both the faradaic charge-transfer reactions occurring across the solid/liquid interface and the recombination processes within the semiconductor. For any particular system, a complete analysis in terms of the chemical identification of all the interposed molecules and their role would be an arduous, if not impossible task. Present knowledge of the catalytic and dielectric properties of the species existing in surface atomic layers and in the inner double layer region adjacent to semiconductor electrodes is particularly limited. We are restricted therefore to trying to identify the key factors affecting the output performance of cells and to devise experimental approaches that might ultimately provide a complete description of the photoelectrochemical processes in molecular detail.

This work attempts to model a semiconductor/molten salt electrolyte interphase, in the absence of illumination, in terms of its basic circuit elements. Measurement of the equivalent electrical properties has been achieved using a newly developed technique of automated admittance measurements and some progress has been made toward identification of the frequency dependent device components (1). The system chosen for studying the semiconductor/molten salt interphase has the configuration n-GaAs/AlCl$_3$: 1-butylpyridinium chloride (BPC) melt/vitreous C, with the ferrocene/ferricenium ion redox couple as the liquid phase charge carrier. Photoelectrochemical cell electrolytes can be divided broadly into two classes, aqueous and nonaqueous. If aqueous mixtures are included within the former classification, the non-

0097-6156/81/0146-0343$05.00/0

aqueous types can be subdivided further into those based upon organic or inorganic solvents and molten salt electrolytes, each of these requiring some means of ionic conduction and/or a suitable redox system. The rationale for studying the semiconductor/molten salt interphase is that in general, from corrosion considerations, nonaqueous electrolytes appear to be more attractive than the aqueous for use with the traditional semiconductor materials. Inherent advantages of developing molten salt electrolyte systems for cells to produce electricity from light have been discussed elsewhere (2).

Experimental

Materials. Single crystals of Sn-doped n-GaAs of orientation (100) were obtained from Laser Diode Lab, Inc. with the following quoted properties, N_D = 2.3 x 10^{17} cm^{-3}, ρ = 0.0071 Ωcm, μ = 3900 $cm^2 V^{-1}s^{-1}$. Ohmic contact was achieved with thermally evaporated Ge-Au alloy annealed in N_2 at 400°C for 15 minutes. A Teflon coated Cu wire was attached with silver epoxy cement and the single crystal and the Ohmic contact masked with a nonconducting epoxy cement. Surface preparation of the crystals was sequentially a 30 sec etch in 6M HCl, a distilled water rinse, a 1 minute immersion in H_2SO_4:H_2O_2:H_2O (3:1:1 vol.), a distilled water rinse, and a final rinse with absolute ethanol. The wet assembly was then placed immediately into the ante-chamber of the drybox to be vacuum dried. The counter electrode comprised a plate of vitreous C (Atomergic Chemetals, Inc.) and the reference electrode for all experiments consisted of 0.5 mm Al wire (Alfa Ventron Puratronic grade, m4N8) immersed in 2:1 molar ratio $AlCl_3$: BPC melt. The preparation of 1-butylpyridinium chloride and melts have been described in earlier publications, e.g (3).

Apparatus and Technique

Circuitry used to obtain capacitance-voltage data for Mott-Schottky plots was similar to that used previously (2). Linear voltage ramps (5-10 mVs^{-1}) and ∿10 mV peak/peak ac signals were added to a PAR 173 Potentiostat/Galvanostat from a PAR 179 University Programmer and Ithaca Dynatrac 3 lock-in amplifier, respectively. The voltage output from the current follower of a PAR 179 Digital Coulometer was connected to the signal input of the LIA and the capacitance-voltage curves were displayed on a PAR RE 0074 X-Y recorder. These measurements were made on cells thermostated to 40 $\pm$ 1°C within a drybox. Most measurements were made in absolute darkness or in cells carefully screened with Al foil to exclude light.

A schematic of the automatic network analysis system used for photoelectrochemical cell measurements is shown in Figure 1. The ac signal of 20 mV p/p amplitude is applied with a Hewlett-Packard 3320B frequency synthesizer to the cell and to the reference signal input of a HP 3570A automatic network analyzer. A high speed current-voltage converter (Teledyne Philbrick 1435) and current booster with a flat frequency response from dc to 1 MHz is used to buffer the analyzer. Cell admittances can be calculated from the measured $G(s) = (G_i + sC_i)/(G_f + sC_f)$ in which G_i and G_f are the conductances, C_i and C_f the capacitances of the cell and the feedback components, respectively, and s equals $j\omega$. The desktop computer (HP 9825A) communicates via the IEEE-488 standard interface bus with the synthesizer and network analyzer to sweep frequency, to collect gain and phase data, to calculate capacitance and conductance values and then to store this information on flexible discs. In these experiments, the bias was adjusted manually using a low noise power supply until the required potential drop was obtained between the reference and working electrodes. Measurements were made in two frequency ranges from 5-50 Hz and 50-100 KHz. Details concerning the calibration of the system and comparison of results obtainable to the data accessible from impedance bridge techniques have been outlined already for metal-insulator-semiconductor (MIS)/semiconductor-insulator-semiconductor (SIS) solar cells (1). Dummy cell calibration curves gave errors nominally < 3 percent for gain and moderate phase angles.

Results and Discussion

Voltammetric Investigations. Linear sweep voltammetry provides a simple, sensitive and rapid means of obtaining a current-voltage profile of the semiconductor/electrolyte interphase. Figure 2 illustrates the dark I-V curves at a n-GaAs electrode in basic 0.8:1 ($AlCl_3$: BPC respectively) and acidic 1.5:1 molar ratio melts. In basic melts, a peak at ~-0.6 to -0.75 V increases with illumination and by scanning further in the positive direcion past the rest potential at -0.48 V (Figure 3). Tentatively, this reduction peak may be assigned to a homogeneous arsenic chloride species at the electrode surface because a similar peak is obtained at a C electrode by the addition of $AsCl_3$ to the melt (Figure 4). An anodic peak in Figure 4 at +0.47 V appears only after the cathodic reaction at -0.4 V but, probably due to filming, repeated scans cause the waves to distort and diminish. The electrochemistry of gallium chloroaluminate species in a high temperature $AlCl_3$: NaCl: KCl eutectic has been studied (4). The standard electrode potentials are reported to be $Ga(III)/Ga(I)$ = 0.766 V, $Ga(III)/Ga(0)$ = 0.577 V and $Ga(I)/Ga(0)$ = 0.199 V versus $Al(AlCl_3$: NaCl: KCl eut.) reference. Extremely low electroactivity was indicated at a C electrode after the addition of $GaCl_3$ to the basic room temperature melt but the peaks at circa +0.2 V and +0.8 V in the voltammograms of n-GaAs may be tentatively assigned to compounds of $Ga(I)$ and $Ga(III)$, respectively, either homogeneous or surface bonded.
Both gallium and arsenic compounds exist in several oxidation

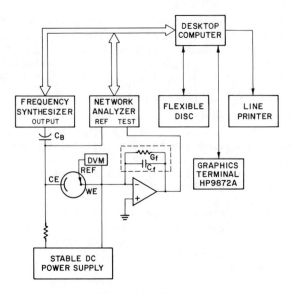

Figure 1.   Computer-controlled network analysis system for photoelectrochemical cell admittance measurements

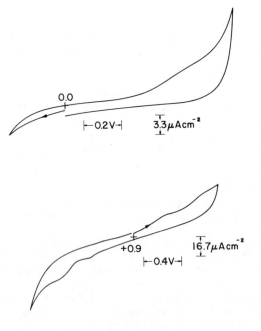

Figure 2.   Cyclic voltammograms of n-GaAs in 0.8:1 (upper) and 1.5:1 (lower) melts at 50 mVs⁻¹ and 20 mVs⁻¹, respectively, nonilluminated, 40°C

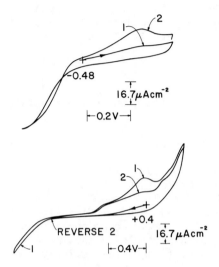

Figure 3. *Voltammograms of* n-GaAs *under low illumination (~ 1 mW cm⁻²) in 0.8:1 (upper) and 1.1:1 (lower) melts at 10 mVs⁻¹ and 20 Vs⁻¹, respectively, 40°C*

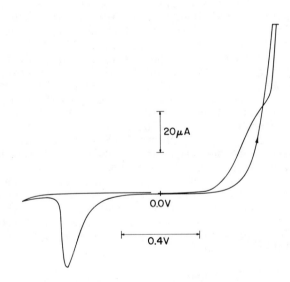

Figure 4.   *Cyclic voltammogram (initial scan) of AsCl₃ (0.136 mL) added to 0.95:1 melt (12.12 g), v = 100 mVs⁻¹, C electrode*

states and the probability of complex equilibria with halo-species
and the degree of irreversibility of the electron-transfer processes makes absolute correlations of potential peaks with definitive
species difficult. An essential problem too, is to distinguish
between the trace amounts of dissolved products from corrosion of
the semiconductor and chemical states that may be present in the
surface atomic layer.

Electrolyte Composition Effects

The major equilibrium process controlling the solvent acidity
is given by equation (1),

$$2 \; AlCl_4^- \;\; \rightleftharpoons \;\; Al_2Cl_7^- + Cl^- \qquad (1)$$

From a potentiometric model for the chloraluminate species present
in these ionic mixtures, an equilibrium constant has been derived
using the mole fraction scale, $K \sim 7.0 \times 10^{-13}$ at 40° (3). A pCl^-
can be defined analogous to pOH^- in aqueous media, $pCl^- = -\log_{10}Cl^-$.
Mixtures with an excess of $AlCl_3$ will be Lewis acids containing
the electron deficient $Al_2Cl_7^-$ ion with some $AlCl_4^-$ ion, whereas
when the molar ratio becomes < 1:1 $AlCl_3$: BPC_2 respectively, the
melt will be termed basic and contain donor Cl^- ions together
with $AlCl_4^-$ ion. Surface interactions between the substrate material and the electrolyte can affect significantly the performance
of cells (5) and these are of two main types (i) solvent acid-base
induced ionization of surface compounds, and (ii) the accumulation
of solute species in the inner double layer region at the semiconductor surface (specific or superequivalent ion adsorption,
organic adsorbates, etc.). Such a classification is not clear cut
because the phenomenological definition of specific ion adsorption
at metal electrodes (6) does not discuss the chemical state of the
species, the essential criterion being that an excess of the species must be present at the potential of zero charge. Different-
iation of these processes requires a knowledge of the chemical
forces involved in the equilibria or the adsorption (electrostatic
plus chemical for ions), the extent of ionization or the adsorption isotherm, and the surface potential parameters associated
with either process. Quantitative expressions of these effects
are restricted in the literature. An explanation of the flat-band
shift found in aqueous solutions arises from a consideration of
surface equilibria of the type,

$$-MOH + OH^- \;\; \rightleftharpoons \;\; -MO^- + H_2O \qquad (2)$$

An additional potential drop occurs across the Helmholtz layer,
(7), given by,

$$\Delta\phi_{DL} = const - (2.3RT/F)\{[pH] + \log(f_{MO^-} \cdot X_{MO^-})\} \; (3)$$

in which $\Delta\phi_{DL}$ is the potential drop across the double layer and f_{MO^-} and X_{MO^-} respectively are the activity coefficient and mole fraction of ionized groups in the surface. A linear relation between the flatband potential, V_{fb}, and the pH should be expected as long as the contribution from the second term is parenthesis is minimal. Gryse, Gomes, Cardon and Vennik (8) have discussed the effect of the double layer potential drop.

Figure 5 contains a summary of the Mott-Schottky plots obtained at 1 KHz and different electrolyte compositions ranging from 0.8:1 to 1.75:1 molar ratios. The intercepts were calculated from the least squares gradients taken in the voltage regions where faradaic processes are least significant, namely 2.0 to 0.6 V for the acidic and 1.0 to -0.5 V for the basic melts. Faradaic processes become apparent as hysteresis in the C-V plots and these effects give rise to considerable scatter in data, particularly near the 1:1 molar ratio composition. Least squares analyses of these data give a shift of the flat-band with pCl⁻ of 0.066 V (standard deviation, σ = 0.009 V), or approximately (2.3 RT/F) V at 40°. This ΔV_{fb}/pCl⁻ value for the (100) orientation n-GaAs is approximately one-half that obtained with (111) n-GaAs crystals (2), indicating that the crystal surface atom density and type can be a significant factor in the interactions between substrate and electrolyte. Flat-band potential values for (100) and (111) n-GaAs/molten salt interphases and for the (111) n-GaAs/aqueous electrolyte interphase are compared in Table I.

TABLE I

Xtal	Orientation			V_{fb}(NHE)	Ref.
n-GaAs ($\bar{1}\bar{1}\bar{1}$)face	aqueous	pH 2.1	-0.95V		(9)
"		pH 12	-1.50V		
n-GaAs (111)face	AlCl$_3$:BPC	pCl⁻2	-1.25V		(2)
"		pCl⁻12	+0.10V		
n-GaAs (100)face	AlCl$_3$:BPC	pCl⁻2	-0.80V		This work
"		pCl⁻12	-0.20V		This work

Typically in aqueous electrolytes the flat-band potential variation per pH unit is (2.3 RT/F) V (7, 9-15) but in the high temperature AlCl$_3$:NaCl molten electrolytes, Uchida and coworkers (16, 17) have reported flat-band shifts of 2(2.3RT/F) V per pCl⁻ unit for Sb-doped SnO$_2$ and n-TiO$_2$ semiconductors.

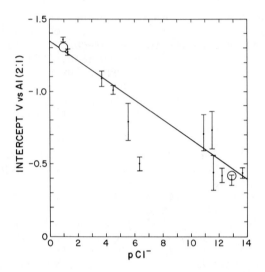

*Figure 5.   Variation of Mott–Schottky intercept with pCl⁻ for (100) orientation
n-GaAs, 40°C. Circles denote intercept values from automated admittance measurements. Bars signify standard deviation of least-squares straight line.*

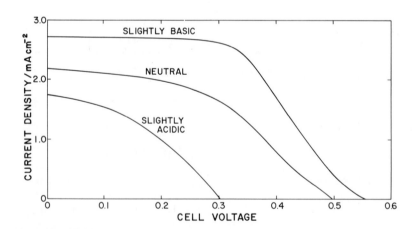

*Figure 6.   Power curves for n-GaAs 0.124M ferrocene cell, electrolyte unstirred,
illumination ~ 100 mW cm⁻²*

The consequences of quite small changes in the surface charge can be significant to device performance. Figure 6 illustrates the illuminated current-voltage behavior of a (100) n-GaAs electrode in a melt containing the ferrocene/ferricenium redox couple. At the exact neutral point, $V_{fb} \sim -0.91$ V, or 1.04 V negative of the redox potential of the ferrocene couple. As the band gap of n-GaAs is approximately 1.44 V, a negative shift of the band edges of 0.4 V would improve the overlap between the redox couple energy levels and the valence band edge. Thus, it may be possible to "tune" photoelectrochemical cells for optimal performance by deliberately varying the extent of the surface charge with suitable adsorbates. These photoeffects have been discussed in greater detail elsewhere (18).

The magnitude of the errors in determining the flat-band potential by capacitance-voltage techniques can be sizable because (a) trace amounts of corrosion products may be adsorbed on the surface, (b) ideal polarizability may not be achieved with regard to electrolyte decomposition processes, (c) surface states arising from chemical interactions between the electrolyte and semiconductor can distort the C-V data, and (d) crystalline inhomogeneity, defects, or bulk substrate effects may be manifested at the solid electrode causing frequency dispersion effects. In the next section, it will be shown that the equivalent parallel conductance technique enables more discriminatory and precise analyses of the interphasial electrical properties.

Automated Admittance Measurements

Some of the experimental techniques that have been developed for studies of MIS solar cells, for example, might be applied successfully to photoelectrochemical cells. Features common to electrochemical solar cells and metal-oxide-semiconductor (MOS), MIS, or SIS devices have been reviewed in a theoretical comparison of their underlying modes of operation (19). An essential starting point for device analysis is to be able to model equivalent circuit that accurately represent and reflect the material properties and cell behavior in both the static and, hopefully, the operational modes. Tomkiewicz (20) has applied a relaxation spectrum analysis technique to the n-TiO$_2$/aqueous electrolyte interphase and he could assign passive elements to two space charge layers with different doping levels. In the light of previous work which has demonstrated (21) that the equivalent parallel conductance technique is more sensitive than capacitance measurements for deriving suface state information in MIS devices, we have made a series of these measurements on a photoelectrochemical cell. For our initial evaluations, the interface was not illuminated and no redox couple was added to the electrolyte. Both the capacitance and conductance methods derive equivalent information as functions of the applied voltage and frequency, however, the inaccuracies that arise in the treatment of data generally are larger using capacitance measurements (21). Figure 6 illustrates an equivalent circuit for the semiconductor/molten salt interphase. The capacitance and resistive elements associated with the large area counter electrode and the cell geometry have been ignored in this analysis.

Mathematical treatments for equivalent circuit of this type for a single level and a continuum of surface state levels have been presented elsewhere (e.g., 10, 20, 21, 22). The admittance of the reduced circuit after removing the influence of R_{bulk} ($=R_{diel} + R_{elec}$) will have the form:

$$Y = G + jB \qquad (4)$$

where
$$\frac{G}{\omega} = \frac{1}{\omega R} + \sum_{i=1}^{n} \omega C_i \tau_i / (1 + (\omega \tau_i)^2) \qquad (5)$$

and
$$B = \omega C_{sc} + \sum_{i=1}^{n} \omega C_i / (1 + (\omega \tau_i^2) \qquad (6)$$

The unknown parallel RC elements that comprise the impedance z in Figure 7 cause a series of maxima in the G/ω vs. ω data at $\omega\tau = 1$, from which the majority carrier time constant τ of a single level interface state, $R_s C_s$, or any general RC time constant may be obtainable. Superposition of curves occurs if the individual time constants are closely spaced.

Figure 8 illustrates the data points and theoretical curve fits for the (100) n-GaAs/0.8:1 melt interphase. To identify the space charge capacitance and the bulk resistance elements, the frequency independent capacitances were derived from the maxima in log (G/ω) vs. log(ω) plots, e.g., Figure 8, as $(G/\omega)_{max} = C_s/2$ and conductances from the maxima in log(ωC) vs. log ω plots e.g., Figure 9. The procedural steps for modeling were (a) firstly, establish the limiting low frequency conductance $G_{dc} = 1/R_{far}$ and, as described above, the R_{bulk} C_{sc} parameters, (b) model a curve for the circuit obtained from the basic elements, (c) substract theoretical curve from experimental data to determine difference maxima and, finally, (d) sucessive trial and error insertion of RC values constituting z to attain best curve fit to experimental data. In Figure 8, the circuit element $R_{far} = 0.83$ MΩ determines the degree to which ideal polarizability has been achieved i.e., the absence of corrosion processes and faradaic losses due to electroactive impurities or electrolyte decomposition (cf. n-TiO$_2$/ aqueous interphase (22), $R_{far} = 10$ KΩ). Addition of a large pseudo-capacity of R_{far} did not alter the slope of the low frequency simulated portion of the curve. A value of $C_{far} \simeq 0.1$ μF causes a peak at low frequencies (<100 Hz). No such peaks were observed in the present data. The influence of decreasing the electrolyte conductance or increasing overpotentials at the

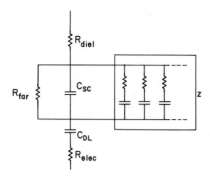

Figure 7. Equivalent circuit for interphase: (R_{diel}) resistance of semiconductor; (R_{elec}) electrolyte resistance, (R_{far}) faradaic resistance; (C_{sc}) space charge capacitance; (C_{DL}) double-layer capacitance; and (z) parallel impedances associated with surface states, faradaic reactions, etc.

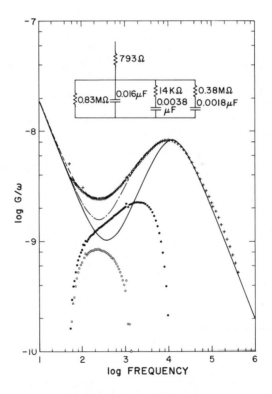

Figure 8. n-GaAs in 0.8:1 melt at 0.0 V bias, area 0.06 cm². Experimental (G/ω) values are crosses: (———) the theoretical curve for R_{far} and single R_{bulk} C_{sc} circuit; (· – · –) curve insertion of high frequency parallel RC; and (– – –) final curve fit. Lower points represent the difference between theoretical and experimental curves.

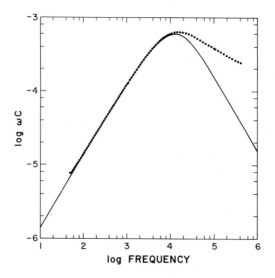

Figure 9.   n-GaAs in 0.8:1 melt, 0.0 V bias. Experimental ωC values (●) and theoretical curve (———)

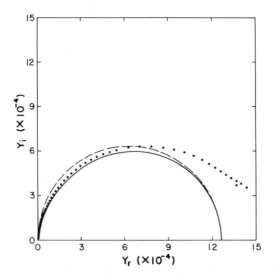

Figure 10.   Cole–Cole plot for n-GaAs in 0.8:1 melt, bias 0.0 V: (●) experimental data; (———) theory for circuit insert of Figure 8; (– – –) single RC hemisphere

secondary electrode is to lower the frequency of the maximum associated with the space charge capacitance and its series resistance, R_{bulk}. Another method for comparison of experimental data and derived circuit components is to examine the Cole-Cole plots (23). Figure 10 portrays the imaginary and real admittances for the experimental data and it is apparent that the total reduced circuit fits more closely than the theoretical model for a single RC_{SC} element. Dispersive distortion of the curves is evident about 10 KHz in experimental data possibly caused by the presence of fast surface states whereas it appears above 100 KHz in a dummy cell of similar format.

It is necessary, however, to include two RC elements for z in order to obtain a good fit to the experimental data. The insert in Figure 8 describes the complete circuit in terms of its passive elements. A higher frequency RC element in z of as yet undetermined physical origin, appeared both at basic (0.8:1) and acidic (1.5:1) interphases with (100) n-GaAs. If it arises from a surface state, an order of magnitude estimate of its density from the expression $N_{SS} = C_S/e$ is 4×10^{11} cm^{-2}. The lower frequency RC element in the basic melts can be associated with the faradaic process seen in cyclic voltammetry at circa -0.7 V. This is shown in Figure 11, where the magnitude of its effect has increased. The influence of trace faradaic processes in the acidic melts also is manifest at frequencies 50-500 Hz. A more detailed discussion of these experimental procedures and theoretical modeling is available (24). Finally, the frequency independent capacitance values C_{SC} obtained at different bias values have been used to obtain values by Mott-Schottky analyses for the flat-band potentials in the 0.8:1 and 1.5:1 melts. These values are shown in Figure 5 and correspond well to the data obtained from slow voltage sweep capacitance measurements. Obviously, the equivalent parallel conductance method can provide a detailed picture of the frequency effects and better indicate which frequencies are most reliable for extracting information for specific interphasial phenomena.

Abstract

Linear sweep voltammetry, capacitance-voltage and automated admittance measurements have been applied to characterize the n-GaAs/room temperature molten salt interphase. Semiconductor crystal orientation is shown to be an important factor in the manner in which chemical interactions with the electrolyte can influence the surface potentials. For example, the flat-band shift for (100) orientation was (2.3RT/F)V per pCl^- unit compared to 2(2.3RT/F)V per pCl^- for (111) orientation. The manner in which these interactions may be used to optimize cell performance is discussed. The equivalent parallel conductance method has been used to identify the circuit elements for the non-illuminated semiconductor/electrolyte interphase. The utility of this

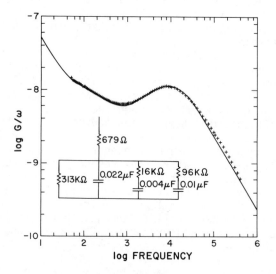

Figure 11.   Illustration of influence of irreversible faradaic process on (G/ω) data, 0.8:1 melt, bias: 0.65 V ((✕) experimental data; (———) theoretical curve for equivalent circuit insert)

technique in providing information on surface states, trace faradaic processes and reliability of Mott-Schottky analyses, is demonstrated using the n-GaAs/AlCl$_3$-Butylpyridinium Chloride interphase as a representative example.

Acknowledgement

Support funds were provided by The Solar Energy Research Institute under Contract No. XP-9-8002-9.

Literature Cited

1. Smith, P.; DuBow, J.D.; Singh, R.; Emery, K.; Proc. 14th Photovoltaic Specialists Conf., San Diego, CA, January 1980.

2. Singh, P.; Rajeshwar, K.; DuBow, J.B.; Job, R. J. Am. Chem Soc., 1980, 102, 4676.

3. Gale, R.J.; Gilbert, B.; Osteryoung, R.A. Inorg. Chem, 1979, 18, 2723.

4. Anders, U.; Plambeck, J.A. Can. J. Chem., 1969, 47, 3055.

5. Rajeshwar, K.; Singh, R.; Singh, P.; Gale, R.J.; DuBow, J.; J. App. Physics (In Press).

6. Mohilner, D.M., "Electroanalytical Chemistry", Vol. 1, Ed. Bard, A.J., Marcel Dekker, 1966, 241.

7. Gerischer, H., "Physical Chemistry. An Advanced Treatise", Vol. 1Xa/Electrochemistry, Ed. Eyring, H., Academic Press, 1970, 463.

8. DeGryse, R.; Gomes, W.P.; Cardon, F.; Vennik, J. J. Electrochem. Soc., 1975, 122, 711.

9. Cardon, F.; Gomes, W.P. J. Phys. D: Appl. Phys. 1978, 11, L63.

10. Dutoit, E.C.; Van Meirhaeghe, R.L.; Cardon, F.,; Gomes, W.P. Ber. Bunsenges Phys. Chem., 1975, 79, 1206.

11. Dutoit, E.C.; Cardon, F., Gomes, W.P. ibid. 1976, 80, 475.

12. Madou, M.J.; Cardon, F.; Gomes, W.P. J. Electrochem. Soc., 1977, 124, 1623.

13. Horowitz, G. J. Appl. Physics, 1978, 49, 3571

14. Lamasson, P.; Gautron, J. C.R. Acad. Sc. Paris, Ser. C. 1979, 288, 149.

15. Tomkiewicz, M. J. Electrochem. Soc., 1979, 126, 1505.

16. Uchida, I.; Urushibata, H.; Toshima, S. J. Electoanal. Chem., 1979, 96, 45.

17. Uchida, I.; Urushibata, H.; Akahoshi, H.; Toshima, S; J. Electrochem. Soc., 1980, 127, 995.

18. Singh, P., Ph.D. thesis, Colorado State University (1980).

19. Kar, S.; Rajeshwar, K.; Singh, P.; DuBow, J. J. Solar Energy, 1979, 23, 129.

20. Tomkiewicz, M. J. Electrochem. Soc., 1979, 126, 2220.

21. Nicollian, E.H.; Goetzberger, A. The Bell System Technical Journal, 1967, XLVl (6), 1055.

22. Sze, S.M., "Physics of Semiconductor Devices", Wiley-Inter-science, 1969.

23. Cole, R.H.; Cole, K.S. J. Chem. Phys., 1941, 9, 341.

24. Smith, P. M.Sc. thesis, Colorado State University (1980).

RECEIVED November 19, 1980.

Analysis of Current–Voltage Characteristics of Illuminated Cadmium Selenide–Polysulfide Junctions

JOSEPH REICHMAN and MICHAEL A. RUSSAK

Research Department A01/26, Grumman Aerospace Corporation, Bethpage, NY 11714

Comparisons of experimental I-V characteristics with those predicted by theoretical models (1) are commonly made to analyze the effects of illumination on semiconductor junction devices. These comparisons have not normally been made for semiconductor-electrolyte (S-E) junctions, most likely due to the lack of suitable theoretical models.

In this paper the calculated I-V characteristics using a model that includes depletion region recombination (2) are compared with experimental results for single-crystal CdSe electrode/polysulfide electrolyte junctions. In particular the predicted effects of redox ion concentration and light intensity on the I-V characteristics are analyzed and compared with experimental data.

Theoretical Discussion

Models. The most commonly used model to analyze the I-V performance of S-E junctions is that of Gartner (3) as given by the following

$$I_g = qF[1 - exp(-\alpha W)/(1 + \alpha L)]$$

where F is the absorbed monochromatic solar flux, α is the absorption coefficient, W is the depletion width, and L is the minority carrier diffusion length. This equation shows a weak dependence of current on voltage through the functional dependence of depletion width on the square root of potential drop across it. This equation often gives a good fit to experimental data at short circuit conditions (the redox potential) and at reverse bias (anodic polarization for an n-type semiconductor electrode). However, in the forward bias range (cathodic polarization), the region of interest for photovoltaic performance analysis, the Gartner model predicts a decrease in current considerably smaller than is observed experimentally. This is due to the neglect of the various recombination mechanisms that contribute to the decrease in photocurrent with increasing

0097-6156/81/0146-0359$05.00/0

photovoltage (or decreasing band bending). The Gartner model, originally derived for a Schottky barrier in reverse bias, only accounts for recombination of carriers generated in the bulk of the semiconductor. Other mechanisms that can contribute to photocurrent loss are:

1. Diffusion of depletion region generated minority carriers to the bulk and subsequent recombination
2. Recombination in the depletion region
3. Recombination at the interface
4. Opposing dark current.

Recombination in the depletion layer can become important when the concentration of minority carriers at the interface exceeds the majority carrier concentration. Under illumination minority carrier buildup at the semiconductor-electrolyte interface can occur due to slow charge transfer. Thus surface inversion may occur and recombination in the depletion region can become the dominant mechanism accounting for loss in photocurrent.

Mechanisms 1 and 2 are included in the model that is used here for comparison with experimental data. Interface recombination and dark current effects are not included; however, the experimental data have been adjusted to exclude the effects of dark current. To include the additional bulk and depletion layer recombination losses, the diffusion equation for minority carriers is solved using boundary conditions relevant to the S-E junction (i.e., the photocurrent is linearly related to the concentration of minority carriers at the interface). Using this boundary condition and assuming quasi-equilibrium conditions (flat quasi-Fermi levels) (4) in the depletion region, the following current-voltage relationship is obtained.

$$I = q\nu p_W \exp(q\ \phi_b/kT)$$

where ν is the rate constant for charge transfer, ϕ_b is the potential drop across the depletion region and p_W is the hole concentration at the depletion edge. To determine p_W the following equation is used

$$P_W = \left[\frac{-K + (K^2 + 4AC)^{\frac{1}{2}}}{2A}\right]^2$$

where $A = q\nu\exp(q\phi_b/kT) + qL/\tau$, $C = I_g + q\nu p_0\exp(q\phi_{bo}/kT)$ and $K = \pi kT(2\varepsilon\varepsilon_0/q\phi_b)^{1/2}/4\tau$.

In the above, L is the hole diffusion length, τ is the lifetime, p_0 is the equilibrium hole density, and ϕ_{bo} is the equilibrium band bending voltage. These equations are good approximations when ϕ_b is not too small and are equivalent to that given in (2) where the exchange current parameter is used instead of the charge transfer rate constant. More accurate equations that apply to the extire range of potential have been used in the analysis given here (5). The equations however are somewhat complex and therefore are not presented in this paper.

Predicted Results. The effects of charge transfer rate constant, ν, and illumination intensity were selected for analysis and comparison with experiment. Predicted results based on the model described above showing the effects of ν on the I-V characteristics are given in Figure 1. Using the values of the parameters given in the caption, this figure shows that as ν gets smaller, losses in current due to depletion region recombination occur at higher values of band bending. This is due to the higher concentration of minority carriers at the interface that is rquired to sustain a given photocurrent as ν decreases. This leads to higher recombination rates in the depletion region as the band bending is reduced.

The effect of varying light intenity on the I-V characteristics are shown in Figure 2. Because the curves are normalized to the incident intensity, they should superpose each other if the recombination losses were linearly dependent on intensity. However, our model predicts that recombination losses in the depletion region is proportional to the square root of intensity. Thus losses due to depletion region recombination become relatively less significant at higher intensities and consequently, for a given band bending, higher collection efficiencies result.

For the parameters used here, the additional recombination due to diffusion from the depletion region to the bulk do not become significant until the band bending becomes small.

Discussion of Experiment

Experimental Procedure. Experimental I-V data for a CdSe single crystal electrode in polysulfide electrolyte was used to compare with predictions of the model that includes depletion region recombinations. The crystal was obtained from Cleveland Crystal Corp., Cleveland, Ohio, and prepared for measurements according to procedures found in the literature (6). Electrodes composed of Na_2S and sulfur in 1M NaOH was used with Argon gas bubbling through the cell during measurements. Voltammetric measurements were made with a PAR model 173 potentiostat and Model 175 universal programmer using a scan rate of 10 MV/sec. Spectral measurements were made using the output of a tungsten-halogen lamp. Monochromatic intensities were obtained by using narrow bandpass filters. Intensities were measured using a Elppley thermopile detector.

Electrode Characterization. To minimize the number of adjustable parameters of the model, the optical and electronic properties of the CdSe crystal electrode were meaured. Diffuse

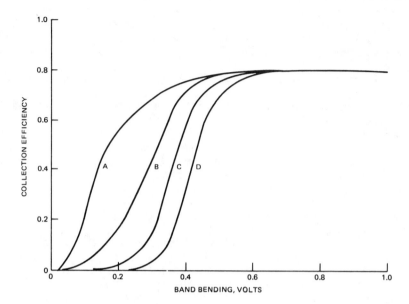

Figure 1. Calculated effect of rate constant on I–V characteristics (parameters: lifetime $\tau = 10^{-9}$ s; $N = 1.5 \times 10^{17}$ cm^{-3}; $L = 1$ μm; $\alpha = 3 \times 10^4$ cm^{-1}) ((A) 1000; (B)100; (C) 10; (D) 1 cm/s)

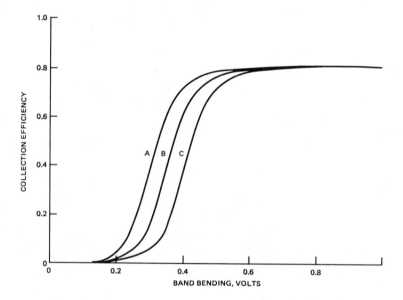

Figure 2. Calculated effects of intensity on I–V characteristics (parameters as in Figure 1 with $v = 10$ cm/s) ((A) 10 mA/cm²; (B) 1.0 mA/cm²; (C) 0.1 mA/cm²)

reflectance measurements as a function of wavelength from the electrode surface immersed in the electrolyte compartment were obtained. The absorption coefficient as a function of wavelength was obtained from reflectance and transmittance measurements of CdSe thin films using conventional methods. The carrier density was obtained from measurements of resistivity and Hall voltage using the Van de Pauw method (7). The flatband potential was obtained by determining the potential at which a photoresponse is observed using a high intensity chopped light source. Additionally, Mott-Schottky plots were also made for additional verification of flat band potential and carrier density. The results of these measurements are summarized in Table I.

Measurements of photocurrent as a function of wavelength at short circuit conditions (-0.8 V rel SCE) were made. The data was reduced to internal quantum efficiency by using the measured values of reflectance, correcting for electrolyte absorption, and measuring the incident intensity. The results are shown in Figure 3 together with a best fit to the data using the Gartner model (Equation 1). The depletion width was calculated using the measured values of carrier density and flat band potential. For a diffusion length of 1 μm and the measured values of the absorption coefficients, a good fit to the data was obtained, as seen from Figure 3.

The I-V characteristics were obtained at varying monochromatic intensities using a chopped beam at a wavelength of 650 nm. The internal quantum efficiency (or collection efficiency) was then determined, as a function of band bending voltage, using the reflectance and transmittance data. This was done by first subtracting the dark current from the photocurrent as obtained from the chopped light response.

Comparison of Theory & Experiment

To compare the effects of intensity, I-V characteristics were obtained for the CdSe/polysulfide junction at intensities of 3.7 and 0.37 ma/cm^2 at a wavelength of 650 nm. The results using the above data reduction method, are shown in Figure 4 together with calculated fits to the data using the model that includes depletion layer recombination . The adjustable constants were carrier lifetime and charge transfer rate constant. Other parameters were obtained from the previous measurements. As can be seen from Figure 4 reasonably good fits to the data are obtained for values of band bending >0.2 V. The predicted decreasing importance of depletion region recombination with increasing intensity is observed experimentally. At low values of band bending, effects not included in the model, such as interface recombination and changes in potential drop across the Helmholtz layer may become important and explain the discrepancy.

Table I. CdSe Single Crystal Electrode Characterization

PARAMETER	MEASUREMENT	VALUES
ABSORPTION COEFFICIENT (α)	TRANSMITTANCE & REFLECTANCE	5×10^3 cm^{-1} TO 7×10^4 cm^{-1}
REFLECTANCE	INTEGRATING SPHERE	0.10-0.12
CARRIER DENSITY (N_D)	HALL METHOD MOTT-SCHOTTKY	1.5×10^{17} cm^{-3} 1×10^{17} cm^{-3}
FLATBAND POTENTIAL (V_{fb})	CHOPPED LIGHT MOTT–SCHOTTKY	-1.60 vs SCE -1.47 vs SCE

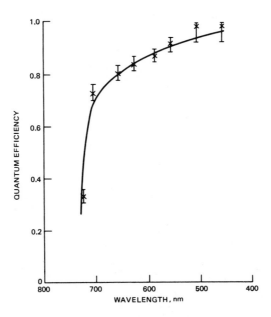

Figure 3. Experimental collection efficiency vs. wavelength with fit to data using Gartner model (measured values of τ and N, and L = 1 μm were used to fit data)

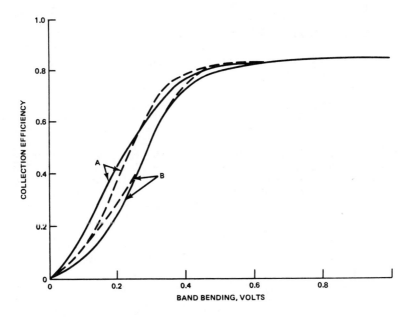

Figure 4. Comparison of theory with experiment: effects of intensity (parameters as in A of Figure 1 with v = 225 cm/s) ((——) experimental; (– – –) calculated; (A) 3.7 mA/cm² intensity; (B) 0.37 mA/cm² intensity)

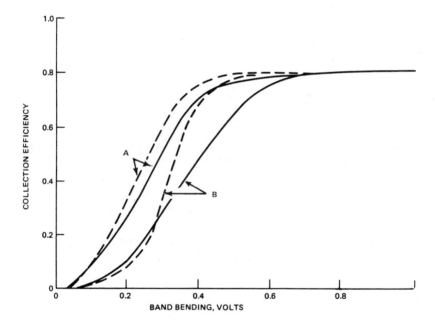

Figure 5.   Comparison of theory with experiment: effect of redox ion concentration (parameters to fit data were as in Figure 1 with v = 225 cm/s for A and v = 22.5 cm/s for B) ((———) experimental; (– – –) calculated; (A) 2.5M Na$_2$S/1M S/1M KOH; (B) 0.25M Na$_2$S/0.1M S/1M KOH)

The effects of charge transfer rate constant were determined by varying the concentration of the redox ions. The results are shown in Figure 5 for electrolyte of composition 2.5 M Na_2S/1 M S/1 M kOH and 0.25 M Na_2S/0.1 M S/1 M KOH. The calculated curves were obtained by assuming the rate constants differed from one another by an order of magnitude to correspond to the redox ion concentration change. For the higher concentration electrolyte, a good fit was obtained. At the lower concentration, however, the fit was poor with the predicted result showing a much more rapid current variation. This may be due to an instability in the surface layer due to buildup of sulfur (8) which would then make ν a function of current density. As the current decreases due to recombination , the sulfur layer gets thinner. This increases ν and thereby reduces the recombination losses making for a more gradual decline in photocurrent. This effect is expected to be more significant at the lower redox ion concentration.

Summary

The observed decrease in photocurrent with decreasing band bending (increasing photovoltage) for CdSe/polysulfide junctions have been accounted for using a model that has depletion region recombination as the major loss mechanism. This is based on reasonably good fits of predicted behavior to experimental I-V data for well characterized electrodes where light intensity and redox ion concentration variations were studied. The significant departure from predictions of the low redox ion concentration I-V data can be ascribed to nonlinear boundary conditions. This appears due to the dependence of the charge transfer rate on current density caused by sulfur buildup on the surface.

Acknowledgements

The authors thank C. Creter and J. DeCarlo for their assistance in obtaining much of the experimental data. Partial support for this study was provided by SERI under Contract XP-9-8002-8.

Abstract

The current-voltage (I-V) characteristics of semiconductor/electrolyte junctions under illumination are analyzed using a model that has depletion region recombination as the dominant loss mechanism. The effects of intensity and charge transfer rate constant on the I-V characteristics are studied theoretically and experimentally using single-crystal CdSe electrodes in polysulfide electrolyte. Reasonably good fits to the data were obtained using a minimum number of adjustable model parameters except at small band bending and for the low redox ion

concentrtion.   These deviations are ascribed to effects not
included in the model.

Literature Cited

1.   Hovel, H.; "Semiconductors and Semimetals, Vol 11, Solar
      Cells;" Academic Press; NY, 1975.
2.   Reichman, J.; "The Current-Voltage Characteristics of
      Semiconnductor-Electrolyte Junction Photovoltaic
      Cells;" Appl Phys Ltr, 36, p 574, 1980.
3.   Gartner, W.W.; Phys Rev; 116, p 84, 1959.
4.   Many, A., Goldstein, Y. and Grover, N.B.; "Semiconductor
      Surfaces;" J. Wiley and Sons, NY, p 156, 1965.
5.   Reichman, J.; to be published.
6.   Miller, B. and Heller,, A.; J. Electrochem Soc; 124,
      p 694, 1978.
7.   Van der Pauw, L.J.; Philips Research Reports; 13, p 1,
      1958.
8.   Cahan, D., Manassen, J. and Hodes, G.; Solar Cell Matls;
      1, p 343, 1979.

RECEIVED October 3, 1980.

Stability of Cadmium-Chalcogenide—Based Photoelectrochemical Cells

DAVID CAHEN, GARY HODES, JOOST MANASSEN, and RESHEF TENNE

The Weizmann Institute of Science, Rehovot, Israel

Cd-chalcogenides (CdS, CdSe, CdTe) are among the most studied materials as photoelectrodes in a photoelectrochemical cell (PEC) (1,2,3,4). Interest in such PEC's stems from the fact that, in aqueous polysulfide or polyselenide solutions, a drastic decrease in photocorrosion is observed, as compared to other aqueous solutions, while reasonable conversion efficiencies can be attained. An important consideration, from the practical point of view, is that thin film polycrystalline photoelectrodes can be prepared, by various methods, with conversion efficiencies of more than half of those obtained with single crystal electrodes and with better stability characteristics than those obtained with single crystal based PEC's (1,4,5).

Recently we showed that PEC's using alloys of Cd(Se,Te) as photoelectrodes can be prepared with stability characteristics much better than those obtained for CdTe based cells and with similar conversion efficiencies (6,7). Those efficiencies could be improved considerably by several chemical and photoelectrochemical surface treatments.

Here we will consider mainly the short-circuit current (SCC) output stability characteristics of both small grain, thin film CdSe and Cd(Se,Te)-based PEC's as well as those of single crystal CdSe-based cells, where different crystal faces are exposed to the solution, after they have undergone any of a series of surface treatments. These studies show a strong dependence of the output stability on solution composition, on real electrode surface area, on surface treatment, on crystal face and on crystal structure (for the Cd(Se,Te) alloys) (1,2,7).

We suggest a predominantly kinetic explanation for these phenomena, involving the fine balance between the rate of the photocorrosion reaction on the one hand and that of the regenerative redox reactions on the other hand.

Because we are interested, ultimately, in practical devices, our aim is to try to attain long term stability (months-years).

0097-6156/81/0146-0369$05.00/0

Therefore unsatisfactory stability by our definition does not necessarily reflect short term behaviour (minutes–hours).

Experimental

Single crystals of CdSe were obtained from Cleveland Crystals Inc., as well as from W. Giriat (IVIC-Caracas), and were n-type with typical resistivities of 3–15 Ω-cm ($\sim 10^{16}$ cm^{-3} donor concentration). Crystal orientation was checked by X-ray diffraction. Electrodes were prepared from them as described elsewhere (1). Polycrystalline CdSe and Cd(Se,Te) electrodes were prepared by painting a slurry of the powder on a Ti substrate and subsequent annealing (5). The Cd(Se,Te) powders were prepared from CdSe and CdTe by cosintering (7). Solution preparation has been described before (1). Except for the outdoors tests, which used 2-electrode cells with CoS counterelectrode (8), all stability tests were done under potentiostatic control at SCC, using Pt counter and reference electrodes. A stabilized quartz-I$_2$ lamp provided illumination. Constant temperature was maintained using a thermostatted water bath surrounding the PEC.

Etching. The polycrystalline electrodes were etched, before use in polysulfide solution, in 3% HNO$_3$ in conc. HCl for ~ 4 secs. Single crystal CdSe electrodes were polished, before use, down to 0.3μ, with alumina polish. Their various etching treatments were as follows:
1. Chromic acid etch only; ~ 10 secs. in CrO$_3$:HCl:H$_2$O (6:10:4 w/w).
2. Aqua regia etch; ~ 20 secs. in aqua regia.
3. Aqua regia/chromic acid etch; ~ 10 secs. in solution of "1", after normal aqua regia etch.
4. Photoetch; 5 secs. in HNO$_3$:HCl:H$_2$O (0.3:9.7:90 v/v), under $\sim$AM1 illumination, as photoelectrode in PEC with carbon counterelectrode, after aqua regia etch (Tenne and Hodes, Appl. Phys. Lett., in press).

Results

During the last few years we have carried out several outdoors test of PEC's containing thin film polycrystalline CdSe photoelectrodes and CoS counterelectrodes. Figure 1 shows the results of one such test carried out during 1979, with an initially high efficiency cell. It is clear from the figure that the drop in efficiency after 2 months is due mainly to parameters affecting the SCC. More recent tests, using several of the improvements that can be obtained from the results discussed below, showed CdSe cells of a similar type to be stable within 10% for 6 months.

The solution used in the cell of figure 1 was chosen in Dec. 1978 and its most important characteristic is the [S]/[S$^=$] (added quantities) of 0.5. Figure 2 shows the considerable importance of this ratio on the output stability of CdSe-based PEC's. We see

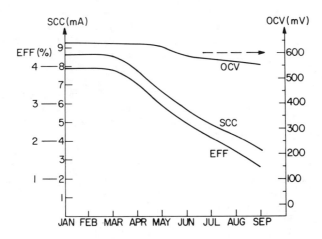

Figure 1. Results of an 8-month outdoors test of PEC containing a 0.8-cm² thin film, painted CdSe photoelectrode (not photoetched), CoS counterelectrode, and 1M KOH, 2M S⁼, 1M S, 1mM Se solution. (OCV) open-circuit voltage; (SCC) short-circuit current; (EFF) solar conversion efficiency (~ AM1.5). Between measurements the cell operated on maximum power (68 Ω load). No appreciable change in fill-factor occurred during the test.

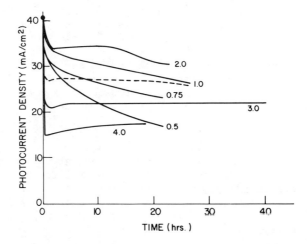

Figure 2. Time dependence of short-circuit current of polycrystalline painted CdSe photoelectrodes as a function of solution composition. All solutions contained 0.8M KOH and 0.8M S⁼. S/S⁼ ratio indicated next to plots. Dashed line shows behavior in 3.2M S solution at 50°C. All other experiments done at 35°C, potentiostatically. Light intensities adjusted to obtain identical initial current densities.

that the $[S]/[S^=]$ of figure 1 is one of the least favourable ones
and that a ratio of 2 would yield much better results. If we compare the output stability of polycrystalline and single crystal
CdSe cells we find the polycrystalline CdSe cells to be much more
stable (figure 3). Figure 3B furthermore stresses the importance
of the initial current density in determining the output stability
of such cells. (Note that the initial current densities of figures
2 and 3 correspond to an illumination intensity equivalent to 3.5
-5xAM1 solar intensity, and as such these experiments are accelerated life-time tests. Such tests may be much more severe than
the factors 3.5-5 would imply).

     For comparison, figure 3B shows results not only for the rela-
tively most stable face of a single crystal (2) and a polycrystalline thin layer prepared by pasting (5), but also for a pressed
pellet of CdSe (2). This last electrode shows stability behaviour
intermediate between that of the two other electrodes. This behaviour is evident also in the decrease in output stability, which
is less steep than that of the single crystal electrode.

     In figure 4 we show how the output stability of a cell using
a single crystal electrode with a specific crystal face exposed to
the solution can vary depending on the surface treatment given to
the electrode, before use. For comparison we include the stability
behaviour of the, least unstable, (1120) face, which was given the
most favourable surface treatment (photoetch). Figure 5 provides
a close look at the faces after the various surface treatments.
(The completely featureless chromic acid etched-face is not shown).
The progressively increasing surface area (from A-C) is evident.
The nature of the small ($\sim 0.1\mu$) holes obtained after purposely
photocorroding the electrode (termed "photoetching") is presently
under investigation.

     Although figures 2-5 indicate how we may improve the output
stability of CdSe-based photoelectrochemical cells (and, in some
cases their conversion efficiency, though never much beyond 4% for
thin-film polycrystalline based ones), a practical device should
have a considerably higher solar conversion efficiency together
with an output stability at least equal to that obtained for the
best CdSe-based cells. CdTe with a bandgap of 1.45eV (at RT) as
compared to ~ 1.7eV for CdSe would be a logical choice for attaining
this goal but, unfortunately, cells using n-CdTe are not stable in
polysulfide solutions under "solar illumination" conditions. (Such
cells are claimed to be stable in polyselenide and polytelluride
solutions (9)). Also addition of tens of millimolar quantities of
Se to polysulfide solutions was shown to improve the output stability of CdSe cells (2). Here we are, for practical reasons,
interested in solutions containing no, or only minimal quantities
of Se and Te because of the much lower toxicity and lower air-
sensitivity of pure polysulfide solutions).

     Figure 6 shows, however, that alloys of $CdSe_x Te_{1-x}$ can have
optical bandgaps comparable to that of CdTe, even if x>0.5 (10).
CdSe and CdTe form homogeneous alloys over the whole composition

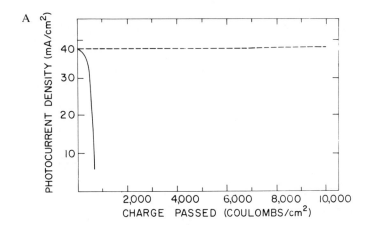

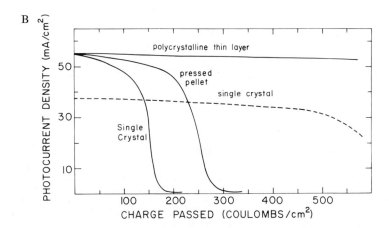

Figure 3. (A) Dependence of short-circuit current on total charge passed for painted polycrystalline (– – –) and single crystal (———, (11$\bar{2}$0) face) CdSe photo-electrodes. Solution contained 1M each KOH, S$^=$, and S. (B) Same as 3A but at higher initial current density. Behavior of pressed pellet CdSe electrode (2) included as well. Other conditions as for 3A (but 5mM Se added to the solution) and for Figure 2. The behavior of the single crystal from 3A is shown, too, for comparison. (Data obtained at Bell Labs.).

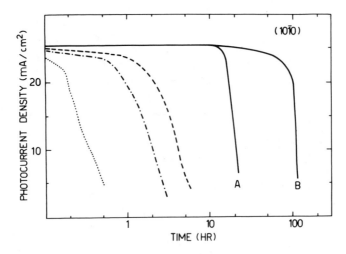

Figure 4.   Logarithmic plot of time dependence of short-circuit current for single-crystal CdSe photoelectrode, (10$\bar{1}$0) face exposed to solution: (· · ·) after chromic acid etch; (– · – ·) after aqua regia/chromic acid etch; (– – –) after aqua regia etch; (——) after aqua regia/photoetch; (A) (10$\bar{1}$0); (B) (11$\bar{2}$0). Further conditions as in Figure 3A.

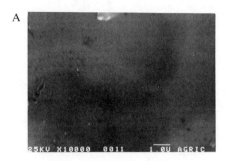

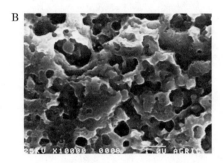

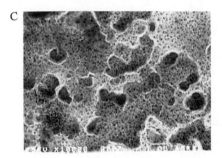

Figure 5. Scanning electron microscope pictures of single-crystal CdSe after several surface treatments. (0001) face, Cd-side, shown. (11$\bar{2}$0) and (10$\bar{1}$0) faces behave similarly. (A) After aqua regia/chromic acid etch; (B) after aqua regia etch; (C) after aqua regia/photoetch.

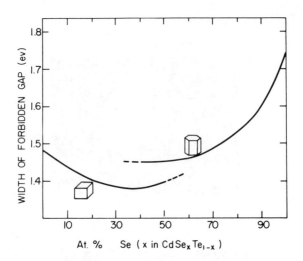

Figure 6.   Direct optical bandgap of Cd(Se,Te) as a function of Se content and crystal structure. Data for well-annealed samples, at room temperature. (After Ref. 10) (7).

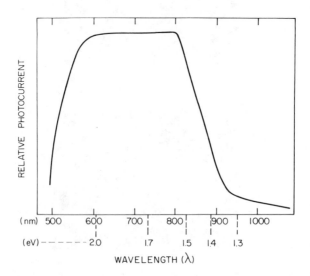

Nature

Figure 7.   Spectral response of polycrystalline, thin-film, painted $CdSe_{0.65}Te_{0.35}$ photoelectrode in 1M KOH, $S^=$,S solution. Corrected for photon density wavelength dependence (6).

range (as do CdS and CdSe (10,11) and CdS and CdTe (10)). The alloy can have either the hexagonal, wurtzite, structure or the cubic, sphalerite, one, depending on its composition and the conditions of preparation (10,12). Table I shows that the lattice parameters of Cd(Se,Te) alloys depend on the alloy stoichiometry in a rather linear fashion, i.e. Vegard's law is obeyed (although deviations of up to one percent are observed).

Table I

$$Cd\ Se_x\ Te_{1-x}$$

X	1-x	Hex(%)	$a(\overset{\circ}{A})$	$c(\overset{\circ}{A})$	Cubic(%)	$a_o(\overset{\circ}{A})$
1.00	0.00	100	4.31	7.02	0	(6.05)
.70	.30	100	4.38	7.165	0	–
.65	.35	>95	4.40	7.18	<5	6.23
.63	.37	>90	4.42	7.20	<10	6.24
.60	.40	70 (±10)	4.43	7.23	30 (±10)	6.26
.57	.43	30 (±10)	4.44	7.24	70 (±10)	6.27
.55	.45	<5	4.44	7.25	>95	6.28
.50	.50	0	–	–	100	6.30
0.00	1.00	0	(4.60)	(7.50)	100	6.48

Crystallographic data for $CdSe_xTe_{1-x}$, from X-ray powder diffraction. Stoichiometries from relative amounts of starting materials, and from Vegard's law. Percentages of cubic and hexagonal structures estimated from powder diffraction intensities. Lattice dimensions in parentheses are literature data.

Figure 6 shows moreover that the two structures can coexist over a range of stoichiometries. The spectral response of a PEC containing $n-CdSe_{.65}Te_{.35}$ as photoanode is shown in figure 7. An effective optical bandgap of ∿1.45eV, in agreement with figure 6, is observed. Such cells can, under optimal conditions, attain solar conversion efficiencies of 8% (6). Here we will consider their output stability as a function of stoichiometry of the photoelectrode or, more precisely, as a function of the crystal structure of the photoelectrode. Figure 8 illustrates the pronounced differences in stability between cells with electrodes having different stoichiometries. The strong decrease in stability which occurs when the stoichiometry is altered only slightly, suggests that another factor besides composition, plays a role. As indicated on the figure this factor is the crystal structure. Electrodes that have a predominantly hexagonal structure are much more stable than those having mainly the cubic structure. This is

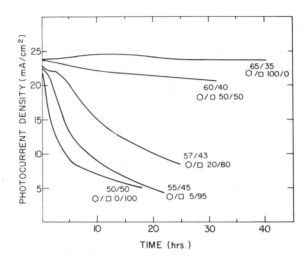

Figure 8.   Time dependence of short-circuit current for several Cd(Se,Te) electrodes in 2M KOH, S⁼, and S. Further details as in Figure 2. See Table I for determination of stoichiometry (upper numbers, Se/Te) and crystal structure (lower numbers) (7).

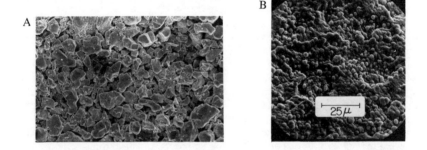

Journal of the Electrochemical Society

Figure 9.   Scanning electron microscope pictures of painted Cd chalcogenide films: (A) Cd(Se₀.₆₅Te₀.₃₅); (B) CdSe (3× larger magnification than A). Both surfaces were etched before the experiment (5).

true also at a fixed stoichiometry. Cells with $CdSe_{0.60} Te_{0.40}$
photoelectrodes could be prepared with a higher hexagonal content
than that shown in the figure, and then showed a stability closer
to that found for the $CdSe_{0.65} Te_{0.35}$ – containing cell.

The surface morphology of such alloy electrodes is shown in
figure 9A and compared to that of CdSe electrodes (figure 9B),
prepared by the same "pasting" method (5). The increased grain
size of the alloy electrodes can be attributed to the double
annealing process needed for their preparation. A single annealing
step (of the slurry on the Ti-substrate) suffices for preparation
of the CdSe electrodes. As we shall see below, this increased
grain size can have a detrimental effect on output stability.

Discussion

The results presented above for CdSe-based polysulfide cells
(figures 1-4) can be understood in terms of the competition between regenerative redox reactions ($S^=$ oxidation) and photocorrosion of the semiconductor as illustrated schematically in figure
10. This scheme is based on the model for deactivation of CdSe
and CdS polycrystalline – and single crystal electrodes, proposed
previously (1,13). This model involves the above mentioned two
possible reaction paths open to a photocorrosion reaction produces
zerovalent Se or S, and Cd^{2+} ions, whose rapid reprecipitation
leads to the formation of a finely polycrystalline top layer of
CdS (1,14, Cahen and Vandenberg, submitted) and, in the case of
CdSe, to S/Se substitution (1,2,14). Although one can argue that
the deactivation of CdSe is due to the formation of a chemically
different (CdS) layer on it, by introducing a barrier for holes
on their way to the surface, figure 11 shows that CdS behaves
rather similar to CdSe. This suggests that the surface restructuring alone, e.g. differences in grain size, and, possibly, a
different doping level of the top layer of CdS, as compared to
the bulk, may suffice to destabilize the electrode, by introducing
a new resistance to current flow. The results shown in figure 1
can be understood in the light of these findings, as we would
expect the SCC to be affected mostly by this increased resistance.
The explanation for the results of figure 2 can be found in the
extreme left-hand process of figure 10, namely the dissolution of
S from the surface. Previously we showed how the solution composition can influence the transient photocurrents in $CdSe/S_x^=$ PEC's,
in a manner which can be correlated with the S-dissolving ability
of the solution (15). An optimum solution composition was found
for a [S]/[$S^=$] ratio of 1.5-2. Figure 2 shows that the same ratio
exerts an optimal stabilizing influence on the cell. This can be
rationalized by considering the effect of elemental S on the
electrode surface. Such sulfur will inhibit the flow of holes to
$S_n^=$ from the solution (or, limit the availability of adsorbed $S_n^=$ at
the surface) and thus increases the probability of photocorrosion.
(A more electrochemical way of looking at this is to consider the

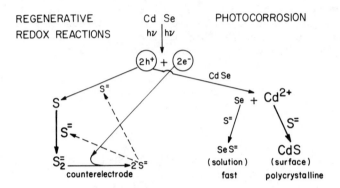

Figure 10. Scheme of competing reactions at illuminated CdSe surface. The process on the extreme left is the main rate-limiting step, while the one on the right can lead to photoelectrode corrosion.

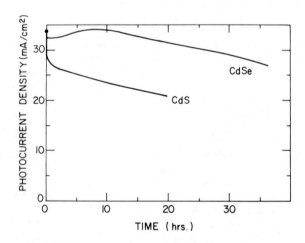

Figure 11. Time dependence of SCC for thin-film polycrystalline CdS and CdSe photoelectrodes (*1M KOH, 2M S⁼, and 0.2M S solution*). Note unfavorable [S]/[S⁼] (see *Figure 2*), chosen here to minimize solution absorption. Further details as in *Figure 2.*

shift in potentials caused by the presence of a, rather non-conducting, sulfur layer on the surface. This reasoning leads to the same conclusions). It is quite likely that, initially, the photocorrosion reaction will affect electrode performance only minimally. However, because photocorrosion leads to restructuring of the electrode top layer, which causes a decrease in hole flow to the surface as well, it is, in a sense, an autocatalytic reaction, at least once it has proceeded beyond a critical point (a critical effective thickness of the restructured top layer). This autocatalytic nature of the destabilization process can be seen in figure 3 for the single crystal electrode. It is clear then from the above that photoelectrode stability will depend on the relative rates of the two competing processes (13). Increasing the illumination intensity may favour the photocorrosion reaction at constant maximal rate of sulfide oxidation and sulfur removal (constant solution composition). On the other hand an increase in solution temperature increases the regenerative redox reaction rate as well as the rate of sulfur removal, and, at constant illumination intensity will lead to increased stability (figure 2). (In the Result's section we sounded a cautionary note regarding the use of accelerated life time tests. In view of the autocatalytic nature of the degradation process, and the fact that at low current densities an electrode may function well, even though some surface restructuring takes place (1), this caution can be understood. In simple terms: 5,000 C/cm^2 at 4xAM1 tells us only that at AM1 the electrode will be stable during passage of at least this amount of charge. It may very well be stable for 50,000 C/cm^2, however!).

Also the results of figures 3 and 4 can be understood in terms of this competition. The regenerative redox reaction ($S^=$ oxidation) will, as any electrochemical process, depend on the real surface area, because this will determine the true current density. The higher the real surface area the lower the true current density and the more the regenerative redox reaction will be favoured over the photocorrosion process, which depends mainly on the hole flux to the surface. Figure 9B shows the rather small grain size of our thin film CdSe electrodes. Therefore, at equal geometric current densities, the true current density passing through the polycrystalline thin film electrode. Figure 3B furthermore shows that an electrode with true surface area intermediate between these, does indeed show a stability behaviour between those of the single crystal and thin film electrode. The results of figure 4 are now clear as well, especially in the light of the SEM pictures of figure 5. The various surface treatments influence, inter alia, the true surface area of the electrodes and thus their stability behaviour. Presently we are studying the effects of these surface treatments further, as, for example, the improved conversion efficiency of photoetched thin film polycrystalline electrodes (6) cannot be explained by this surface area effect. Clearly also the previously observed (2),

and here confirmed differences in stability between electrodes
with different crystal faces exposed to the solution, cannot be
attributed to surface area effects, as the various surface treatments all had similar effects on the three crystal faces studied:
$(10\bar{1}0)$, (0001) and $(11\bar{2}0)$. A possible reason for the differences
in stability between these faces (which we found to have decrea-
sing stability in the order $(11\bar{2}0)$, $(10\bar{1}0)$ $\sim$ (0001) for all surface treatments) may be found in the number of bonds that has to
be broken to dislodge a Cd ion from the surface. In the $(11\bar{2}0)$
plane there are two in-plane bonds and one into the bulk phase;
for the $(10\bar{1}0)$ plane there is only one in-plane bond and one or
two into the bulk phase; the (0001) plane has no in-plane bonds
but 1 or 3 into the bulk phase. In this, admittedly, simplified
picture the $(11\bar{2}0)$ plane is the stable one because 3 bonds need
to be broken to disrupt it. Alternatively one can look at the
possibility for strong surface-solute interaction, which can influence the relative energetics of the two competing processes
and which, in its turn, will be dependent on the number of non-
bonded electrons available at the surface. Future experiments
will try to clarify these hypotheses through the use of other
crystal faces.

We now turn to the strong crystal structure dependence of
the output stability of Cd (Se,Te)-based PEC's. In figure 12 we
compare the two crystal structures that can be adopted by the
alloys. Both space-filling and stick-and-ball models are shown,
because the close relationship between the two structures can be
best appreciated from the space-filling models, while the Cd and
chalcogen coordination is seen clearest in the stick-and-balls
model. (The hexagonal phase is essentially a close-packed ABAB
hexagonal stacking of tetrahedrally coordinated Cd atoms, and
the cubic phase has ABCABC stacking. The first coordination shell
is identical in both forms). Several possible explanations can
be forwarded to explain this dependence. For example, because
the output stability of hexagonal CdSe depends on the crystal face
exposed to the $S_n^=$ solution, preferred orientation of the, more
stable, $(11\bar{2}0)$ plane, which is known to be a preferred cleavage
plane of the wurtzite phase could explain the improved stability
of the hexagonal phase over that of the cubic one. (The sphalerite
(110) plane is crystallographically equivalent to the wurtzite
$(11\bar{2}0)$ one). Comparison between observed X-ray powder intensities
and calculated ones (on the basis of the known crystal structures)
shows indeed some preferred orientation in these pasted electrodes,
but rather of the wurtzite $(10\bar{1}0)$ plane, and of the crystallographically similar sphalerite (111) one.

Probably we are dealing here with an intrinsic difference
between these two structure types. Their solid state chemistry
shows the wurtzite structure to be the more rigid one (in terms
of substitution, for example). This could imply a difference in
Cd-X (X=S,Se,Te) bond strengths. These bond strengths are,
according to Pauling's ionicity concept, influenced by the

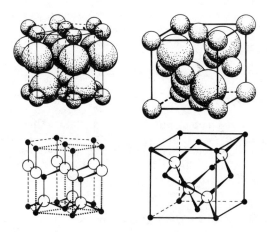

Figure 12.   Space filling (top) and stick-and-ball (bottom) models of wurtzite (left) and sphalerite (right) structures

effective ionic charges on the atoms and these charges can be calculated from piezoelectric coefficients. For ZnS (the only case
where these constants are well-known for both structure types) Zn
is found to have a higher effective ionic charge in the hexagonal
structure, implying stronger bonding, than in the cubic one. Also,
from figure 6 we see that the cubic form has a consistently lower
optical bandgap than the hexagonal one (over the range where they
can coexist). This we can associate, be it somewhat loosely, with
a weaker interaction between Cd and X (in all II-VI compounds that
adopt these structures, this difference in bandgaps is found).

Since the stability of Cd chalcogenide photoelectrodes in polysulfide solution depends on the relative probabilities of self-
oxidation,which leads to breaking a Cd–X bond,and interfacial charge transfer,a small change in bond strength (only 1-1.5 Kcal/mole,
from the differences in bandgaps,i.e. 2-3% of the total bond dissociation energy) may influence the semiconductor stability pro-
foundly if it is assumed that the difference in activation energies
for the cleavage of the Cd–X bond,in the two structures,is of the
order of 1-1.5 Kcal/mole too. This can be pictured schematically as
follows:

$$\left(h^+\right) \quad X \xrightarrow[f_1]{} \quad Cd \xrightarrow[f_2]{} \quad S_n^=$$

Here f_1 and f_2 represent the bond strength of the Cd–X bond,
dictated by bulk crystal structure, and that between the surface
Cd and the probably adsorbed, polysulfide species from solution,
resp. The larger f_1, the more likely it is that the photogenerated hole,h^+, will react with the polysulfide species, i.e. the
lower the probability of self-oxidation.

From this explanation for the observed differences in output
stability, we may infer that the decomposition potential of any
CdX depends on its crystal structure. In view of the results presented here and in ref. 2, it is likely that this potential is a
function of exposed crystal face as well. Further insight in those
dependences, development of optimal surface treatments, and use of
electrodes with optimal grain size, may all be of help in devising
strategies to obtain even more stable $CdX/S_n^=$ photoelectrochemical
systems.

Acknowledgements

Partial support of the U.S. Israel Binational Science Foundation, Jerusalem, Israel, and Ormat Turbines Ltd., Yavneh, Israel,
is gratefully acknolwedged. DC thanks Bell Labs. for hospitality
and   A. Heller, B. Miller and M. Robbins for providing some of
the electrodes used for Figure 3B.

Abstract

The output stability of CdS, CdSe and to some extent,
Cd(Se,Te)/polysulfide photoelectrochemical cells can be explained

in terms of a competition between the semiconductor self-oxidation and the desired polysulfide oxidation. This stability is shown to be dependent on solution composition, real electrode surface area, crystal structure and crystal face. Optimal solutions have maximal sulfur dissolution ability. For CdSe increase in the electrode surface area causes an improvement in output stability, thus explaining the observed differences between polycrystalline and single crystal electrodes, and between single crystal electrodes which are given different surface treatments. Among these, a deliberate photocorrosion process is found to lead to greatest stability. Variations in output stability between single crystal faces may be connected with differences in the number and nature of bonds to surface atoms. Cd(Se,Te) electrodes have reasonable stabilities only if they have a predominantly wurtzite structure. The inferior behaviour of such electrodes with the cubic structure is explained in terms of small differences in bond strengths, which affect the probabilities of the above-mentioned reactions.

Literature Cited

1. Cahen, D.; Hodes, G.; Manassen, J. J. Electrochem. Soc., 1978, 125, 1623.
2. Heller, A.; Miller, B. Adv. Chem., 1980, 184, 215; Miller, B.; Heller, A.; Robbins, M.; Menezes, S.; Chang, K.C.; Thomson Jr., J. J. Electrochem. Soc., 1977, 124, 1019.
3. Ellis, A.B.; Kaiser, S.W.; Wrighton, M.S. J. Amer. Chem. Soc. 1976, 98, 6855.
4. Russak, M.A.; Reichman, J.; Witzke, H.; Deb, S.K.; Chen, S.N. J. Electrochem. Soc., 1980, 127, 725.
5. Hodes, G.; Cahen, D.; Manassen, J.; David, M. J. Electrochem. Soc., 1980, in press.
6. Hodes, G. Nature (London), 1980, 285, 29.
7. Hodes, G.; Manassen, J.; Cahen, D. J. Amer. Chem. Soc., 1980, in press.
8. Hodes, G.; Manassen, J.; Cahen, D. J. Electrochem. Soc., 1980, 127, 544.
9. Ellis, A.B.; Bolts, J.M.; Kaiser, S.W.; Wrighton, M.S. J. Amer. Chem. Soc., 1977, 99, 2839;2848.
10. Tai, H.; Nakashima, S.; Hori, S. Phys. Stat. Sol. A., 1975, 30(a), K115.
11. Noufi, R.N.; Kohl, P.A.; Bard, A.J. J. Electrochem. Soc., 1978, 125, 376.
12. Stuckes, A.D.; Farrell, G. J. Phys. Chem. Solids, 1964, 25, 477.
13. Cahen, D.; Manassen, J.; Hodes, G. Sol. En. Mater., 1979, 1, 343.
14. Gerischer, H.; Gobrecht, J. Ber. Bunsenges. Phys. Chem., 1978, 82, 520.
15. Lando, D.; Manassen, J.; Hodes, G.; Cahen, D. J. Amer. Chem. Soc., 1979, 101, 3969.

RECEIVED October 3, 1980.

Photoeffects on Solid-State Photoelectrochemical Cells

PETER G. P. ANG and ANTHONY F. SAMMELLS

Institute of Gas Technology, IIT Center, 3424 South State Street, Chicago, IL 60616

Photoeffects in solid-state photovoltaic devices have been demonstrated for a variety of interfaces, e.g., at p-n junctions and Schottky barriers. For such devices a series of highly-controlled fabrication steps are necessary since high-purity semiconductor materials are required to reduce electron-electron hole recombination reactions at trap sites. Greater simplicity in solar cell fabrication can, however, be achieved with liquid-junction photoelectrochemical cells (PEC). For such devices a junction is formed merely by introducing a semiconductor electrode into a redox electrolyte. Illumination of, for example an n-type semiconductor, can result in oxidation of a redox species via its reaction with electron holes photogenerated in the semiconductor space charge region. Solar energy conversion efficiencies achieved with such liquid-junction devices have recently been approaching those of conventional solid-state photovoltaic devices.

Photoelectrochemical devices have the inherent advantage that either electricity or useful chemical species, potentially available for later electrochemical discharge, can be produced at the interface, whereas only direct electricity generation occurs with solid-state photovoltaic cells. High solar energy conversion efficiencies have been accomplished by narrow band-gap materials which can utilize a large fraction of the solar spectrum. However, in many cases such materials can manifest photoanodic corrosion effects with time, particularly when the redox species equilibrium potential is close to the semiconductor decomposition potential. Approaches to stabilize those semiconductors have included the evaluation of both organic and inorganic nonaqueous electrolytes. Most work is, however, at an early stage. Extending the thought of identifying electrolytes having higher stability and lower nucleophilicity has led us to consider the utility of solid electrolytes for photoelectrochemical devices. No work has been previously reported on the evaluation of solid-state photoelectrochemical systems. It is difficult to conceive a regenerative solid-state PEC because of the ionic specificity of most solid electrolytes. This has led us to consider semiconductor/solid electrolyte interfaces, where an immobile redox species might be present, either in a solid electrolyte lattice

0097-6156/81/0146-0387$05.00/0

position, or as a separate phase, in intimate contact with the solid electrolyte. Illumination of a semiconductor/solid electrolyte interface containing such a redox species might result in either oxidation or reduction of this species, depending upon whether the semiconductor is a photoanode or a photocathode. Because the redox species is immobile, such a device would store the energy produced. To some extent the viability of this approach will, in part, be dependent upon the accommodation of a concentration of mobile cations somewhat greater than that required for acceptable ionic conductivity across such a solid-state cell. This is necessary to preserve electroneutrality in the proximity of the immobile redox species. From crystallographic considerations, beta-alumina has a particularly suitable lattice structure for this concept. Here, the a and b axes are periodically separated by Al-O-Al bonds, between which the mobile sodium ions migrate. A significant excess of sodium ions can be accommodated within this material. In the course of this work solid electrolytes evaluated have included beta-alumina, lithium iodide, and rubidium silver iodide. Unfortunately, the sodium ion conductivity of beta-alumina at room temperature is too low to be of utility in solid-state photoelectrochemical devices. Any photopotential generated would probably be insufficient to overcome resistance losses within the solid electrolyte. Because of the high ionic conductivity of the rubidium silver iodide at ambient temperature, experimental emphasis was placed on this material.

Rubidium silver iodide, $RbAg_4I_5$, has a conductivity of 0.27 ohm^{-1} cm^{-1} at ambient temperatures [1] and has found application in galvanic cells [2]. Solid electrolytes have also been synthesized through compound formation between AgI and various substituted ammonium halides [3-5] and also between AgI and sulfonium iodides [6]. The single crystal structures of certain unique phases have been investigated [7-11] to determine the structural features that permit fast ion diffusion in solids. It was observed that the iodide ions form face-sharing polyhedra (generally tetrahedra) with the silver ions situated within these polyhedra. On the average there are two or three vacant sites per silver ion, thus permitting ready diffusion through the faces of the polyhedra.

Photoeffects at the semiconductor/$RbAg_4I_5$ interface were investigated with the objective of identifying the utility of this solid electrolyte material in solid-state photoelectrochemical devices with storage.

Experimental

Solid-state cells were fabricated using $RbAg_4I_5$ (Research Organic Inorganic, Sun Valley, California), silver powder (Cerac, Inc.) and PbI_2 and/or iodine powder. Layers of the appropriate mixtures were prepared inside a commercial KBr die and pressed at 2000 kg/cm^2 using a Carver laboratory press. The resulting cell (about 2mm thick) was sandwiched between a counterelectrode and a transparent semiconductor working electrode. Current collection to the working electrode was made by a transparent conducting glass. The conducting

glass electrode was tin oxide-coated Corning 7059 glass with a
thickness of 0.032" and a standard resistivity of 100 ± 30 ohms per
square. A small silver wire reference electrode was inserted inside the cell via a small hole.

The photoelectrochemical properties of the semiconductor/solid
electrolyte interface was investigated using a Wenking ST-72 potentiostat and an Oriel 150-Watt xenon light source. Monochromatic
light was obtained by using an Oriel Model 7240 monochromator. The
slit opening of the monochromator was set at 1.5mm, thus giving a
spectral bandpass of 10 nm. The electrode potential was measured
using a Wenking PPT-70 electrometer. A Tacussel type GSTP2B pulse-
sweep generator was used to control the potentiostat. The current-
voltage relationships of the electrodes were recorded on a Hewlett-
Packard 7046 X-Y recorder. An n-type polycrystalline CdSe was made
by thermal vacuum evaporation of CdSe powder (99.999%, Cerac, Inc.)
on the conducting glass. This was accomplished by using an Edwards
306 thin-film evaporator at a pressure of 10^{-6} Torr. The electrode
was annealed in air at 400°C for 30 minutes and cooled down slowly.

Results and Discussion

Photoactivity of $RbAg_4I_5$. Before evaluating photoeffects at
the semiconductor-solid electrolyte interface, it is first necessary
to characterize any photoactivity present in the solid electrolyte
itself. Rubidium silver iodide can be expected to possess some photosensitivity by analogy to the photovoltaic effect demonstrated
by the binary silver halides (12,13). This was performed by evaluating a cell of configuration Cond. glass/PbI_2 + $RbAg_4I_5$/$RbAg_4I_5$/Ag +
$RbAg_4I_5$/Cond. glass. Solid-state cells based upon this solid electrolyte usually consist of a silver anode and a complexed iodine
cathode (14). This cell was mounted between two transparent conducting glass electrodes. Upon illumination of the Cond. glass/PbI_2 +
$RbAg_4I_5$ interface, this side developed positive photopotentials of
between 3 and 50 mV. However, after initially passing both anodic
as well as cathodic currents through this interface, larger photopotentials could be achieved. During passage of anodic current at
this interface, dark areas appeared on the electrolyte surface. This
might be explained by the formation of free iodine at this interface.
Larger photopotentials were realized at these darker areas. For
example, a potential of 450 mV in the dark and 850 mV under illumination was found versus the silver reference electrode. This is
shown in Figure 1. However, when the light beam was not focused on
the darkened spot, the photopotential was only about 5 mV. Large
photopotentials were still detected at wavelengths longer than 4750 Å,
when a sharp-cut filter was used in the light path. The response
of the cell under this condition is shown by the second deflection
shown in Figure 1. The action spectra of the cell will be described
later in more detail.

Photoexcitation of the $RbAg_4I_5$ can be expected to result in the
formation of electron-hole pairs. The photoelectrons will diffuse
from the surface into the bulk because of their much greater drift
mobility compared to the photoholes. A net positive space charge

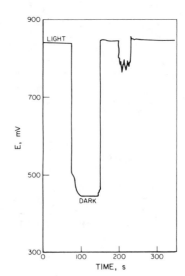

Figure 1. Photopotential in the cell: cond. glass/PbI_2+$RbAg_4I_5$/$RbAg_4I_5$/Ag + $RbAg_4I_5$/cond. glass. The cond. glass/PbI_2 + $RbAg_4I_5$ interface was illuminated with 200 mW/cm^2 xenon light.

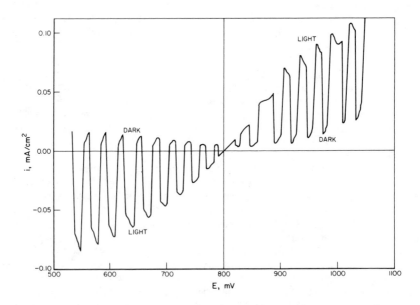

Figure 2.   Current–voltage characteristics of the cell: cond. glass/PbI_2 + $RbAg_4I_5$/ $RbAg_4I_5$/Ag + $RbAg_4I_5$/cond. glass under chopped illumination (200 mW/cm^2) at 25°C (voltage scan rate: 25 mV/s)

will probably be formed near the illuminated electrode due to this
higher concentration of photoholes. These photoholes can be expected
to oxidize any free silver present at this interface. These processes can lead to a positive photovoltage at the illuminated electrode, which can be correlated to Dember-type photovoltage (13,15).

The current-voltage characteristics of the cell were analyzed
by sweeping the cell voltage using a potentiostat. The cell voltage
was measured as the voltage between the illuminated side versus the
non-illuminated silver side. Under chopped xenon light illumination,
cathodic as well as anodic photocurrents were recorded, as shown in
Figure 2. These photocurrents, as well as the photopotential in
Figure 1, indicate that photoexcited charge carriers are produced
in addition to the silver ions present in the dark. This correlates
well with the Hall effect of $RbAg_4I_5$ in the dark and under illumination
that has been previously reported (16). Hall signals in the dark
are caused by the Ag^+ ion carriers. Under illumination the sign of
the Hall voltage, however, is reversed. From this, and by virtue
of the magnitude and temperature dependence of Hall mobility, it was
concluded that the Hall signals under illumination was primarily
contributed by photoelectrons.

If we assume that the dark areas are formed by photolytic iodine,
then the positive photopotentials in Figure 1 can be expected to
lead to larger cathodic currents. Here the iodine will be reduced
to iodide. Under continuous illumination, it has been shown that
cathodic photocurrents can be drawn for many hours. When the potential of the illuminated interface was made anodic, iodide will be
oxidized to iodine on the surface. Photogenerated charge carriers
may reduce the internal resistance of the cell, thereby increasing
the effective current density upon illumination, as shown in Figure 2.
However, at large cell voltages, the electrolyte will decompose according to (14): $2 RbAg_4I_5 \rightarrow 4 Ag + 2 RbI_3 + 4 AgI$, with the con-
current equilibration $RbI_3 \rightleftarrows RbI + I_2$. Iodine generated by such
mechanisms may be responsible for the dark areas on the electrolyte
surface. An increase in the anodic dark current at voltages higher
than 900 mV in Figure 2 can be correlated with this process. The
experiment, however, was performed with no compensation for resistance and polarization losses within the solid electrolyte cell.
Hence, the voltages for the inception of electrochemical reactions
may differ from values expected from a reversible electrode.

It is of interest to investigate the photosensitivity of $RbAg_4I_5$
solid electrolyte doped with free iodine. Here the surface of the
pressed electrolyte was coated with solid iodine for a few seconds
until it became light brown. The cell was assembled between transparent conducting glass current collectors. The potential of the
iodine-treated surface was measured against a silver reference electrode. Due to the high activity of the iodine, the potential was
661 mV in the dark. This surface immediately developed a large positive photopotential of about 100 mV upon illumination with the xenon
light. The current-potential characteristics of this surface was
essentially the same as the one shown in Figure 2. However, an increase in the dark current in the anodic direction was noticed. The
dependency of the photopotential of this cell upon the wavelength

of the illuminating light is shown in Figure 3. Below 300 nm, essentially no photopotential was detected. This was not surprising since the glass and the tin oxide layer on the conducting glass electrode would absorb light with wavelengths below this value. The band-gap of SnO_2 is within this range, i.e., 3.8 eV. This would correspond to a wavelength of about 326 nm. The conducting glass had no photoactivity when it was illuminated in liquid electrolytes. In Figure 3 the photoresponse peak of this solid electrolyte system occurred at a wavelength of about 425 nm. The photopotential decreased at longer wavelength. The peak at 425 nm can be correlated to the 425 nm band in the optical absorption spectrum of $RbAg_4I_5$. The spectrum actually showed four bands at 485, 425, 375, and 335 nm (17). The 425 nm band was attributed to a forbidden internal transition in the Ag^+ which becomes permitted because of the tetrahedral coordination of the Ag^+ ion in $RbAg_4I_5$. The photopotential, as well as the anodic and cathodic photocurrents were seen up to wavelengths approaching 700 nm.

The photopotential was also measured as a function of light intensity. At 200 mW/cm^2 xenon light illumination, a photopotential of about 100 mV was measured. The photopotential decreased upon attenuation of the light intensity with neutral density filters. A plot of the photopotential versus the relative light intensity is shown in Figure 4. The photopotential is proportional to the logarithm of the light intensity, as observed in the Dember photovoltage of AgBr (13).

CdSe/Solid Electrolyte Junction. It has been demonstrated in the previous section that $RbAg_4I_5$ is photosensitive. Our primary goal in this study was actually to see photoactivity at a semiconductor/solid electrolyte interface. An n-type CdSe was chosen for this study since a thin-film, partially transparent electrode could be fabricated by thermal vacuum evaporation on a transparent conducting glass. A cell of the configuration CdSe on Cond. glass/I_2 + $RbAg_4I_5$/ $RbAg_4I_5$/Cond. glass was fabricated. A small quantity of iodine was introduced onto the solid electrolyte surface as redox species. The CdSe electrode was then placed in contact with this surface. The potential of the CdSe was measured versus a silver wire reference electrode. The potential in the dark was 650 mV and it moved to about 300 mV under 200 mW/cm^2 xenon light illumination. However, the actual light intensity reaching the CdSe/solid electrolyte interface was significantly lower. This rapid negative photopotential is shown in Figure 5. This phenomenon is analogous to the behavior of n-CdSe in liquid-junction photoelectrochemical cells.

The dependency of the CdSe electrode potential upon the wavelength of the light is shown in Figure 6. The spectrum was not corrected for variations in light intensity by the monochromator. The mean intensity at 625 nm was 70 μW/cm^2 with variations at other wavelengths estimated as ±25 μW/cm^2. No photopotential was seen below 300 nm. Between 300 and 375 nm, a positive photopotential was detected. This could have the same origin as the photopotential of the Cond. glass/$RbAg_4I_5$ interface shown in Figure 3. A large nega-

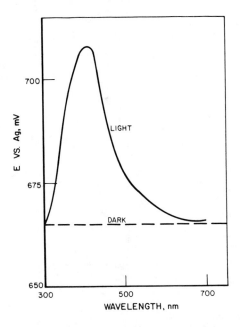

Figure 3.   Potential of cond. glass/I_2 + $RbAg_4I_5$ interface vs. Ag as a function of wavelength

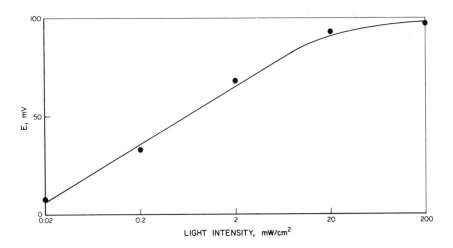

Figure 4.   Photoresponse of cond. glass/I_2 + $RbAg_4I_5$/cond. glass vs. light intensity

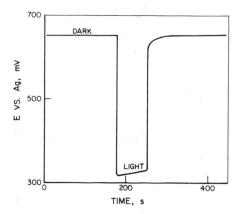

Figure 5.   Photopotential of CdSe/I$_2$ + RbAg$_4$I$_5$/RbAg$_4$I$_5$/cond. glass, back-illumination with 200 mW/cm^2 xenon light

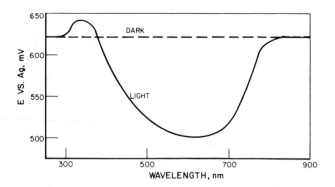

Figure 6.   Potential of CdSe in contact with I$_2$ + RbAg$_4$I$_5$ vs. Ag as a function of the wavelength of the light

tive photopotential can be seen in Figure 6 at wavelengths longer than 375 nm, up to about 800 nm, which corresponds closely to the 1.7 eV band-gap of CdSe. Thus, the negative photopotential can be assigned to the CdSe/solid electrolyte interface. The photoresponse between 300 and 500 nm in Figure 6 might very well be a superposition of the positive photopotential of the Cond. glass/$RbAg_4I_5$ interface and the negative photopotential of the CdSe/$RbAg_4I_5$ interface. The magnitude of the negative photopotential was largest at 625 nm. The dependency of this photopotential upon the intensity of the incident light was also analyzed. Figure 7 shows the negative photopotential as a function of incident optical power. The relative scale at 10° corresponds to 70 µW/cm², where the photopotential was −85 mV. With approximately 100 mW/cm² white xenon light, photopotentials up to −300 mV have been detected. At the very low light intensity of 0.7 µW/cm² at 625 nm, photopotentials of −2 mV were still observed. At moderate light intensities the photopotential is proportional to the logarithm of the light intensity.

The current-potential characteristics of the CdSe electrode were studied under chopped illumination. The result is shown in Figure 8. The open-circuit potential versus Ag was 350 mV under xenon light. When the potential of the CdSe electrode was scanned in the positive direction, anodic photocurrents flowed. At high potentials the photocurrent increased, but the dark current increased as well. When scanned in the negative direction, a small cathodic current flowed due to reduction of iodine. Initially, small anodic photocurrent deflections were seen under chopped illumination near the resting potential. However, at higher cathodic polarization the photocurrent became cathodic. A cathodic photocurrent has already been demonstrated by $RbAg_4I_5$ as shown in Figure 2. Bringing the potential more cathodic than the equilibrium potential of silver resulted in a large cathodic current corresponding to reduction of silver ions to silver. When the potential was brought back to the anodic direction, an anodic current flowed due to reoxidation of the silver. Usually, the photopotential decreased when the electrode is brought to near 0 mV versus silver. This could be due to plating of silver on the electrode surface. However, upon reoxidation of the silver the photopotential was restored. The decomposition of $RbAg_4I_5$ to either silver or iodine species may create limitations to the use of this solid electrolyte for photoelectrochemical cells.

Conclusion

We have demonstrated photoactivity at solid electrolyte/Cond. glass interfaces. The positive photopotentials could, however, be a Dember-type photovoltage. Cathodic photocurrents passed when the cell was discharged. Illumination may yield iodine on the illuminated side with some reduced species (probably silver) in the bulk of the electrolyte.

When CdSe was in contact with the solid electrolyte containing iodine redox species, negative photopotentials and anodic photocurrents were detected. This is analogous to the behavior of n-type semiconductor in liquid-junction solar cells. In the case of solid

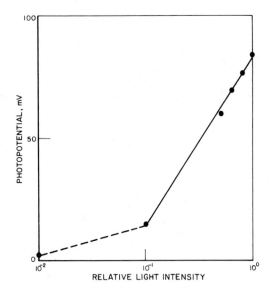

Figure 7.   Dependence of the photopotential of CdSe/I_2 + RbAg$_4$I$_5$ at 625 nm on light intensity ($10°$ corresponds to 70 μW/cm^2)

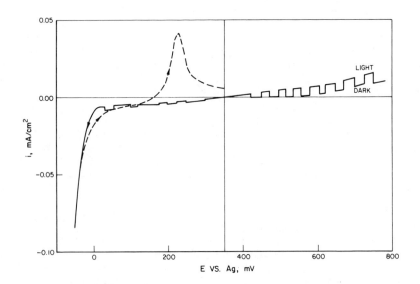

Figure 8.   Current–potential characteristics of CdSe/I_2 + RbAg$_4$I$_5$ interface under chopped illumination (voltage scan rate: 40 mV/s)

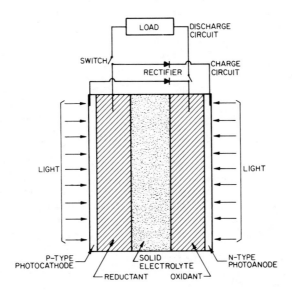

Figure 9.   Photoelectrochemical cell with a solid electrolyte containing reductant and oxidant on each side (the charging circuit contains rectifiers to prevent self-discharge in the dark)

electrolytes, the redox species are immobile. This opens up the possibility of constructing a solid-electrolyte photoelectrochemical cell with a storage capability as shown in Figure 9. The solid-state electrolyte device would consist of three phases in intimate contact where in the ideal case the crystal lattice structure of the reductant, oxidant, and solid electrolyte phases would be similar to facilitate rapid movement of ions between each phase. The reductant and oxidant portions of the device would contain selected redox species either introduced into fixed immobile lattice sites in the solid electrolyte (in the ideal case) or dispersed and in intimate contact with the solid electrolyte in each of the respective compartments. The reductant and oxidant electrodes would possess both ionic and electronic conductivity, whereas the separating solid electrolyte would possess only ionic conductivity. After separation of electrons and holes in the n-type semiconductor, the following general reactions may occur during charge: At oxidant electrode —

$$nC^+ + M^{+n}_{1(\text{lattice})} + h^+ \rightarrow M^{+(n+1)}_{1(\text{lattice})} + (n-1)C^+$$

At reductant electrode —

$$(n-1)C^+ + M^{+m}_{2(\text{lattice})} + e^- \rightarrow M^{+(m-1)}_{2(\text{lattice})} + nC^+$$

where: M_1 represents a redox species which is oxidized by an electron hole h^+; C^+ represents a mobile ion conductor; and M_2 represents a redox species which is reduced by an electron e^-.
    During discharge, the following reaction may occur:

$$M^{+(n+1)}_{1(\text{lattice})} + M^{+(m-1)}_{2(\text{lattice})} \rightarrow M^{+n}_{1(\text{lattice})} + M^{+m}_{2(\text{lattice})}$$

The basic requirement of the redox oxidant in contact with n-type semiconductors is that it has an equilibrium potential more negative than the decomposition potential of the semiconductor and more positive than the lower edge of the semiconductor conduction band. The basic requirement of the reductant electrolyte is that its redox equilibrium potential be negative of the oxidant electrolyte and more positive than the lower edge of the semiconductor conduction band. More work will be necessary at characterizing solid electrolyte/semiconductor interfaces with those solid electrolytes available before satisfactory solid-state devices capable of photocharge can be realized.

Acknowledgment

This work was sponsored by the Solar Energy Research Institute for the United States Department of Energy.

Literature Cited

1. Raleigh, D.O. J. Appl. Phys., 1970, 41, 1536.

2. Owens, B.B.; Sprouse, J.S.; Warburton, D.L. 25th Power Sources Conference, 1972, 8.

3. Owens, B.B. J. Electrochem. Soc., 1970, 117, 1536.

4. Owens, B.B.; Christie, J.H.; Tiedeman, G.T. J. Electrochem. Soc., 1971, 118, 1144.

5. Owens, B.B., "Advances in Electrochemistry and Electrochemical Engineering," Delahay, P. and Tobias, C.S., Eds., Vol. 8, New York: Wiley Interscience, 1971.

6. Christie, J.H.; Owens, B.B.; Tiedeman, G.T. The Electrochem. Soc. Extended Abstracts, 1974, No. 20, p. 53, Fall meeting, New York.

7. Geller, S. Science, 1967, 157, 310.

8. Geller, S.; Lind, M.D. J. Chem. Phys., 1970, 52, 5854.

9. Geller, S. Science, 1972, 176, 1016.

10. Geller, S.; Owens, B.B. J. Phys. Chem. Solids, 1972, 33, 1241.

11. Geller, S.; Skarstad, P.M.; Wilber, S.A. J. Electrochem. Soc., 1975, 122, 332.

12. Masters, J. J. Electrochem. Soc., 1970, 117, 1378.

13. Tan, Y.T.; Trautweiler, F. J. Appl. Phys., 1969, 40, 66.

14. Owens, B.B.; Oxley, J.E.; Sammells, A.F., "Applications of Halogenide Solid Electrolytes," in Topics in Applied Physics, 1977, Vol. 21, Solid Electrolytes, Geller, S., Ed., Berlin: Springer-Verlag.

15. Dember, H. Physik. Z., 1932, 33, 207.

16. Kaneda, T.; Mizuki, E. Phys. Rev. Letters, 1972, 29, 937.

17. Radhakrishna, S.; Hariharan, K.; Jagadeesh, M.S. J. Appl. Phys., 1979, 50, 4883.

RECEIVED October 3, 1980.

INDEX

INDEX

403